Design Process

This book introduces systematic design process for product and engineering design projects by adopting a design model and the use of several design methods. Starting with a product idea normally outlined by the senior management as a design brief, it guides to plan the design process, define the problem, generate and choose a near-optimal or optimal solution and complete the embodiment, all under a systematic design process model. The main strength of this book is its provision of several worked examples in the use of several design methods at all stages of the design process.

This book explains how to:

- Start with the design brief and define the problem by eliciting and refining stakeholder requirements.
- Establish the functional representation of the product as a function tree or function structure.
- Create conceptual solutions using 12 different conceptual design methods.
- Evaluate and prove that the proposed conceptual solutions are of high grade before choosing one for further development, using the decision matrix method and Pugh's controlled convergence method.
- Use the embodiment design method by Pahl and Beitz to develop the embodiment design for the chosen concept.

It is primarily written for senior undergraduate and graduate students in the fields of industrial engineering, production engineering, manufacturing engineering, mechanical engineering and aerospace engineering.

Design Process
A Hands-on Approach

Sangarappillai Sivaloganathan

CRC Press is an imprint of the
Taylor & Francis Group, an **informa** business

Designed cover image: Sangarappillai Sivaloganathan

First edition published 2025
by CRC Press
2385 NW Executive Center Drive, Suite 320, Boca Raton FL 33431

and by CRC Press
4 Park Square, Milton Park, Abingdon, Oxon, OX14 4RN

CRC Press is an imprint of Taylor & Francis Group, LLC

Library of Congress Cataloging-in-Publication Data
Names: Sivaloganathan, S. (Sangarappillai), author.
Title: Design process : a hands-on approach / Sangarappillai Sivaloganathan.
Description: First edition. | Boca Raton, FL : CRC Press, 2024. |
Includes bibliographical references and index.
Identifiers: LCCN 2024001771 (print) | LCCN 2024001772 (ebook) |
ISBN 9781032560540 (hbk) | ISBN 9781032778204 (pbk) |
ISBN 9781003484950 (ebk) | ISBN 9781032778235 (eBook+)
Subjects: LCSH: Product design.
Classification: LCC TS171 .S545 2024 (print) | LCC TS171 (ebook) |
DDC 658.5/752—dc23/eng/20240321
LC record available at https://lccn.loc.gov/2024001771
LC ebook record available at https://lccn.loc.gov/2024001772

ISBN: 9781032560540 (hbk)
ISBN: 9781032778204 (pbk)
ISBN: 9781003484950 (ebk)
ISBN: 9781032778235 (eBook+)

DOI: 10.1201/9781003484950

Typeset in Sabon
by codeMantra

To my sweet parents,
darling wife Vathana and lovely children Purany and Kumaran
for their undiminishing love, care and support
and
To my wonderful teachers
in particular to late Mr N. Mahadevan
at Chavakacheri Hindu College, and
late Dr M.V. Mendis at University of Moratuwa
Sri Lanka
and
To my student gems at
Brunel University, London and
at United Arab Emirates University

Contents

PART I
Defining the problem

PART 3
Complete design projects

About the author

Sangarappillai Sivaloganathan was a curriculum-based engineer, like most of the contemporary engineering graduates. He graduated as a Mechanical Engineer from the University of Sri Lanka in 1976. The young man who until his mid-teens spent his life with bulls, bullock carts and a rackety bicycle went for training in the casting, forging, machining and millwright shops of the Sri Lankan Railways. Following this, after working in the Sri Lankan Cement Industry, Siva joined the University of Aston in Birmingham and obtained his Master's in Production Technology in 1982. Birmingham, which housed a third of the British Industry at that time, had a major impact in Siva's life. Siva returned to Sri Lanka and became the commissioning engineer and later the operations manager for Lanka Cement Ltd. He joined City University, London in 1985 and obtained his PhD in Mechanical Engineering in 1991. During this time, he joined the activities of the EPSRC Engineering Design Centre and later became the centre's Senior Research Assistant. He joined Brunel University in 1995 as a faculty member and continued to work there until he moved to the United Arab Emirates University in 2011. During his stay at Brunel University, he has supervised nine PhDs, published more than 60 papers in reputed journals and conferences, and convened and functioned as chair for two international conferences in Design. He was the founding course director for the successful MSc program in Advanced Engineering Design. Dr Siva moved to the United Arab Emirates University in 2011. As the program coordinator, he restructured the struggling Master of Engineering Management program. He teaches Product Development in the program from the time he joined UAEU. He presented his teaching philosophy to the ASEE annual conference in 2014 and got the best paper award in the Engineering Management division.

Foreword

I am delighted to write a note on this book whose author is a dear friend, and I am curious because I have been a keen observer of his teaching activities in Product Development in the Master of Engineering Management program over the past 12 years. Dr. Sangarappillai Sivaloganathan, known to us as Dr. Siva, started his involvement with design when he was the Senior Research Assistant at the EPSRC Engineering Design Centre at City University in London and continued his work at Brunel University before coming to us at UAEU in August 2011. While at Brunel University, he published several articles in design in several journals and conferences, was the Convening Chair of two international conferences in Design and was the founding Course Director of the successful MSc program in Advanced Engineering Design.

As the Assistant Dean for Research and Graduate Studies in the College of Engineering at UAEU, I had the opportunity to work closely with Siva when he took over the leadership for our Master of Engineering Management program from September 2011 and led a major structural change to the program. The new structure of the program was designed to focus on Product Development, which he taught, as the main theme and supported by other courses on engineering technology and business. Siva presented the program and its teaching product development methodology at the annual conference of the American Society of Engineering Education in 2014 and received the Best Paper Award from the Engineering Management Division. Later in my new role as the Vice Dean and now Dean of the College of Graduate Studies, I continued working closely with Siva who led the program to obtain accreditation by CAA (Commission for Academic Accreditation). Most of the times, I have participated as an evaluator of the hands-on projects in the Product Development course, which is one of the final and most challenging courses in the program. The project requires students to apply a systematic design process and a hands-on approach to a real problem and deliver a high-quality solution within a limited time frame. The projects are evaluated based on their technical feasibility, design and economic viability, social desirability, and environmental sustainability. Student's enthusiasm to design and manufacture a prototype with limited budget and time has been outstanding. This is an essential skill for engineering managers, as they often have to deal with resource constraints and time pressure in real-world projects. This activity helps students develop their technical, problem-solving, teamwork and entrepreneurial skills. Many impressive and innovative prototypes were created by students in this activity.

The book covers the systematic design process from start to finish. It provides a comprehensive and detailed description of the systematic design process which starts at the design brief and reaches its final end as the definition of a product that can be manufactured. The hands-on experience is provided by the numerous examples with solutions that relate to actual solutions developed by students, to illustrate how the design process can be applied to various real problems. Moreover, this book encourages a hands-on approach by providing opportunities for students, to design and manufacture prototypes, test and validate solutions, and present and communicate results. This book also features complete case studies that demonstrate how real-world projects have been designed and managed using a systematic design process. This book makes systematic design concepts and methods, accessible and understandable to a wide range of readers, regardless of their prior knowledge and experience.

The special feature of this book is its usage of simple but real examples to explain and illustrate systematic design process. This book demonstrates how simple examples can help to clarify, grasp and consolidate the principles of problem definition, identification of design criteria and constraints, generation and evaluation of alternative solutions, and selection and implementation of the best solution. Siva has used his personal experience as a practicing engineer for a decade and as a professor and researcher in design, for over three decades, to explain the concepts of the design process through hands-on examples. Enjoy the reading. Gain a wealth of experience in systematic design process.

Prof. Ali Al-Marzouqi
Dean, College of Graduate Studies
United Arab Emirates University
PO Box 15551
Al Ain
United Arab Emirates

Preface

Systematic design process divides the process into constituent components and combines and utilizes the results of the components to achieve the desired result. In this approach, no specific parts are overlooked and, more importantly, definite starting point and finishing point are defined. It suits well to my background, going from junior to senior schools and then graduating as a mechanical engineer, with sound theoretical and limited application knowledge. During my career, I took systematic design as the topic of study. In this book, I have presented the experiences in systematic design, from the experiments and design projects that have been undertaken over the past 30 years. Like the peasant who thought the water he enjoyed in an Oasis is worthy for consumption by the king, I believe that these experiences are worthy of learning by the design community. I submit them for consumption and apologize for any shortcomings.

I had a wonderful maths teacher whose speciality was in the choice of examples. He always chooses very simple examples to explain complex and thought-provoking topics that he teaches new to the class. His examples are always independent and specific to the new topic he is introducing and hence do not require a lot of background knowledge. Because of the simplicity, we all leave the class with contentment and confidence. Our inspiration would be very high, and as a result, we entered into student competitions, among ourselves in the class as well as with students from other colleges in the region. At that time, we were one of the best schools in the country. In this book, I endeavoured to bring simple and independent examples to explain the applications of various design methods, under a systematic five-stage design model.

The challenges facing the designer can be identified into two classes: (i) defining the problem well and (ii) identifying a near-optimal or optimal conceptual design solution from the unknown solution space. Part 1 of this book describes and provides experiences in the problem-defining process. Part 2 describes and provides experiences in various design methods to generate conceptual design solutions, methods to evaluate them and choose the best solution from a sample of design solutions. Also it describes how to embody the chosen conceptual design and give it a physical form. Part 3 describes and provides experiences in some simple examples which integrate some methods from Parts 1 and 2. A five-stage systematic design model is used as the backbone for this book. 'Experience is what you experience' is the maxim that is followed here, and examples are used at every stage to reinforce the content. Some achievements by students are given in between to provide a feel to the activities in the tasks and to keep a foot on the ground.

Part 1 starts with Chapter 1 giving a brief introduction on design to comprehend the design task and an example to get to the bottom of the process in full. Chapter 2 explains how the design process for any product can be planned in a systematic design process. Design brief is the document defining 'what the product is', originated by the senior management and given to the design team. This is the origin of the design process and is described in Chapter 3. The design team reads and assimilates the design brief and decides what kind of information from the stakeholders can enable them to design and deliver a more satisfying product. Often checklists are employed to assist the design team on the selection of the types of requirements. Obtaining requirements is described in Chapter 4. Functional description is the first attempt by the design team to describe the intended product in the functional domain. Chapter 5 describes the two approaches: (i) function tree modelling and (ii) function structure modelling. The needs or requirements as obtained and presented may not be measurable. To specify or to quantify, the requirements are translated into metrics using the need-metric matrix. This is described in Chapter 6. In Chapter 7, the design brief, functional representation and the metrics to a greater extent are used to draw the specifications. The functional description and the specifications define the problem well and form the product concept.

Part 2 made up of Chapters 8–10 describes and provides experiences in (i) the generation of a subset of designs that contains near-optimal and/or optimal designs (ii) evaluation and establishment that they are highly valuable designs and (iii) choosing a conceptual design for further development. Chapter 8 describes several design methods for generating conceptual designs. Chapter 9 describes the evaluation and selection of a concept from the sample of designs generated. It adopts two methods: (i) the decision matrix method and (ii) Pugh's controlled convergence method for the evaluation and selection of the preferred design. Chapter 10 describes embodiment design and is developed using the method developed by Pahl and Beitz. As the design process progresses towards Chapter 10, more and more geometric and engineering knowledge are needed. Part 3 describes and provides experiences through simple, but complete examples using the Design Model and Desig Methods described in Parts 1 and Part 2.

This book summarizes the experiences, from simple experiments in using new techniques in design, from 1990 at City University Engineering Design Centre, at Brunel University from 1995 to 2011 and at United Arab Emirates

University from 2011. Several hundreds if not thousands of students have been participating in these experiments during these periods. My most sincere thanks are due to them. There are several colleagues who have been contributing and helpful towards this endeavour. I choose three of them to mention here. Professor Yousef Haik, Vice Chancellor Research, University of Sharjah, was a constant inspiration for me to embark and stay in this hard task. Professor Ali Masrouki, Dean, Graduate School at United Arab Emirates University, was a staunch ally for all my experiments at UAEU, since 2011. He was very supportive and provided the Foreword for this book. Professor Khalifa Hareb, Department of Mechanical Engineering at UAEU, was a friend, colleague and partner in all my mechatronic experiments. A great appreciation and sincere thanks are due to them. Several students have contributed and helped in this endeavour. Five students, Maitha Obaid Alshamsi, Latifa Ali Suhail Aldhaheri, Mariam Sultan Khalfan Alnuaimi, Maitha Ali Sarhan Alshamsi and Fatima Ali Alaryani, whom I call the Pilot Group, were helping me for the past 3 years. They were the test benches for the writing of various chapters and topics. Their feedback helped me enormously to improve the presentation of this book. A very special appreciation and thanks are due to them.

I would like to put on record my sincere appreciation and thanks to Taylor and Francis for their immense contribution for the success of this book. Three unsung hero and heroines helped immensely to structure the thoughts and their presentation in this final form of the book. They are Mr Reen Gauravjeet Singh, Ms Mehnaz Hussain and Ms Assunta Petrone (codeMantra). The book and I have benefitted a lot by their principled work with undivided attention. Special thanks and appreciation to you.

I hope this book that provides step-by-step description of systematic design process employed to convert an abstract brief into the definition of a physically realizable product will be helpful to beginners and students studying the design process.

Dr Sangarappillai Sivaloganathan
Mechanical Engineering and Aerospace Department
United Arab Emirates University

Chapter 1

Introduction

1.1 INTRODUCTION

Design Process or design activity is the subject of this book, and introduction to the book is the aim of this chapter. *Following a systematic design process with, a design model and use of design methods, is the technique adopted and advocated for design throughout this book.* Model questions and answers are integrated in each chapter to provide the hands-on experience. This chapter explains the nature and definition of design, the systematic design process and a glimpse of it through an example design, developed using the systematic design process. It also explains the contents of the remaining chapters which describe components of systematic design process.

Life without designed objects is unthinkable. Designed objects are the result of discoveries aimed to alleviate difficulties. Design activity, aimed at facilitating life, is at the epitome of all engineering endeavours and design has been present, in some form or other, throughout the ages. The subject is vast and has several facets, and several researchers are actively engaged in researching design. Early designers identified solutions to problems using their skillset and imagination, and in this method, called 'Conventional Design', the design process is not transparent. 'Systematic Design', developed in the recent past, is transparent and divides the design process into constituent components and integrate their outputs to find the solution to the design problem. The description of the design process as a sequence of its components (stages or activities) is called the design model and the tools and techniques used at different stages are called design methods.

A design stage model with (a) requirements (b) product concept (c) solution concept (d) embodiment design and (e) detail design as its stages, is chosen for use in this book. Two common challenges (a) defining the problem well and (b) finding the optimal or a near-optimal solution from the unknown solution space, are faced by all design teams. Part 1 of this book describes how the first challenge, defining the problem well, is handled. Part 2 of the book describes how a near optimal design is developed, evaluated and chosen. Part 3 provides some complete examples. For ease of understanding the design activity is divided into two classes, product design and

DOI: 10.1201/9781003484950-1

engineering design. Further how the development of the activity model for a specific product, incorporating activities and appropriate design methods at various design stages, is explained in `design process planning'.

1.2 NATURE OF DESIGN

Humans always have needs for goods and services to carry on with their lives. Characterization of a societal need is generally called a 'Design Problem'. Such a design problem generally will have several possible solutions, and these possible solutions as a collection is called the solution space. A design problem will have many requirements, and acceptable solutions meet several of these requirements while the optimal and near-optimal solutions meet most of the requirements. Two unique characteristics of a design problem are: (i) the problem is ill-defined and (ii) the solution space is unknown. The challenge is to define the problem well and then identify the optimal or near-optimal solution from the unknown solution space in an efficient way. In this book, the design model adapted lends itself to (i) define the problem well in the first two stages, using several design methods and (ii) establish a representative solution space (assumed to contain the optimal and near-optimal design solutions), from which a preferred solution is chosen and developed to the full. The adopted design model is explained in Section 1.4.

1.2.1 Definition of design

There is no universally accepted definition for design yet, and there are several definitions. Ulrich and Eppinger [1] define 'Product Development' as the set of activities beginning with the perception of a market opportunity and ending in the production, sale and delivery of a product. In the 60s, a committee to the Science and Engineering Research Council (SERC) of the UK [2] described 'Engineering Design' as the use of scientific principles, technical information and imagination in the definition of a mechanical structure, machine or system to perform pre-specified functions with the maximum economy and efficiency. The Accreditation Board for Engineering and Technology (ABET) definition [3] states that 'engineering design is the process of devising a system, component, or process to meet desired needs'. It is a decision-making process (often iterative), in which the basic science, mathematics and engineering science are applied on imagination to optimally convert resources to meet a stated objective.

1.2.2 Product design and engineering design

Any physical object (the raw material) in the real world will obey the laws of physics. When its structure is subjected to change, to meet some societal needs, in many situations, it will continue to behave in the same way as it behaved before the change. In such a situation, other requirements such as easy to make and use, and aesthetically pleasing appearance, take priority. In this book, such designs are categorized as product design. There are several situations where the designed artefact should explicitly adhere to, scientific and engineering principles and technical information, as described by the SERC committee and ABET. They require engineering analysis. These designs are categorized as Engineering Design in this book. In many situations, the product design approach and the engineering design approach mix together, as in the example described in Section 1.6.

Product Design in this book is defined as the process of developing a product from the identification of a societal need to the complete definition of the physically realizable product without involving engineering analysis. The considerations and outcomes here are the geometric definitions.

In the same way, *Engineering Design* in this book is defined as the process of developing a product from the identification of a societal need to the complete definition of the physically realizable product with the use of engineering analysis. In addition to geometric definitions, the considerations and outcomes ensure that the engineering characteristics are within acceptable limits.

1.3 SYSTEMATIC DESIGN PROCESS

Systematic design process aims to define the problem well and then generates candidate concepts, and chooses one for further development. It divides the process into constituent components and completes the components in sequence, to achieve the full design. The division can be based on either (i) the intermediate stages (mile stones) the design process goes through or (ii) the activities that are carried out during the design process. Description of the design process in this way is called the design model. In the systematic design process 'Design Methods' are various tools and techniques that are used at different stages of the design process. Typical examples can be (i) customer interviews, (ii) observation of an adjacent product in action or use and (iii) focus group interviews. All these are design methods to establish customer requirements. Similarly, morphological analysis, C sketching and gallery method are design methods for conceptual design. Computer experiments and meta modelling are examples of design methods at the detailed design stage. There are several design methods that the designer can choose from.

In short, the systematic design process needs a Design Model and, a collection of chosen appropriate design methods. This is the essence of design process planning explained in detail in Chapter 2.

1.4 DESIGN MODEL

In this book, the design stage model for systematic design, proposed by Haik et al. [4], is followed. The first two stages in the design model defines the problem. The third stage, solution concept stage, generates a representative 'solution space with sample designs' from which a design solution is chosen. This representative solution space is presumed to contain the optimal and near-optimal solutions at the beginning. Later, evaluation establishes that the sample contains high-grade designs, before choosing one from them. This chosen concept is further developed as completed design. Thus, the last three stages address the challenge of choosing a near-optimal or optimal solution. Figure 1.1 illustrates the design model.

- *Requirements*: The genesis of the design process is the design brief that comes from the senior management or a customer to a design company. It describes 'what the product is (Goal)'. Every product is developed to perform certain functions. The design team thinks about the functions that can be expected from the product and forms a mental picture that would give the indication of inputs that the customers or stakeholders only can provide. This is about the performance of the product and as such, is related to the functions. The design team in the end has two inputs: (i) from the company's senior management stating 'what the product is', through the design brief and (ii) from the stakeholders stating 'what the product should do' from their use point of view, termed as customer requirements or customer needs. The design team establishes a list of prioritized requirements by carrying out some surveys among the stakeholders.
- *Product concept*: Product concept consists of two items: (i) the functions the product would perform or how the product behaves

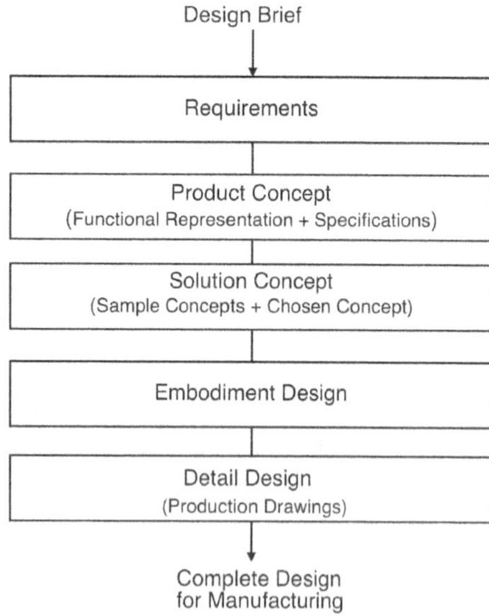

Figure 1.1 Systematic design model.

and (ii) the specifications that define the product or set boundaries to it. A Metric, a Value and a Unit form a single specification, and product specifications consist a list of several single specifications. Specifications are drawn from the two outputs of the requirements stage: the design brief and the prioritized customer requirements.

The combined Requirements and Product Concept stages in the design model define the problem well.

- *Solution concept*: Any product is a collection of components harmoniously integrated in a particular way, to impart a pre-defined behaviour, that delivers the intended functions. Solution concept is the integrated arrangement or the 'scheme' in which the chosen components are arranged to deliver the intended functions safely and efficiently. Depending on the magnitude of the problem, a chosen concept can be reached in several stages, including concept generation, concept evolution, concept prototyping and proving, and evaluation and selection. Conceptual design is a difficult and important part in the design process, and there are several design methods to assist.

  Solution concept stage in general, generates a representative solution space, ensures that the space contains high class designs, and chooses one for further development.
- *Embodiment design*: Embodiment design provides the physical structure to the concept or scheme outlined in the solution concept stage. It defines the architecture incorporating the various function providers, details the arrangement or layout and defines their dimensions if appropriate. It determines the overall dimensions and operating envelope of the product when needed. Finer adjustments to the dimensions may be carried out during the detailed design stage when engineering analyses are carried out. Embodiment design provides the definitive layout and overall dimensions.
- *Detail design*: Detail design comprises two main parts: (i) the details required for manufacturing and (ii) detailed design or the required engineering calculations. These calculations ensure correct functionality and safety in operation or provide the required validating data from applied mathematical calculations. In product design, the first part without detailed design is deemed adequate while engineering design incorporates engineering analysis or detailed design as an integral part.

1.5 DESIGN METHODS AND ACTIVITY DESIGN MODEL FOR A SPECIFIC PROJECT

Activity design model describes the sequence of activities that have to be discharged for a project, and this can be easily established by starting with a stage model and adding the list of suitable design methods needed for each stage of the specific project. In this context, design methods are tools and techniques to achieve a requirement at a design stage. Figure 1.2 shows some example design methods suitable for the design stages.

1.5.1 Planning the design process

Choosing the activities for each stage of a systematic design model is the essence of planning the design process. Once the design brief is received, each project should start with the 'Planning of the Design Process'. This is the task of filling the dotted lines in the planning framework shown in Figure 1.3. As

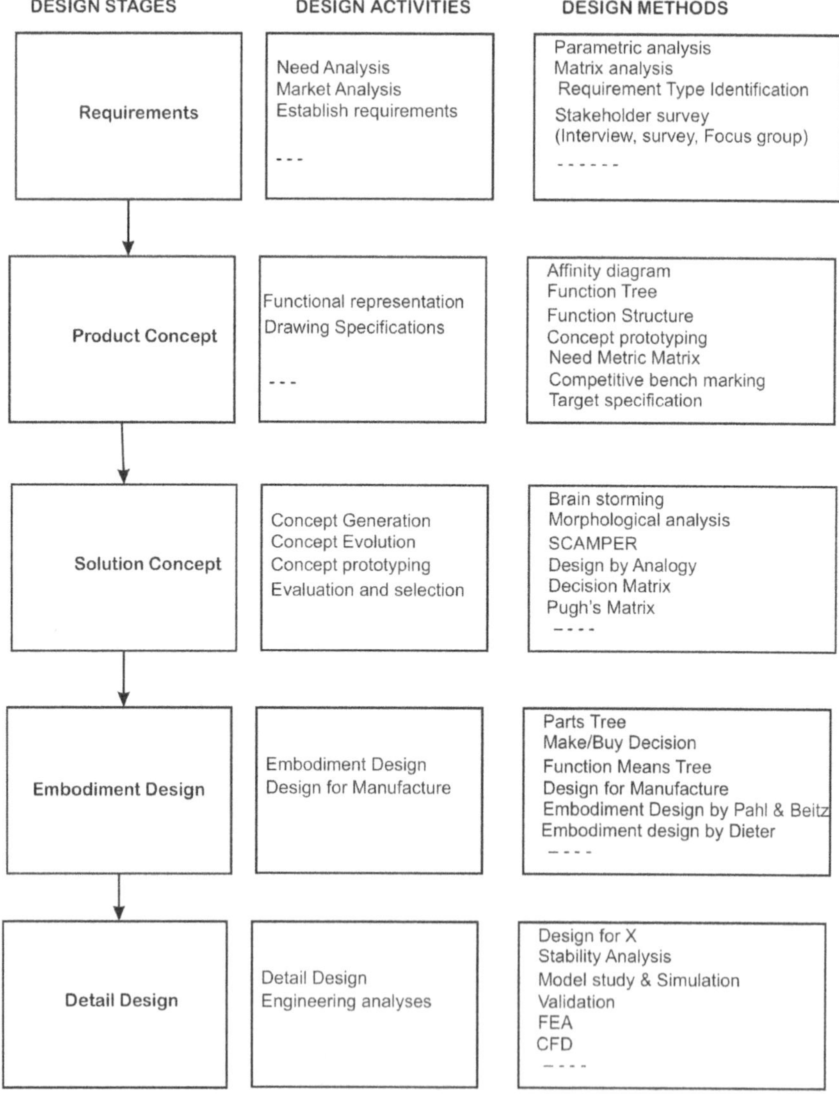

Figure 1.2 Design model, activities at each stage and possible design methods.

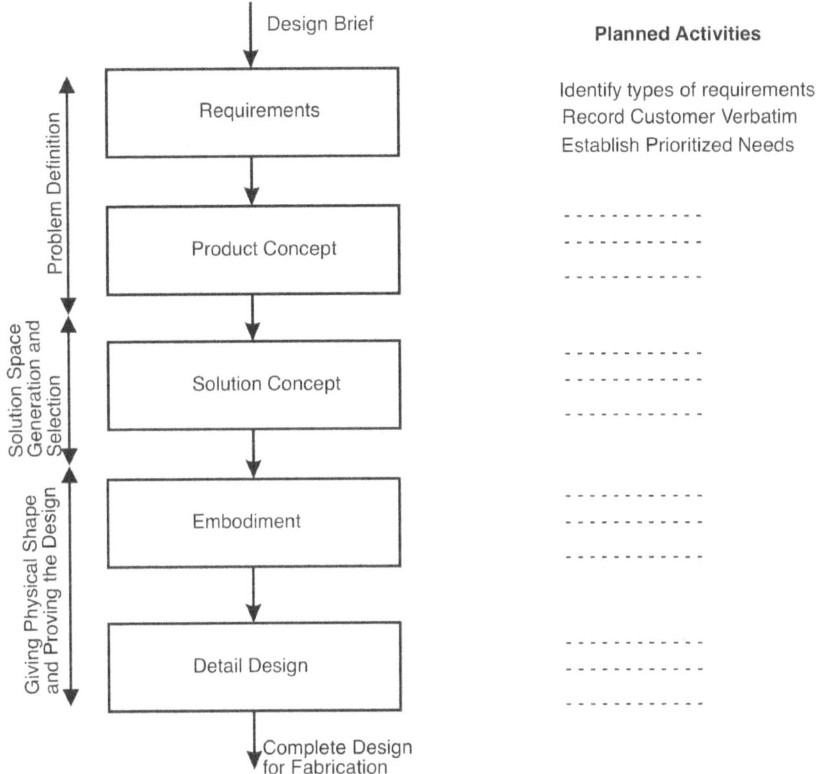

Figure 1.3 Planning the activities at each stage in the design model.

examples of the activities, the activities to be carried out in the requirements stage are given as: (i) identify the types of requirements, (ii) record customer verbatim and (iii) establish prioritized needs, as shown in Figure 1.3.

The procedure for solving a design problem is to plan the design activity model that would address the specific needs of the problem and execute the plan.

Consider an example, to understand the complete solution method, that goes through all five stages of the design model and all associated design methods. The example is given in Section 1.6.

1.6 A DISPLAY BOASTING THE GLORY OF MECHANICAL ENGINEERING: AN EXAMPLE

Designing and building of a product, which requires both product design and engineering design approaches, are described here. It shows how the design model and design methods are used in the systematic design process of a typical product.

The problem: Universities conduct open-days to attract students. Their departments display student projects and artefacts explaining and demonstrating the contents of the disciplines and their excellence. A product intended to (i) boast the glory of mechanical engineering (product design part) and (ii) demonstrate the subject content of machine design courses (Engineering design part) that form a significant part of mechanical engineering is developed in this project. It was designed and built by a group of final year mechanical engineering students under the guidance of the author as their graduation project.

1.6.1 Design brief

See Table 1.1.

Table 1.1 Design brief for a display

Design brief of a display showing the glory of mechanical engineering	
Product description	A relatively light weight display unit, consisting of a transparent or open cuboid of about 1–1.5 m^3 in size as a base that houses a power train made of different machine elements, driven by a motor to rotate a shaft, and the shaft carrying a rotating structure that has multiple views, showing visual and textual forms of events, artefacts and personnel, relating to the glory of mechanical engineering.
Product concept	A two-part item consisting (i) The base, the machine part that houses the power train consisting of motor, chains, gearbox, bearings and the driving shaft and (ii) the display fitted on the shaft that contains and displays items boasting the greatness of mechanical engineering.
Benefits to be delivered	Spread impressive information about Mechanical Engineering field in a uniquely attractive way. Explain interesting details of mechanical engineering. Showcasing parts of Machine Design 1 and 2 courses. Easily moveable display from one place to another.
Positioning and target price	Middle of the range elegant exhibit weighing about 50–70 kg and price in range of about \$750–\$1,000
Target market	Universities, Technical colleges and institutes, Mechanical departments and museums
Stake holders	Engineering students, faculty members and head of Mechanical Engineering department, and engineers
Possible features and attributes	Different shapes of rotary part (diamond, cubes, blocks, etc.). Different achievements, landmarks and personnel.
Possible areas for innovation	Displaying methods (painting, machining and engraving)

Drafted by Dr Sangarappillai Sivaloganathan

1.6.2 Design process plan

'Planning the Design Process' is an important action that should be carried out by the design team at the beginning when they receive the design brief. The five-stage design model is taken as the basis for the plan. The design model is responsible for the following activities

1. Problem definition
2. Establishing a subset of the design space, consisting of plausible conceptual designs (representing optimal and near-optimal designs) and choosing one from them
3. Giving the physical shape to the product
4. Modelling and proving the design whenever necessary and producing a set of production drawings.

Problem definition consists of (i) establishing the type of requirements (ii) recording customer verbatim and (iii) translating the verbatim into prioritized needs. These needs are then converted into metrics and the metrics form the basis for target specifications. A final functional representation is prepared at this stage. The problem definition thus produces two records: (i) specifications and (ii) functional representation.

Conceptual designs can be generated using several different methods. In this example, a morphological chart is prepared first, and from the chart different conceptual designs are generated. A criteria set is used to evaluate the proposed conceptual designs and decision matrix is used for the choice.

Giving the physical shape phase uses non-engineering analysis methods to propose embodiment. This would be sufficient for product design. When Engineering analysis is important, first the embodiment is proposed and the dimensions are derived or proven to be adequate using various engineering calculations. Figure 1.4 schematically illustrates the design model-based design process plan for designing the rotating display.

The input to the design process is the design brief. The needs from the stakeholders are then established by carrying out three activities to produce the end result, prioritized needs. In the next stage the initial thoughts on functions, are finalized to establish the functional representation of the intended product. The needs are converted into measurable metrics using the need-metric matrix. From the list of metrics, specifications are drawn. The problem definition is complete with the production of the functional representation and the target specifications. In the next stage, solution space is created by constructing a morphological chart and conceptual design from this chart. A criteria set is used to evaluate and choose a concept using decision matrix. In this project, the display and base are considered separately. The chosen concepts for the display and base are further established by giving the embodiment. The product design aspects are completed with the entire product, and the CAD model is built for both the display and base. The base design is further proved by carrying out engineering analysis.

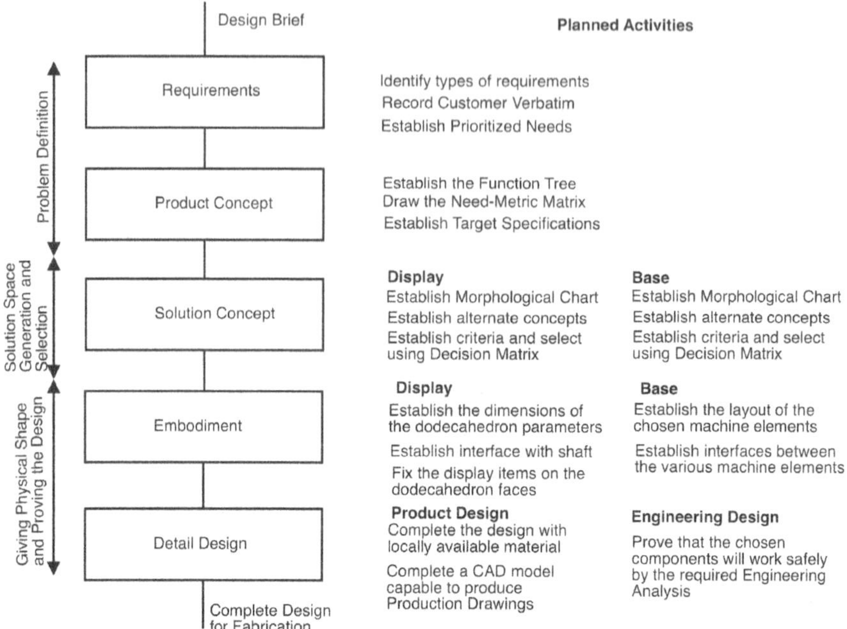

Figure 1.4 Design process plan for the rotating display.

Having thus planned the activities that have to be performed to complete the project, the activities are executed for completing the project. The activities are described in the following sub-sections.

1.6.3 Identification of requirement types

Stakeholder requirements are identified to ensure that the resulting product meets the requirements and expectations of the stakeholders. There are certain areas where knowing the stakeholders' requirements would be very helpful. Experienced designers and companies will know these types of requirements and they will use this knowledge to extract the useful set of requirements from the stakeholders. The experienced designers establish checklists to help young designers. This phenomenon of extracting stakeholder requirements gained prominence after 'Quality Function Deployment' came into use in design. Because of this reason, earlier checklists are more biased toward engineering design requirements and the later ones are more biased towards product design requirements. Two such popular checklists can be attributed to Pahl and Beitz [5] and Ullman [6]. Referring these and any other such lists, the relevant types of requirements are identified. The types of requirements identified for this display are shown in Table 1.2.

Table 1.2 Types of requirements and explanations

Type	*Explanation*
1. Human factors	Appearance, steps for operation, ability to start and stop at will, mobility
2. Functions	Flow of information (types of items for display), operational steps (rate of information flow)
3. Physical requirements	Convenient space for display and foot space, number of items to display, etc.
4. Reliability	Safety hazard assessment
5. Lifecycle Requirements	Easily installable, cleanable, maintainable, etc.
6. Resource concerns	Usage of natural materials (environmental) Machine elements incorporated
7. Manufacturability	Easy to manufacture and assemble

1.6.4 Stakeholder verbatim

The design team decided to pursue the interview method, and the stakeholders of their choice included, the head of the mechanical engineering department, few senior faculty members, few senior mechanical engineering students and few junior students who are about to choose their specialty. The strategy was to record the statements as made by the stakeholders and to scrub them with the entire design team to establish the full list of needs (requirements). The established needs can then be prioritized by the interviewed stakeholders or by the interviewing members reflecting the interview. In this case, the interviewers assigned weights. The listed verbatims are given in Table 1.3.

Table 1.3 Customer verbatim

Type	*Verbatim*
Human factors	Engraving will be better than painting (I like it). I prefer the ability to stop and start the display. It should be easily moveable to different places. The structure is from light and natural material.
Functions	Should bring information on important machines. I like special aircrafts and cars The display should rotate at reasonable speeds (about 3–6) Not too many things to show (4–6 will be nice) Should show many machine elements from courses
Physical requirements	The display should be at eye level The display item should not exceed $1.5\,m^3$ The power train should be spaced well for clear sight.
Reliability and safety	Drive mechanism should be guarded and safe. Should be reliable
Lifecycle Requirements	Should be easily installable Should be easily maintainable Should be easily cleanable
Resource concerns	The structure is from superclass material.

1.6.5 Establishment of prioritized needs

The stakeholders' verbatim is first scrutinized by the design team for information content and clarity. The rewritten verbatim is called 'needs' in this book. To understand the information content, consider the verbatim 'the structure is from superclass material'. In general, the design team has to break this kind of verbatim down into more than one need. In the case of this project, the design team broke it into three needs as shown in Table 1.4. The information content of each individual need should be similar, to have an effective relative importance for them. The importance ratings for the needs can be obtained by referring the needs back to the stakeholders. Alternately the ratings can be assigned by the design team members who conducted the interviews. In this project, it was assigned by the design team who conducted the interviews.

Table 1.4 Verbatim to prioritized needs

No	*Verbatim*	*Need*	*Importance*
1	Engraving will be better than painting (I like it).	Engrave the item displayed	9
2	I prefer the ability to stop and start the display.	Facility for customer to stop and start	7
3	It should be easily moveable to different places.	Easily moveable	8
4	Should bring information on important machines.	Display should include important machines	9
5	I like special aircrafts and cars	Display should have special Airplanes	9
6	The display should rotate at reasonable speeds (3–6 rpm)	Slow rotational speed of display	7
7	Not too many things to show (4–6 will be nice)	Number of displayed items from 4 to 6	7
8	Should show many machine elements from courses	Use more and different machine elements	9
9	The display should be at eye level	Display at eye level	8
10	The display item should not exceed 1.5 m^3	Reasonable occupied volume of the display	6
11	The power train should be spaced well for clear sight.	Spaced laying of the machine elements at base	7
12	Should be reliable	Should be reliable	8
14	Should be easily installable	Easily installable	7
15	Should be easily cleanable	Easily cleanable	8
16	The structure is from superclass material.	Made up of natural and light material	6
		Material is easily obtainable	6
		Material should be environmentally friendly	6
19	Should use standard commercially available components	Should use standard commercially available components	7

1.6.6 Establishment of need-metric matrix

To ensure that a need is properly deployed, a precise and measurable characteristic of the product that will reflect the degree of satisfaction of that particular need is identified. Such a measurable characteristic is called a metric. All needs should be translated as metrics so that the metrics can be kept within the required limits to fully satisfy the customer's need. Each need is considered individually, and a metric is identified for it. In this way, the customer verbatim is translated into easily measurable metrics with units. Each metric is assigned a numeric value to form a specification and the full set of specifications derived from the full set of needs is called the specifications of the product. The process of this translation is facilitated by the need-metric matrix, shown in Table 1.5, where the needs are in the rows and the suitable metrics are identified and placed in the columns. A one-to-one relationship is preferred, but there may be few, one to several relationships. Instead of looking for engineering measures it is worth noting the following when choosing metrics. These are explained further in Chapter 6.

a. In many situations, the need may lead to a physically measurable quantity. Then choose the right units that would assist visualization.
b. The need may be, to have a particular feature present in the product. In these situations, yes/no is a correct metric to see whether the need is fulfilled or not.
c. Similarly, the need may be a particular characteristic that can be completely fulfilled by a few selected provisions. In such situations, the appropriate measure can be the membership in a list or not.

Table 1.5 shows the need-metric matrix for the display of this project. Full explanation of measurement levels and the choice of metrics is given in Chapter 6.

1.6.7 Functional description

This stage started with the formulation of an idea about the product to be designed to meet the expectations given in the design brief. The students looked at various analogous products such as rotating globe and various memorabilia items to comprehend the base part and display part. They soon formed a rough function tree to describe the various functions the product should perform. The function tree is firmed up after establishing the requirements as shown in Figure 1.5.

1.6.8 Drawing the specifications

Different people can easily interpret the needs differently as they may leave too much margin for subjective interpretation. For this reason, a set of specifications that spell out in precise, measurable detail what the product

Table 1.5 Need-metric matrix

	Units	mm	Yes/No	Yes/No	List	List	List	RPM	Count	Count	mm	Cubic metre	Yes/No	hours	Yes/No	Yes/No	List	Importance ratings (Low 1–10)
	Metric	Depth of engraving	On/Off switch	Wheel mounted base	Historical machines	Special planes	Speed of rotation	No of items displayed	No of elements	Height of display	Volume of the display	Visible elements	Good operating hours	Move as assembled	Single wipe cleaning	Material from list	Units from market	
	Needs																	
1	Engrave the item displayed	■																9
2	Facility for customer to stop and start		■															7
3	Easily moveable			■														8
4	Made up of natural and light material				■													6
5	Should display important machines					■												9
6	Display should have special airplanes						■											9
7	Slow rotational speed of display							■										7
8	Number of displayed items from 4 to 6								■									7
9	Use more machine elements									■								9
10	Display at eye level										■							8
11	Reasonable occupied volume of display											■						6
12	Spaced laying of the machine elements												■					7
13	Should be reliable													■				8
14	Easily installable													■				7
15	Easily cleanable														■			8
16	Made up of natural and light material															■		6
17	Material is easily obtainable															■		6
18	Environmentally friendly materials															■		6
19	Use commercially available units																■	7
	Importance of the Metrics																	

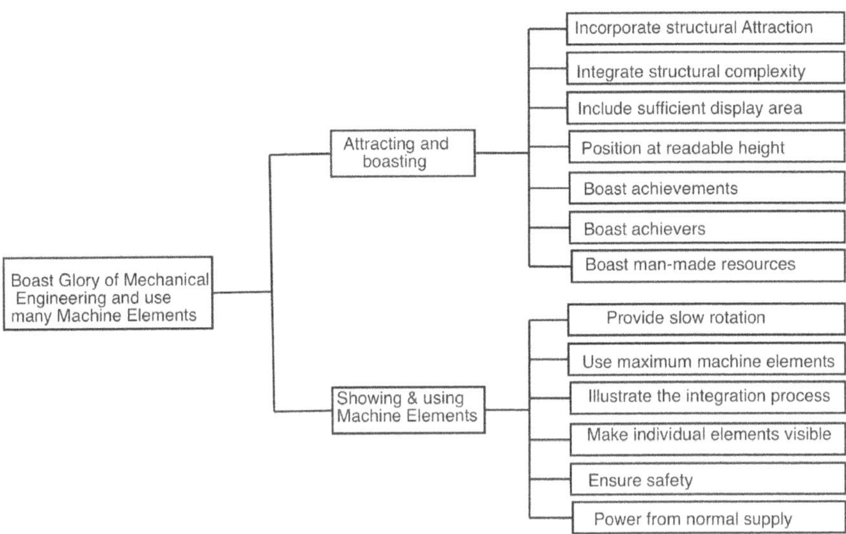

Figure 1.5 Function tree of the rotating display.

is and has to do, is established. Product design specification (PDS) is a set of statements of what the product should be and should do. It is intended to define what the designer should be trying to achieve. It is a specific type of goal statement that clearly and specifically lists the requirements as well as the specific performances required of each requirement. In short, this is the answer to the first challenge, 'Defining the problem well'. It is drawn, based on the design brief, functional description, list of needs and the need-metric matrix. This can be further assisted by the metric-based competitive benchmarking (Table 1.6).

1.6.9 Conceptual design

A scheme can be described as a clever mental plan to put an idea into effect often with a graphic representation. Conceptual design is an outline solution in the form of a scheme to integrate functional subsystems into a complete product. It should have been carried to a point where the means of performing each major function has been fixed, as have the spatial and structural relationships of the principal components [7]. Design concepts are the means for providing functions. They are therefore developed in consideration of the function of the device being designed. In this context, the functional description of the product can be very useful to start conceptual design. For the purpose of this book, conceptual designs are composed of the 'harmonious integration of function providers delivering the functions, set out in the functional description'.

Table 1.6 Specifications of the rotating display boasting the glory of mechanical engineering

Specification	*Details*	*Origin*
Overall weight	50–70 kg	Brief
Overall height	1.5–1.75 m	Needs
Display volume	1.5–1.75 m^3	Brief
Visible, units and functions in base	Yes	Needs
Engrave depth of the item	10–20 mm	Needs
Facility for customer to stop and start	Yes	Needs
Easily moveable with wheels	Yes	Needs
Made up of wood and steel	Yes	Needs
Should display important machines	Choice from list	Needs
Display should have special airplanes	Yes	Needs
Slow rotational speed of display	10–15 rpm	Needs
Number of displayed items	4–6	Needs
Number of machine elements	More than 4	Brief
Spaced laying of the machine elements	1.5–1.75 m^2	Needs
Special inventions list	Yes	Needs
One-time installation	Yes	Needs
Ability to move as assembled	Yes	Needs
Lockable wheels	Yes	Function tree

Conceptual design of the rotating display is divided into three parts. The rotating part has the design of (i) rotating body and (ii) the exhibits or figures attached to it. The third part is the drive mechanism and the base.

1.6.9.1 Design of the rotating body

The rotating body can take several forms like prisms, cylinders, spheres, bodies with planar faces and several other similar bodies. The requirements identify four to six display items as preferable. The choice therefore should be from solids of various forms. Table 1.7 shows some of the solids considered by the students. These form the conceptual designs for the rotating bodies.

The designs can be explained in the following way:

- *Design (a)*: This design has the top face as an equilateral triangle and the bottom face as a regular hexagon both have the same length of sides. This results in three squares and three equilateral triangles in the sides.
- *Design (b)*: This design has the mirror image of the body in Design (a) about the hexagon attached to it. This results in six squares and six equilateral triangles in the sides.

Table 1.7 Conceptual designs for the rotating bodies

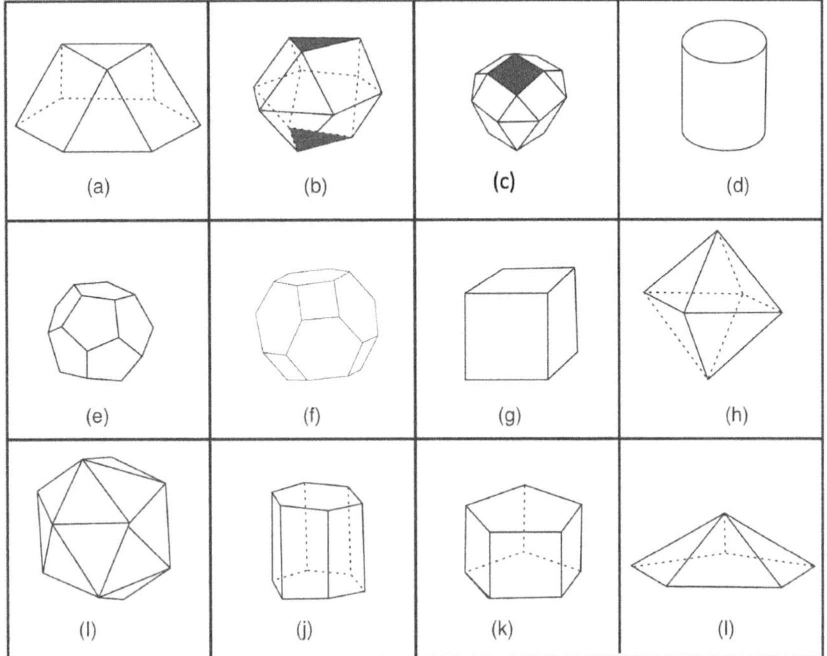

- *Design (c)*: This design has the top face as a square and the intermediate plane as an octagon both have the same length of sides. This results in eight squares and eight equilateral triangles in the sides.
- *Design (d)*: The cylindrical face in this design is divided into sector slices where glorious materials are displayed. The surface can accommodate four to six items easily.
- *Design (e)*: The dodecahedron design is an assemblage of 12 regular pentagons. It has five uniformly placed pentagons at the top and a similar number at the bottom.
- *Design* (f): This truncated octahedron has square top and dropping hexagons. The half of it has a bigger square bottom with four trapeziums in the side.
- *Design (g)*: This cube-shaped design has four faces on the side and one face on top. The axis of rotation can be in different orientations.
- *Design (h)*: A regular octahedron is considered as the design here. It has eight faces shaped like an equilateral triangle.
- *Design (i)*: A regular icosahedron is considered as the design here. It has a body surrounded by 20 equilateral triangles.
- *Design (j)*: A hexagonal prism is considered as the design here. It has six faces for displays.

- *Design* (*k*): A pentagonal prism design is considered. It has five square sides to display.
- *Design* (*l*): A regular pentagonal pyramid is the design considered here. It has a regular pentagonal base and five erected triangular faces joined together at a common vertex point.

  The designs were evaluated with the following criteria and weights using a decision matrix.

1. Uniformity of the surface: 0.15
2. Accommodating capacity: 0.12
3. Attractiveness: 0.08
4. Ease of manufacture: 0.04
5. Ability in giving 3D sense: 0.19
6. Arousing geometric interest: 0.15
7. Recognition as intellectual work: 0.27 (Figure 1.6)

	Criterion 1 0.15	Criterion 2 0.12	Criterion 3 0.08	Criterion 4 0.04	Criterion 5 0.19	Criterion 6 0.15	Criterion 7 0.27	
Design 1	7 / 1.05	7 / 0.84	6 / 0.48	8 / 0.32	6 / 1.14	6 / 0.90	6 / 1.62	6.35
Design 2	7 / 1.05	7 / 0.84	7 / 0.56	8 / 0.32	6 / 1.14	6 / 0.90	6 / 1.62	6.43
Design 3	7 / 1.05	7 / 0.84	7 / 0.56	8 / 0.32	7 / 1.33	7 / 1.05	7 / 1.89	7.04
Design 4	7 / 1.05	7 / 0.84	7 / 0.56	8 / 0.32	6 / 1.14	6 / 0.9	7 / 1.89	6.7
Design 5	10 / 1.5	9 / 1.08	8 / 0.64	7 / 0.28	9 / 1.71	8 / 1.2	10 / 2.7	9.11
Design 6	7 / 1.05	7 / 0.84	7 / 0.56	8 / 0.32	8 / 1.52	7 / 1.05	8 / 2.16	7.5
Design 7	9 / 1.35	9 / 1.08	6 / 0.48	9 / 0.36	5 / 0.95	5 / 0.75	5 / 1.35	6.32
Design 8	9 / 1.35	6 / 0.72	7 / 0.56	8 / 0.32	6 / 1.14	6 / 0.9	7 / 1.89	6.88
Design 9	9 / 1.35	6 / 0.72	7 / 0.56	7 / 0.28	7 / 1.33	6 / 0.9	7 / 1.89	7.03
Design 10	9 / 1.35	9 / 1.08	6 / 0.48	10 / 0.4	6 / 1.14	5 / 0.75	5 / 1.35	6.55
Design 11	9 / 1.35	9 / 1.08	6 / 0.48	10 / 0.4	6 / 1.14	5 / 0.75	5 / 1.35	6.55
Design 12	9 / 1.35	7 / 0.84	6 / 0.48	9 / 0.36	6 / 1.14	5 / 0.75	5 / 1.35	6.27

Figure 1.6 Decision matrix – display body.

The decision matrix indicated that four concepts out of the 12 have scored more than 7 and all others scored more than 6. One scored more than 9. This shows that the designs considered are of good grade and can be treated as a set containing optimal and near-optimal designs. The dodecahedron with a score of 9.11 identified itself as the right choice for the display. This choice set the number of items to be showcased as 5. They are selected in the following way.

1.6.9.2 Items for the display

Items to be considered for exhibiting are proposed by reviewing the history. They are: (i) gear wheels, (ii) steam locomotive, (iii) machine tools, (iv) concord plane, (v) Geneva mechanism, (vi) wind mill, (vii) Isaac Newton, (viii) Albert Einstein, (ix) Robert Hooke, (x) model T car and (xi) composite materials. They were evaluated according to the following criteria using a decision matrix.

1. Easy to machine: 0.20
2. Geometric features: 0.15
3. Importance to Mechanical Engineering: 0.08
4. Beneficial to Mechanical Engineering: 0.15
5. Exciting to students: 0.20
6. Contains sufficient information: 0.07
7. Impact on society and mechanical eng: 0.10
8. Popular among the public: 0.05

Five items chosen for displaying from the decision matrix are

1. Isaac Newton
2. Albert Einstein
3. Composite materials
4. Concorde plane
5. Model T car.

1.6.10 Embodiment of the display

The embodiment design for the rotating display was divided into two halves: (i) Display (the product design part) and (ii) Rotating unit (the engineering design part) as shown in Figure 1.7. In this section, the embodiment of the display part is described. It followed the steps by Pahl and Beitz [5]. To proceed, a good understanding of the geometry of the dodecahedron is needed.

1.6.10.1 Geometry of the dodecahedron

The dodecahedron can be built by adding the roof shown in Figure 1.8b on all six faces of the cube shown in Figure 1.8a. The completed dodecahedron is shown in Figure 1.8c.

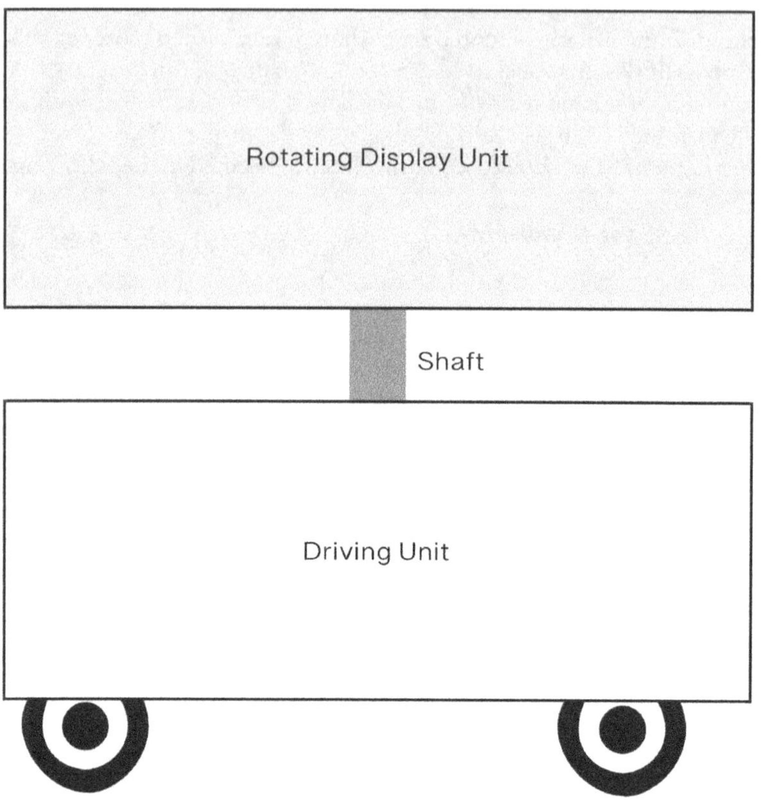

Figure 1.7 Generic structure of the rotating display.

Consider the pentagon APQDV to understand the geometry.

Assuming the side is of length '*a*'. The following can be deduced.

Angle $\angle AVD = \frac{5 \times 180 - 360}{5} = \frac{540}{5} = 108^{c}$ — Angle of a regular pentagon.

From this $\angle VAD = \angle ADV = \frac{180 - 108}{2} = 36^{\circ}$ — Angle of an isosceles triangle.

Therefore, the side of the cube $AD = 2 \times a \cos 36$.

To build the dodecahedron the angle between two adjacent pentagons is needed. The angle between the pentagon APQDV and pentagon BPQCR is one such angle.

1.6.10.2 Calculation of the angle between adjacent pentagons

Consider the triangle QDL in the pentagon APQDV in Figure 1.9.

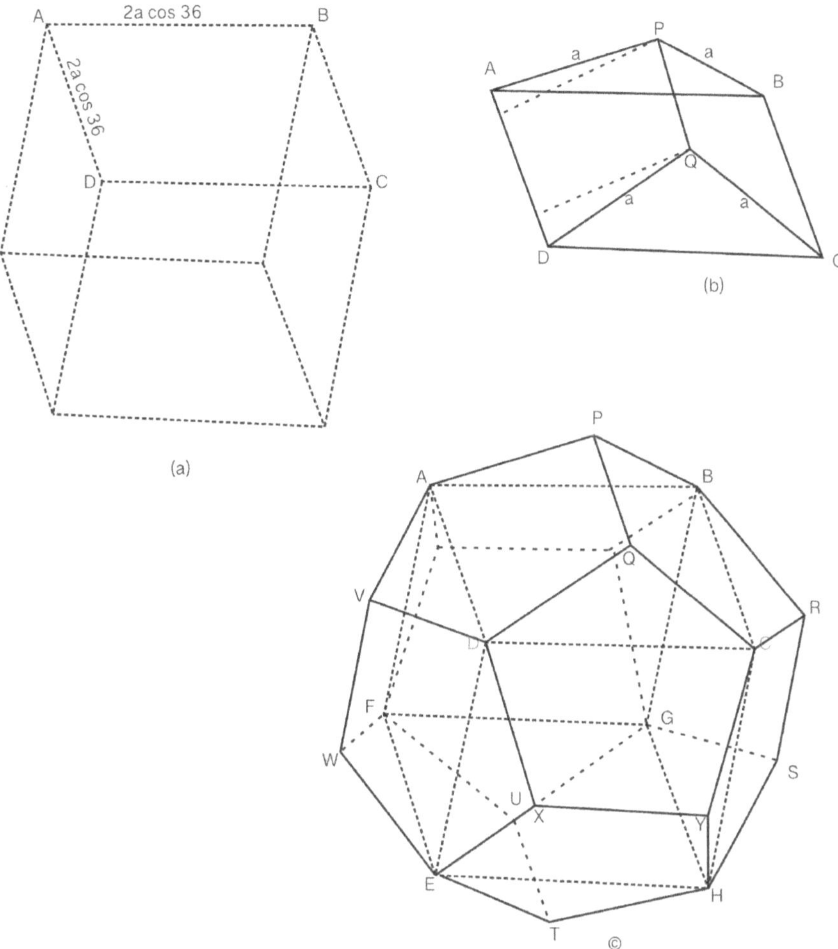

Figure 1.8 Forming the geometry of the dodecahedron.

Angle $\angle LDQ = \angle VDQ - \angle ADV = 108 - 36 = 72°$.

Therefore $QL = a \sin 72$.

Consider the triangle DQC in the pentagon DQCYX.

The side DC is $= 2 \times a \cos 36$.

Therefore, the side LN in the triangle QLN is $= a \cos 36$.

Now consider the triangle LQM as shown in Figure 1.9b. Angle $\angle LQM = 2\theta$ is the angle between the adjacent pentagons. $\theta = \sin^{-1} \dfrac{a \cos 36}{a \sin 72} = 58.28°$.

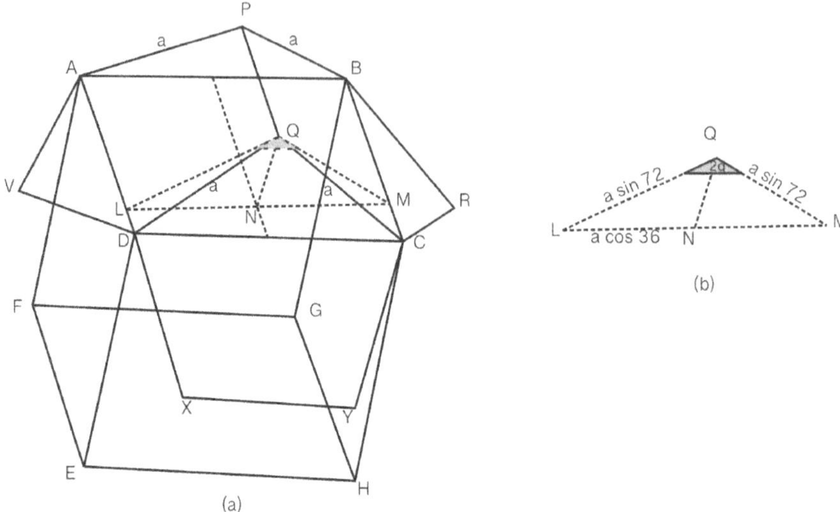

Figure 1.9 Figuring angle between adjacent pentagonal faces.

1.6.10.3 Height of the dodecahedron

A dodecahedron sitting on a horizontal surface has vertices lying in four horizontal planes as shown in Figure 1.10, which cut the solid into three parts. Each of the three parts is equal in volume. Each is one-third of the total.

Considering the heights, the height between the top two planes and the bottom two planes are equal and $= a\cos 26.56$.

The height between the middle two planes is $= a \sin 36 \cos 26.56$.

Therefore, the height of the dodecahedron $= a(2 + \sin 36)\cos 26.56$.

1.6.11 Embodiment design of the system

Embodiment determining requirements are

i. The height should be around 1,500 mm
ii. The maximum dimension of the individual pentagon is around 500 mm to fit in the machine table for machining.

If a is the length of an edge of a pentagon maximum dimension is $= 2 \times a \times \cos 36 = 1.618a$.

Permissible value for $a = \dfrac{500}{1.618} = 309$ mm.

Therefore, it is safe to assume that the side of the pentagon is 300 mm.

If $a = 300$ mm, height of the dodecahedron $= 2 \times 300 \times \cos 26.56 + 300 \times \sin 36 \times \cos 26.56 = (600 + 176.33)\cos 26.56 = 694.4$ mm.

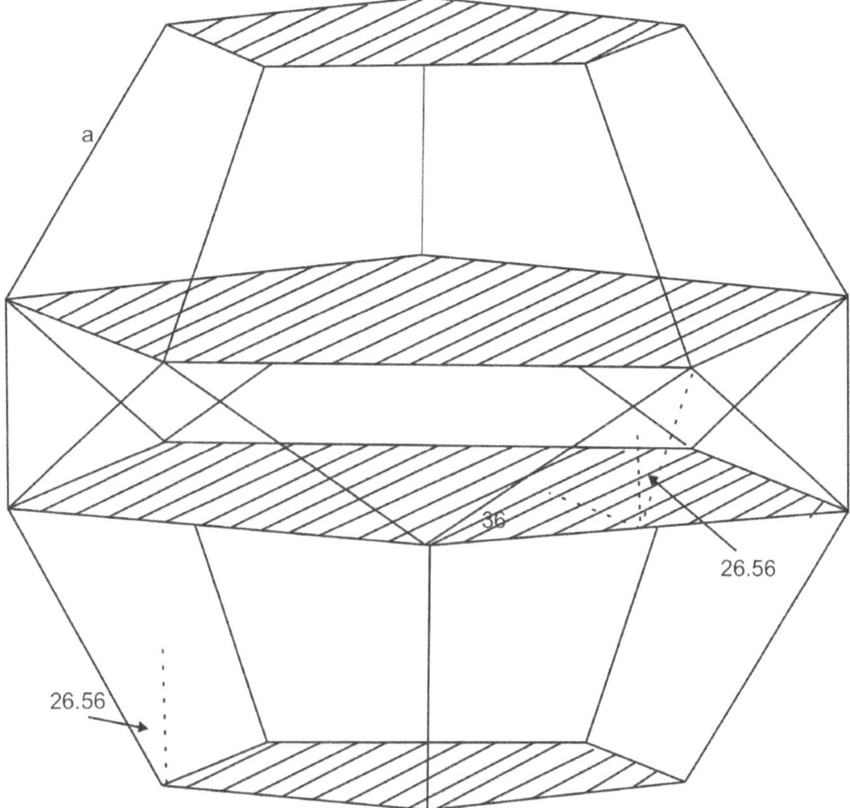

Figure 1.10 Four parallel planes in the dodecahedron.

This means the base can be 550 mm tall, and the shaft between them can be 200 mm.

The dodecahedron is made by joining pentagons, which are made up of timber of 40 mm thick. The angle between two adjacent pentagons is 116.56°. Therefore, the edges have to be chamfered to 58.28°. The manufacturing of it is from the pentagonal piece of timber, as shown in Figure 1.11.

Having designed the pentagons and their interfaces, the dodecahedron was divided into top half and bottom half. As details, wooden plugs were introduced to hold the pentagons at 116.56° when joined. A bracket (called the spider) to sit the assembly on the shaft also was developed and fitted to both top and bottom halves of the dodecahedron. The bottom half is shown in Figure 1.12.

The five items, Isaac Newton, Albert Einstein, Composite Materials, Concorde Plane and Model T car, were modelled in CAD and transported

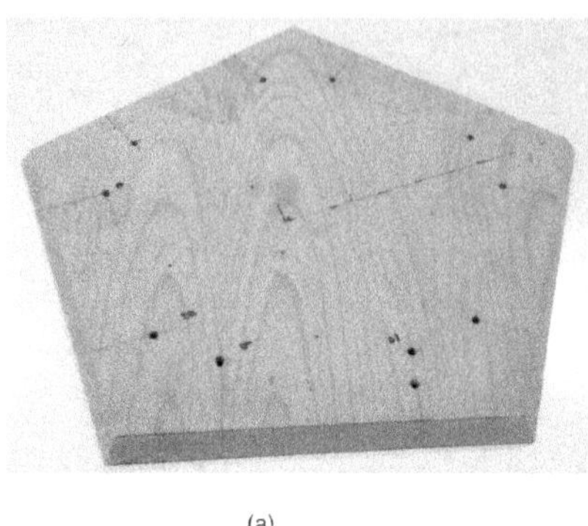

(a)

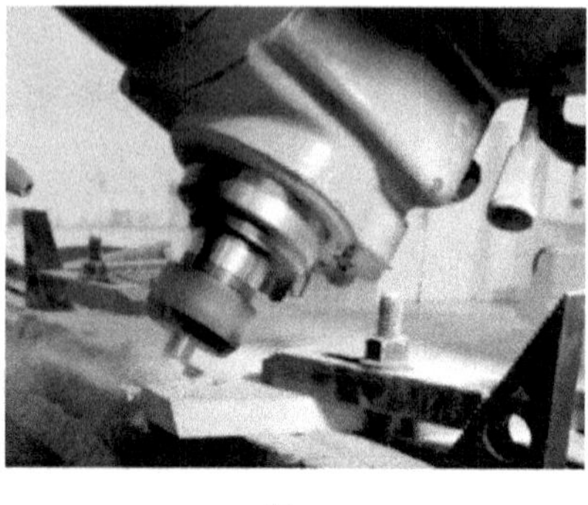

(b)

Figure 1.11 Pentagonal panel and manufacture.

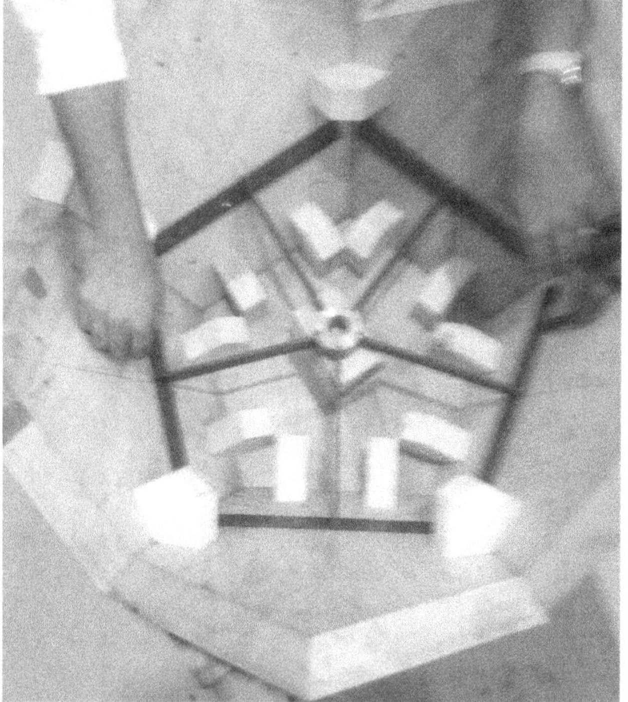

Figure 1.12 The bottom half of the dodecahedron.

to MasterCAM manufacturing software in a neutral data format. The NC code was developed and processed for a Heidenhain controller fitted on an Arrow 2 Cincinnati vertical machining centre. The final product is shown in Figure 1.13.

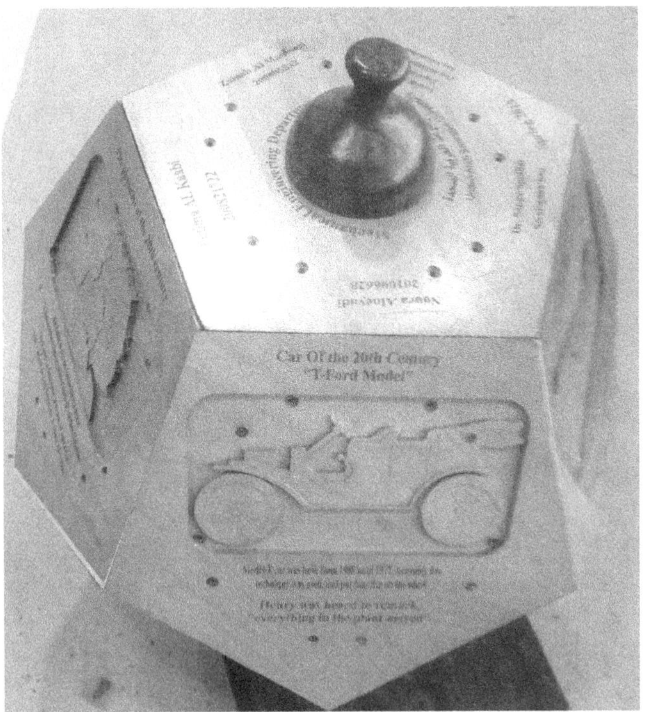

Figure 1.13 Final dodecahedron display.

1.6.12 Conceptual design of the drive train

The base part is required to house a power mechanism to turn the dodecahedron display at slow speeds at the press of a button, and it should demonstrate the use of several machine elements learnt in the machine design 1 and machine design 2 courses. The output should be a vertical shaft that could carry and turn the dodecahedron. Gears for parallel shafts (spur and helical), gears for shafts at right angles (bevel and, worm and wheel), flat and V belt drives, chain drives, shafts, key ways and other similar components can be considered as elements of the Machine Design courses. Similarly, several kinds of motors and control systems could be considered for inclusion. To keep the project complexity under control, a simple induction AC motor running at 1,440 rpm using standard 230 V supply is chosen as the prime mover. It was decided to use machine elements to produce the required height and speed reduction. Three concepts shown in Figure 1.14 are proposed.

Concept (i) shows a normal AC motor directly coupled to a high reduction worm and wheel gear set. It is a simple system. In concept (ii), between the worm gear box and the motor, a reduction belt is introduced. This helps to reduce the speed by about half. Concept (iii) has a further addition of a

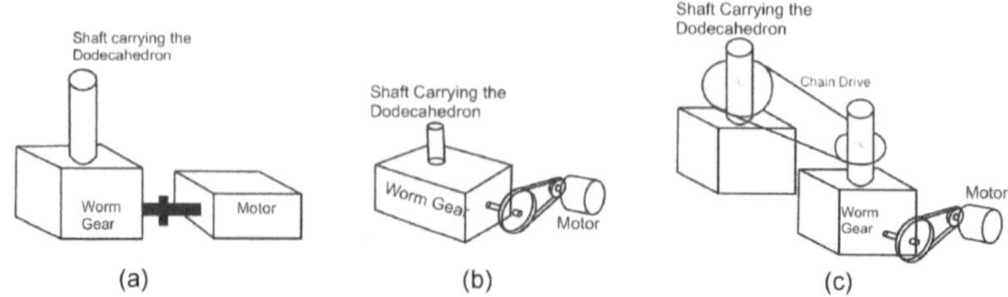

Figure 1.14 Conceptual designs for the drive.

chain drive between the output shaft of the worm gear and the dodecahedron shaft. It can easily achieve very high reduction. After evaluating the designs, concept (iii) was chosen for further development.

1.6.13 Embodiment design for the base

The embodiment determining requirement for the base is stability which means the base should be bigger than the projected area of the dodecahedron. In such a base the motor, gearbox, the shaft, bearings and other components should be positioned with due consideration for safety. The motor, which is the prime mover, a 'V' belt drive to transmit low power at relatively high speed, a worm and wheel gearbox to have a high speed reduction, and change in the shaft direction by 90°, and a chain drive to drive the shaft carrying the dodecahedron are spaced in the base while ensuring clear visibility. The proposed embodiment is shown in Figure 1.15.

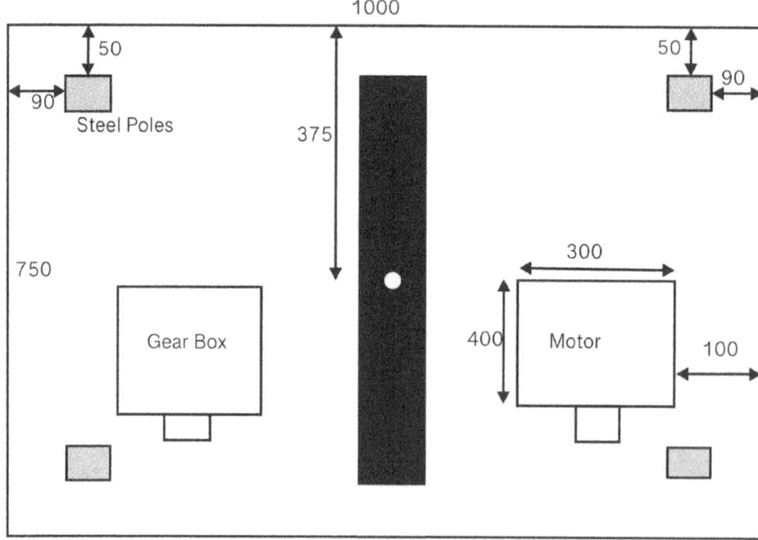

Figure 1.15 Layout for the driving unit and base (plan view).

General approach to 'Engineering Design' is aimed at ensuring that 'The stresses that are imposed on a component are less than the permissible stresses of the component's material'. The following is an excerpt from Mott [8]

> All design approaches must define the relationship between the applied stresses on a component and the strength of the material from which it is to be made, considering the conditions of service. The strength basis for design can be yield strength in tension, compression, or shear; ultimate strength in tension, compression, or shear; endurance strength; or some combination of these. The goal of the design process is to achieve a suitable design factor. N, (sometimes called a factor of safety) that ensures the component is safe. That is, the strength of the material must be greater than the applied stresses.
>
> The sequence of design analysis will be different depending on what has already been specified and what is left to be determined. For example,
>
> **Geometry of the component and the loading are known:** We apply the desired design factor, N to the actual expected stress to determine the required strength of the material. Then a suitable material can be specified.
>
> **Loading is known and the material for the component has been specified:** We compute a design stress by applying the desired design factor, N to the appropriate strength of the material. This is the maximum allowable stress to which any part of the component can be exposed. We can then complete the stress analysis to determine what shape and size of the component will ensure that stresses are safe.
>
> **Loading is known, and the material and the complete geometry of the component have been specified:** We compute both the expected maximum applied stress and the design stress. By comparing these stresses, we can determine the resulting design factor, N, for the proposed design and judge its acceptability. A redesign may be called for if the design factor is either too low (unsafe) or too high (over designed).

This project was executed in a place where structured and regulated supply of raw materials and components with all necessary scientific data and required specifications were difficult if not impossible to find. Therefore, the components were made or purchased with guestimates and the assembled system and components were proven as adequate. This follows the last method as described above.

1.6.14 Detail design

Detail design that incorporates all the components were modelled using the CATIA modelling software. From the created model, a production drawing set was created and used in the manufacturing of the product. The drawing set is not given here, to save space. But few pictures of the sub-assemblies made from the fabricated components and assemblies are shown.

Figure 1.16 shows the base which is the fabrication of the embodiment shown in Figure 1.15. The bearing support structure is the interface between the base and the dodecahedron. The shaft connecting them is shown in Figure 1.17. Section P in the shaft locates and surface A carries the top-spider of the dodecahedron, while section Q locates and surface B carries the bottom-spider. The two spiders are connected to the dodecahedron and the entire load is taken up by the shaft and transferred to the base by the load bearing. Section R is the part that is visible between the base and the dodecahedron. Section S has the step needed to accommodate the load bearing. The shaft ring is attached to this shaft and the shaft transfers the load to the fixed ring in the bearing support structure in the base. Section T has a uniform cross section and accommodates the sprocket wheel and the guiding bearing.

The dimensions and details of the bought-out components are as follows:

1. The motor – 14,440 rpm with 1 kw power drawn from the national grid with 230 V and 50 Hz
2. The motor pulley – Single grove with pitch circle diameter 85 mm
3. The driven pulley – Single grove with pitch circle diameter 150 mm
4. Belt connecting the pulleys — 450 mm between the centre
5. Worm and wheel gearbox – 1 kW power with a reduction of 100–1.
6. Chain drive — driving pulley — number of teeth 18 teeth

Figure 1.16 Embodiment of the base.

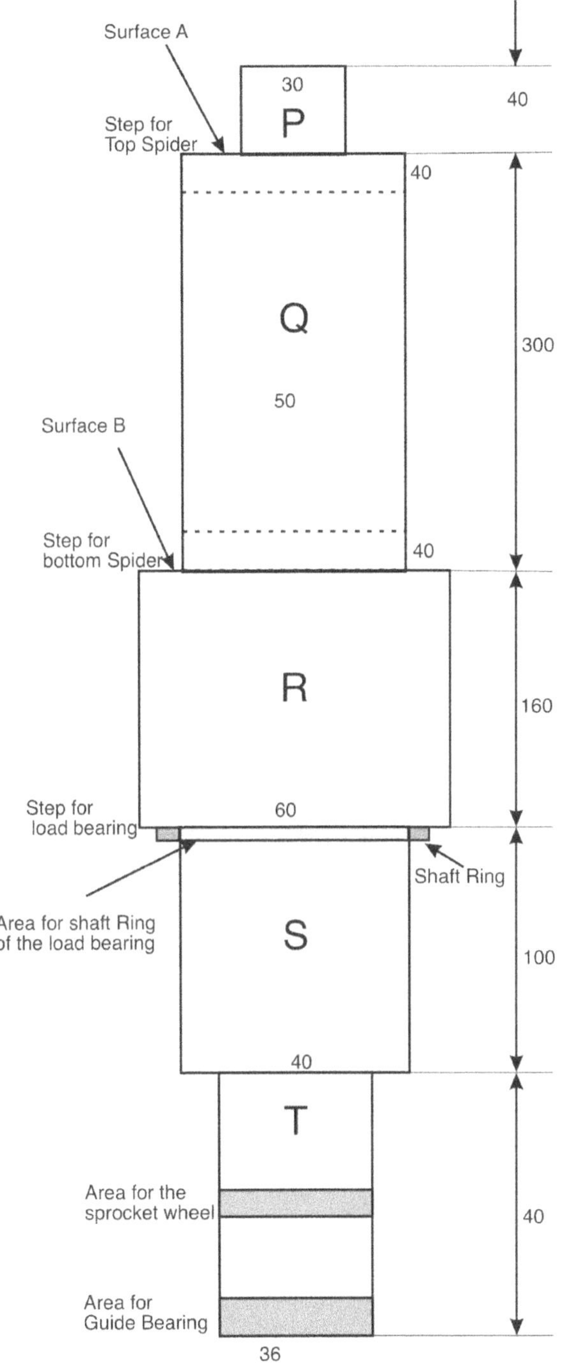

Figure 1.17 Shaft carrying the dodecahedron.

7. Chain drive – driven pulley – number of teeth 39 teeth
8. Chain – centre to centre 280 mm
9. Load bearing – 42 mm load ring, 40 mm shaft ring, 65 mm outer diameter.
10. Guide bearing – 36 mm internal diameter
11. Shaft – Length 640 mm and original diameter 65 mm
12. Bearing Support structure plate – 150 mm wide and 10 mm thick mild steel bar

The completely assembled rotating display boasting the glory of mechanical engineering is shown in Figure 1.18.

1.6.15 Detailed design or the engineering calculations

In the student project situation, the students have to identify available and matching motor, pulleys, gearboxes, chains, sprockets and shaft. The hole diameters, shaft diameters and keyways should match so that they can be assembled as a system. They have found them and incorporated them to form the embodiment design. Thus, the solution method chosen has two

Figure 1.18 Rotating display boasting the glory of mechanical engineering.

parts: (i) a harmoniously matching power train from the motor to the shaft carrying the display using different machine elements and (ii) engineering analysis to prove that the design is safe. This design is of the last type as identified by Mott [8] and explained in Section 1.6.13.

The following calculations have to be performed for the design to be proven to have, sufficient design factor and meet the specifications.

a. Final rotational speed is required to be between 3 and 6 rpm.

Input speed to the worm and wheel = moto speed $\times \dfrac{\text{Diameter of the motor wheel}}{\text{Diameter of the driven wheel}}$.

$= 1440 \times \dfrac{85}{150}$.

Output speed of the worm and wheel $= \dfrac{\text{Input speed}}{\text{Reduction ratio}}$.

$= 1440 \times \dfrac{85}{150} \times \dfrac{1}{100}$.

This is the input speed of the chain drive.

Output speed of the chain drive = speed of the dodecahedron display.

= Input speed × Reduction ratio.

$= 1440 \times \dfrac{85}{150} \times \dfrac{1}{100} \times \dfrac{18}{39} = 3.766$ RPM.

b. The motor takes about 2 seconds to reach the full speed. This can be the time for the display to reach the full speed.

Rotational kinetic energy = ½ moment of inertia * (angular speed)^2.

Moment of Inertia from CATIA = 3.702 kg/m^2.

$$\text{Kineticenergyafter2seconds} = \frac{1}{2} \times 3.702 \times \left(\frac{3.766 \times 2\pi}{60}\right)^2 = 0.2879\ \text{J}$$

$$\text{Energy gain per second} = \frac{0.2879}{2} = 0.14395\ \text{W}$$

As expected, this is small, since the rotational speed is small.

The power of the motor is sufficiently high.

c. The bending strength of the beam

Mass of the display and shaft assembly from CATIA = 43.035 + 8.049 = 51.084 kg

The load acting therefore = 51.084 × 9.8 = 508.5 N (Figure 1.19) From the bending moment diagram maximum bending moment is at the centre.

Maximum bending moment $= \dfrac{1}{2} \times 508.5 \times 375 = 95343.75$ N mm = 95.344 Nm

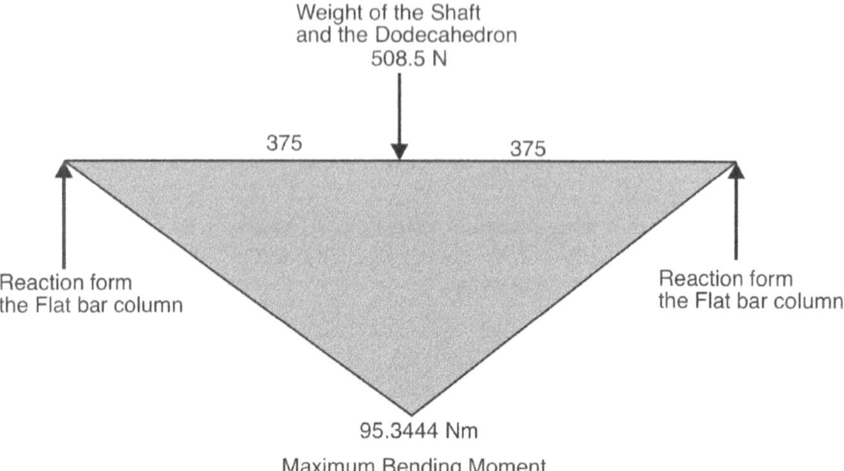

Figure 1.19 Bending moment diagram.

Second moment of area at centre (considering the hole) $= \frac{1}{12}\left((150-65)\times 10^3\right) = 7083\ \text{mm}^4$.

Maximum bending stress therefore $= \frac{95344}{7083} \times 5 = 67.3\text{MPa}$

Assuming the material to be AISI steel 1,040 having a yield strength of 415 MPa, the safety factor available is $\frac{415}{67.3} = 6.2$, which is more than adequate.

d. Adequacy of the shaft

Shaft should be adequate to transmit the torque.

The torque on the shaft is $= \frac{\text{power}}{\text{angular speed}} = \frac{0.14395 \times 60}{3.76 \times 2\pi} \times 10^3$

$= 365.6$ Nmm

This is a very small torque.

1.6.16 Summary of the example

This example demonstrates how the systematic process is used in simple, product design and engineering design problems. As the problem gets bigger more and more activities and design methods can be incorporated at different stages in the design model. One such method could be Finite Element Analysis (FEA) in detailed design. The design process of the display started from the design brief and followed as a single entity through the recording of verbatim, translation into prioritized needs, need-metric matrix development and establishment of target specifications. Then the display, and base and drive, were handled as separate units after that. For the design of the display, morphological charts were developed and, decision matrices were used for the selection of the geometry and the items for displaying. The embodiment of the display was determined using proper geometric calculations using the properties of the dodecahedron. The details for portraying or placing in the display were chosen using a decision matrix. CAD and CAM were used to engrave them using the CNC machine. The conceptual design of the base and drive system was chosen from alternatives using a decision matrix and the layout or embodiment was designed following the procedure described by Pahl and Beitz [5]. Engineering calculations were carried out to ensure that the product would operate safe and sound at the required speed with no stress hot spots.

1.7 STRUCTURE OF THE BOOK

The book has a collection of examples to give experience at various stages and the use of different design methods. It consists of model questions and model answers in the activities needed at different stages of the design process. Chapter 2 describes design process planning which essentially is determining the activities that are needed to be carried out in the conversion of the design brief into the definition of the product, in the form of a set of production drawings. These activities are placed on a Gantt chart to formulate a time plan for the project. Two examples are given for the reader to gain hands-on experience in making design process plans. Chapter 3 explains the design brief and provides examples in the development of design briefs for different kinds of products. Chapter 4 outlines the process of obtaining stakeholder requirements and contains example questions and answers in the entire process of establishing prioritized stakeholder requirements. The first major activity by the design team is 'Function Modelling'. Chapter 5 explains 'Function Tree Modelling', and the important 'Function Structure Modelling' needed for Mechatronic designs. It contains several examples giving hands-on experience in both approaches. Chapter 6 explains how measurements of design properties, at different levels are carried out to facilitate the design process. It introduces need-metric matrix as a tool for defining the needed metrics for measuring the design properties to meet the stakeholder requirements. Need-metric matrices form the basis for drafting the target specifications. Examples are given to provide experience in forming need-metric matrices. Chapter 7 explains target specifications and their derivation. Thus, Chapters 2–7 cover the important sequential steps in the definition of the problem. This forms Part 1 of this book.

Conceptual design is an important stage in the design process. Several design methods have been developed to assist conceptual design process and there is a lot of research work carried out in this area. Chapter 8 explains various design methods used for generating concepts. It also contains example questions and answers, for conceptual design problems. These questions and answers give hands-on experience to the reader. Chapter 9 describes concept evaluation and selection on a multi-factor basis. A decision matrix is constructed using a set of criteria and their importance weights. The proposed concepts are evaluated based on the criteria and the weighted sums of the designs are computed. If the weighted sums are high, that means that the concepts considered are of high grade. Then the concept with the highest score is chosen. In another method developed by Stuart Pugh [9], the developed concepts are progressively evaluated and improved until a preferred design is found. Chapter 10 covers embodiment design. Thus, Chapters 8–10 cover the proposing and identifying a preferred design solution and developing it further to produce the production drawings of a physically realizable product. This forms Part 2 of the book.

Chapter 11 which makes Part 3 of the book describes full case studies from the beginning to the end. It includes the detail design as drawings developed using CATIA models and engineering analyses wherever needed.

REFERENCES

1. Ulrich K.T. and Eppinger S.D., *Product Design and Development*, 3rd edition, Tata McGraw Hill Publications, New Delhi, 2004.
2. Feilden B.R., *Engineering Design: Report of Royal Commission*, 2nd edition, HMSO, London, 1963.
3. *Goals and Priorities for Research in Engineering Design*, A Report to the Design Research Community, American Society of Mechanical Engineers, New York, 1986. https://www.abet.org
4. Haik Y., Sivaloganathan S. and Shahin T.M., *Engineering Design Process*, 3rd edition, Cengage, Noida, 2017.
5. Pahl G. and Beitz W. *Engineering Design*, The Design Council, London, 1984.
6. Ullman D.G., *The Mechanical Design Process*, 4th edition, McGraw Hill Higher Education, New York, 2010.
7. French M.J., *Conceptual Design for Engineers*, Springer, New York, 1990.
8. Mott R.L., *Machine Elements in Mechanical Design*, 4th edition, Pearson Prentice Hall, Upper Saddle River, NJ, 1992.
9. Pugh S., *Creating Innovative Products Using Total Design*, edited by Don Clausing and Ron Andrade, Addison Wesley Publishing Company, Reading, MA, 1996.

Part 1

Defining the problem

Chapter 2

Design process planning

2.1 INTRODUCTION

Design Process Planning is the strategy formulation process, for the conversion of, an abstract description or brief of `what the product is', into the complete definition of a physically realizable product. Systematic design process is used as the basis for this strategy. Haik's [1] `design stage model' with the five stages: requirements, product concept, solution concept, embodiment design and detail design, is adopted for systematic design in this book. The strategy is to introduce `design methods', various tools and techniques, for use at different stages of the design process. The genesis of the design process is the Design Brief from the senior management. The design brief is the trigger for the design process to spring into action. *Design Process Planning consists of establishing the following four constituents*:

a. Activities to be carried out, using design methods or otherwise, at each design stage
b. The deliverables at the end of each stage so that a stage-gate method can be used to manage the design process
c. Project specific knowledge components that have to be collected through literature survey. (It is assumed that the designer has knowledge in the design process and the various design methods)
d. A Time Plan for the activities - A Gantt Chart

This chapter, in sections 2.2 to 2.5 describe these constituents and, explains the four-column schema or framework developed for `design process planning'.

2.2 ESTABLISHING THE ACTIVITIES TO BE CARRIED OUT AT EACH DESIGN STAGE

Establishing the activities needed to complete at each of the design stage in the design model is a suitable technique to follow here. Consider the stages

DOI: 10.1201/9781003484950-3

of the design model to identify the activities that have to be performed in the following way:

1. *Requirements stage*: The first task for the design team is to establish the requirements of the product to meet a societal need. Requirements are elements of the societal need, and they can be several and belong to several types. The possible types of needed requirements can be identified by using the checklist method (details can be seen in Chapter 4). Several methods including questionnaire, one-to-one interview, focus group method and observing a similar product in use by a user, are recommended. The suitable method should be chosen for the project, to establish the requirements. The requirements are then prioritized by giving each requirement a weight from 1 to 10. Thus, the activities to be carried out, for example, can be (i) identifying types of suitable or needed requirements (ii) recording customer verbatim and (iii) establishing prioritized needs.
2. *Product concept stage*: Using the requirements and the design brief, design specifications have to be drawn. In general, a product is designed to perform certain functions, and the design team should try and visualize the functional representation of the product (details can be seen in Chapter 5). The functional representation and the specifications jointly establish the product concept and (i) establishing the functional representation and (ii) establishing the specifications can be example activities in the product concept stage. The activities performed in the first two stages meet the challenge of 'Defining the problem well'.
3. *Solution concept stage*: Solution concept stage proposes various schemes called conceptual designs to get the product to perform the required functions. Conceptual design is made up by integrating function providers, to behave harmoniously and to provide the functions, the product is intended to perform. If all conceptual designs are considered as a whole, they form the 'Solution Space'. It is not possible to construct the entire solution space, and this makes the second challenge, where the design team has to choose a solution concept from this unknown solution space, at the early stage in the design process. The strategy adopted is to construct a subset of the solution space that contains high-grade designs and choose one of them for further development. Several design methods are available for the designer to develop conceptual designs (details can be seen in Chapter 8). In the evaluation and selection process, the individual conceptual designs in the generated sample space are evaluated to establish their grade, and one high-grade design is selected. Two approaches, the 'decision matrix' approach [2] and the 'controlled convergence' approach by

Pugh [3] are employed for this purpose (details can be seen in Chapter 9). These activities of generating a sample conceptual solution space and choosing one from the samples generated, help to meet the second challenge. Typical activities can be (i) choosing a design method and generating sample designs, (ii) improving the designs, (iii) evaluating the designs and (iv) choosing a design.

4. *Embodiment design stage*: Embodiment design is the process of giving physical form and shape for the concept chosen in the above solution concept stage. The method developed by Pahl and Beitz [4] can be used for developing the embodiment design. For product design, few geometric calculations only would be needed to decide the geometric form and shape. This would determine the final design. But for engineering product design, many engineering analyses may be required to determine or prove the dimensions. The required analyses are dependent on the project and the design approach. Following the guidelines to produce the (i) preliminary layout and (ii) definitive layout can be typical activities in this stage.
5. *Detail design stage*: Detail design for 'Product Design' is relatively easy as all required geometric data would have been determined at the embodiment design stage. All what is required is to create a CAD model of the product and its components and generate a set of 'Production Drawings'. However, for 'Engineering Product Design', required engineering analyses should be carried out to ensure that the product and components have engineering properties within acceptable range. Only after ensuring that the properties are within acceptable range, the CAD model can be confirmed and 'Production Drawings' can be produced.

2.3 DELIVERABLES AT THE END OF EACH STAGE FOR MANAGING WITH STAGE-GATE METHOD

The concept of stage-gate system, developed to manage processes, divides them into a number of stages or work stations. Between each work station or stage, there is a quality control checkpoint or gate [5]. The gate, after the activities at each stage are completed, evaluates the results obtained in that stage. A decision, about allowing the project to proceed to the next stage or to stop the project, is made based on the results of the evaluation.

Following the stage-gate approach, a set of deliverables or intermediate outputs can be established for each stage in the design process, and a gate can be established after each stage to evaluate the interim outputs as shown in Figure 2.1. The design brief acts as the initiator for the design process. Activities relating to stage 1, the establishment of prioritized requirements

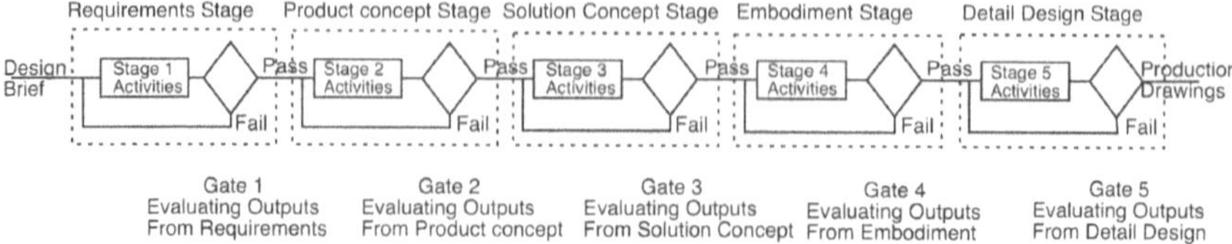

Figure 2.1 Stage-gate system for managing the design process.

and the related details, are carried out and the outputs are fed to the gate. They are evaluated at the gate. If they pass, they are taken for stage 2, the product concept stage, where activities relating to stage 2 are carried out. The outputs are taken to the gate in stage 2. They are evaluated and if they pass, they are passed to stage 3, the solution concept stage, as inputs. In this way, the process passes through the five stages, and the set of production drawings is produced at the end of the fifth stage. If at any stage the outputs fail in evaluation, the process refers them back to the activities block in the same stage for further work and additional resources are allocated if needed.

2.4 REQUIRED KNOWLEDGE COMPONENTS (LITERATURE SURVEY)

The product to be designed may require two kinds of knowledge (i) engineering theory and applications and (ii) available off the shelf materials, components and their properties. Engineering theory like, complex stresses in beams and shafts, fluid flow through pipes, and thermo-fluid combinations can be found in books. This forms part of engineering curriculum, and the main task here is to identify the required theory. Applications of the theory can be found in research papers and case studies. Another important knowledge is the use of specialist software packages like the ones for finite element analysis or mechanism analysis. Deciding the engineering theory, applications and software tools needed, and acquiring them are important parts of the design process. They should be considered during planning.

Materials like sheet metal and extrusions come in certain dimensions only. Choosing these dimensions will reduce the manufacturing effort and cost. In the same way, there are several components such as the chains and sprockets, belts and pulleys, and ball and roller bearings, that are manufactured by specialist manufacturers, at specific dimensions only. It is important that the design team should have a knowledge of these items that have to be purchased for incorporation in the design of the product.

2.5 PLANNING THE THREE COMPONENTS

The three components, planned activities, deliverables and needed knowledge components, described in the preceding sections can be planned by filling the dotted lines in Figure 2.2 so that the plan can be seen in a single chart.

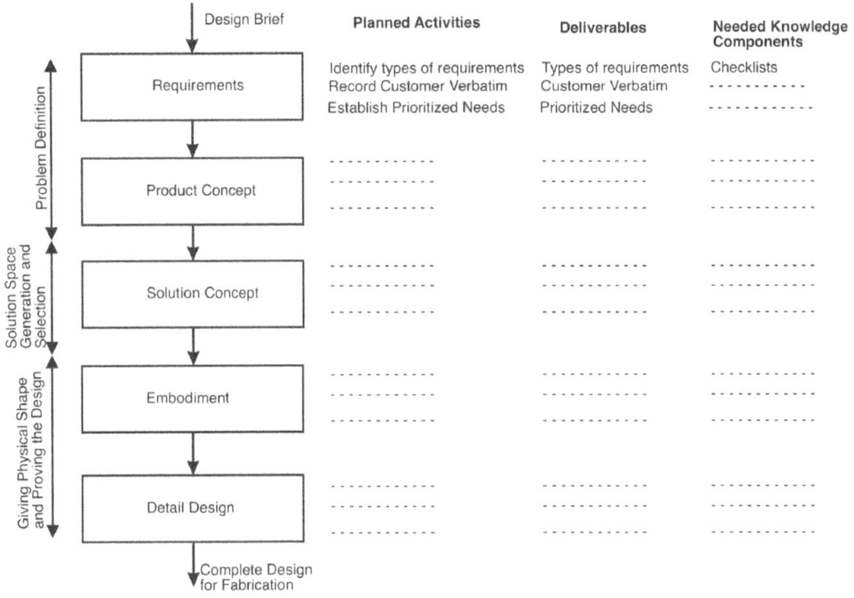

Figure 2.2 Schema of the design process planning.

2.6 TIME PLAN – GANTT CHART

A Gantt chart is a graphical representation of activity against time that assists the creation of a time plan for the design. The activities planned as shown in Figure 2.2 and the time allocated or planned for each of them are detailed in the Gantt chart. In addition to that, the set of interim outputs expected to be produced at the end of each stage is planned on the time scale in the Gantt chart. These sets of outputs enable the management of the design process according to the stage-gate process explained in Section 2.2. Another important requirement that could be incorporated in the time plan is the acquiring of the knowledge components required during the design process. A typical time plan is shown in Table 2.1.

In conclusion, design process planning produces two outputs (i) completely filled schema showing the activities, deliverables and needed knowledge components relating to each stage of the design model and (ii) a time plan showing the above fitted in a time frame. The following examples will

Table 2.1 Typical time plan

Activities	*Semester weeks*													
	1	*2*	*3*	*4*	*5*	*6*	*7*	*8*	*9*	*10*	*11*	*12*	*13*	*14*
Design brief and understand project														
Design process planning														
Determining requirement types														
- - - - - - - - - - - - - - -														
Establish prioritized requirements														
- - - - - - - - - - - - - - -														
Drawing specifications														
- - - - - - - - - - - - - - -														
Developing conceptual designs														
Evaluating the concepts														
- - - - - - - - - - - - - - -														
Embodiment design generation														
Detail design														
- - - - - - - - - - - - - - -														
Preparing production drawings														
Deliverables	Set 1						Set 2		Set 3			S4	Set 5	
Knowledge needed – problem defn.														
Knowledge needed – embodiment														
Knowledge needed – in analysis														
Knowledge of resources in market														

give the experience in producing a 'Design Process Plans' within a systematic design process.

Example 2.1: Design process plan for a single stage scissor lift

A power screw drives the scissor lift shown in Figure 2.3. The four straight legs are connected with bolts at their joints and can rotate freely. The two free bottom ends, and the two free top ends, are connected to wheels with bearings and can travel to and fro horizontally. The power screw drives these ends through the middle bar about which the legs can rotate freely. An AC motor running at 1,440 rpm taking power from the national supply, drives the screw. A two-stage gearbox is needed to reduce the speed by a ratio of 12:1. Plan the design process for this scissor lift.

The design process for this product can be described in the following way. The types of needed requirements can be of the types geometry, kinematics, forces, energy, safety, ergonomics, assembly and maintenance as given by the checklist by Pahl and Beitz [4]. To get the prioritized needs, customers have to be consulted and their verbatim should be recorded. After obtaining the requirements, the design team should visualize the required functions and produce a function tree. With the insights from the design brief, prioritized needs, and the function tree, the specifications of the product are drawn. There are several design methods for conceptual design. In this project, Gallery method is assumed as the choice. A set of evaluation criteria are developed after that, and a Pugh's matrix has to be created for evaluation. From the evaluation, suitable designs can be considered for further improvements, and the most suitable design, after the improvements, can be chosen for further development. For embodiment, the requirements

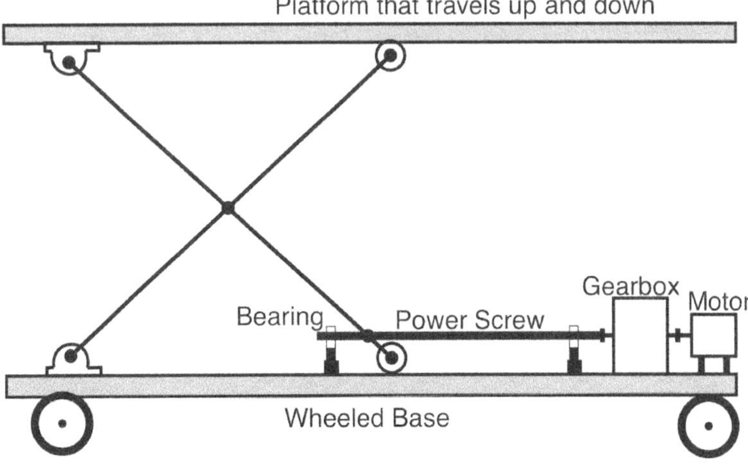

Figure 2.3 Schematic of a single stage scissor lift

should be reviewed, and 'embodiment determining requirements' should be identified in the order of necessity. In the next step, suitable function providers are identified and integrated to form the embodiment design. Since this is an engineering product dimensions have to be decided in consultation with engineering analysis results. Gear wheel dimensions have to be decided based on the power transmission, speed of the wheels, bending stresses and contact stresses. A similar list of analyses is needed for the shaft design and bearing design. The analyses and the embodiment design may be iterative until the design team proves that the lift is functional and all required engineering characteristics are within the acceptable limits. At the end of this process, all required dimensions would be ready. Once all these are done CAD models of the components have to be created by identifying the features in the individual components. A set of production drawings starting with the labelled components in an exploded view and individual drawings for each component with all necessary dimensions is produced as the output of the design process.

The planning part of the process produces two outputs (i) the plan and (ii) activity-time plan. Figure 2.4 shows the plan, and Table 2.2 shows the activity-time plan. In Figure 2.4, the first column shows the design model. The second column from the left shows the activities described above. Interim outputs are established from these activities, which can be monitored by a stage-gate process. The interim outputs are shown in the third column. Specialist subject knowledge

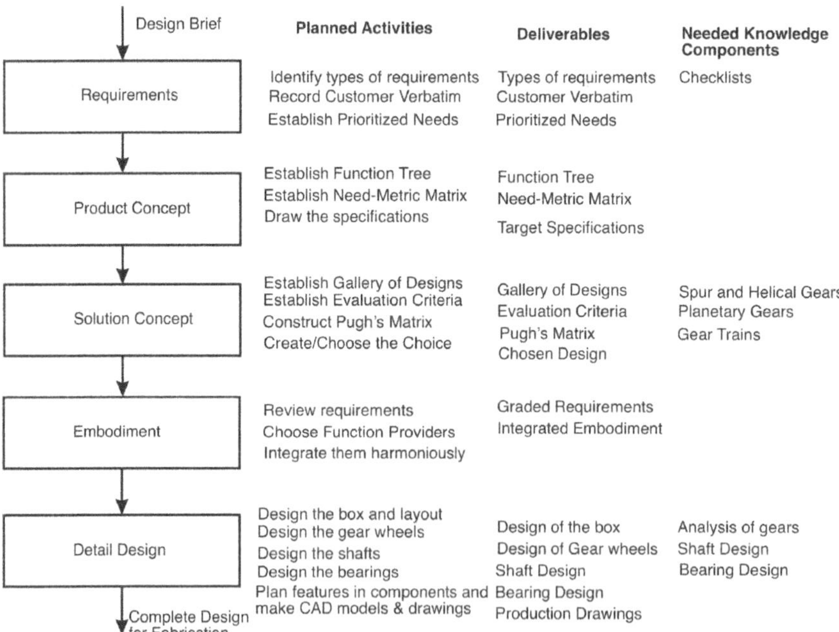

Figure 2.4 Design plan for designing a scissor lift.

Table 2.2 Activity-time plan: Gantt chart

Activity	*Semester 1: Academic year 2022–2023*													
	1	*2*	*3*	*4*	*5*	*6*	*7*	*8*	*9*	*10*	*11*	*12*	*13*	*14*
Design brief and understanding														
Needed requirement types														
Record customer verbatim														
Established prioritized needs														
Establish function tree														
Establish need-metric matrix														
Draw specifications														
Draw a gallery of designs														
Establish evaluation criteria														
Construct pugh's matrix														
Create/choose choice														
Review requirements														
Choose function providers														
Integrate function providers														

(*Continued*)

Table 2.2 (Continued) Activity-time plan: Gantt chart

Activity	*Semester 1: Academic year 2022–2023*													
	1	*2*	*3*	*4*	*5*	*6*	*7*	*8*	*9*	*10*	*11*	*12*	*13*	*14*
Design the box layout														
Design gear wheels														
Design shafts														
Design bearings														
Plan & construct CAD model														
Create production drawings														
Deliverables	Set 1			Set 2		Set 3				Set 4		Set 5		
Knowledge on checklist	Ready													
Knowledge on gear types	Ready													
System design knowledge	Ready													
Gear drive analysis	Ready													
Shaft and bearing design	Ready													

components may be needed at different stages. They are identified in the fourth column. A template of the Planning Schema, based on the design model, can be used for this purpose.

These activities are placed in a time plan as shown in Table 2.2. The deliverables divided into five sets are also placed in the time plan showing the time points at which they should be ready. In the same way, the time plan for the knowledge components also is shown.

Example 2.2: Design process plan for a memorabilia clock

The photograph of a table clock memorabilia, made up of a stand and clock insert, is shown in Figure 2.5. The stand reflects CAD/CAM theory and its application in curve definition and contour milling. The profile of the stand could be defined as a composite Bezier curve or a spline curve. The profile is sketched in a grided paper, and it is broken into straight line and curve segments, for onwards transfer to the computer. If the intention is to produce 1,000 pieces of similar memorabilia establish the Design Process Plan for the stand. It can be assumed that the profile design does not need any engineering analysis. The main focus can be showing the different features of the curves in the profile.

The design brief is the starting point for this design. After understanding it, the design team should plan the types of requirements that can help it to design an appealing product. They may refer the

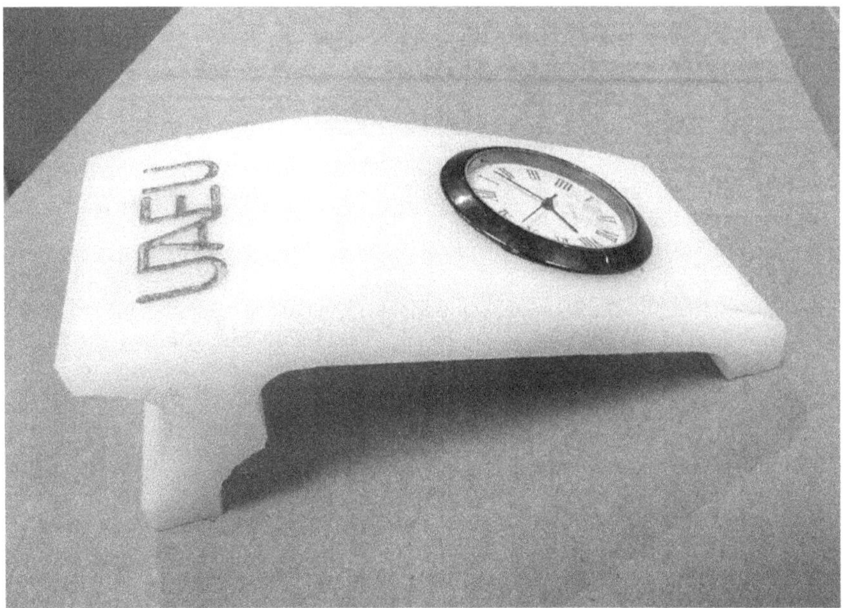

Figure 2.5 Memorabilia clock made up of a machined base and a clock insert.

checklists by Pahl and Beitz [4] or Ullman [6]. Once the types of needed requirements are identified, they can interview the potential stakeholders and record their verbatim. The verbatim is translated into needs with equal information content, and their importance ratings are determined. Thus, the activities in the requirements stage can be

a. Determine the types of needed requirements
b. Record customer verbatim
c. Translate the verbatim into needs of equal task content
d. Obtain the importance ratings of the needs

In the product concept stage, the expected main function is limited to attractive curvy profile for the stand. Therefore, the only activity is to establish the specifications relating to the needs and design brief.

Solution concept stage starts with the sketching of contour profiles and continues with the choice of one from them. The chosen sketch is broken into constituent curve segments with their starting points, finishing points and some intermediate points through which the curve segment passes through. The activities can be

a. Draw a set of selection criteria with weights
b. Make sketches of different profiles
c. Construct a decision matrix and, evaluate and choose a sketch
d. Break the sketch into segments and points

Embodiment design stage gives the geometry and form to the chosen concept. Stability and sufficient space for the clock are the embodiment determining requirements. These will decide the position of the hole for the clock. The activities therefore can be

a. Determine the dimensions for the embodiment determining requirements
b. Create the definitive layout

In the detail design stage, CAD model for the base is created. The contour is made up of several Bezier curve segments, and some of their joints are smooth and some other joints are sharp. The control points of the curves are also firmed up during this stage. They are needed during the manufacturing stage. The activities therefore can be

a. Create the CAD model for the base
b. Create the production drawings

The second column from the left in Figure 2.6 shows these activities. Interim outputs are established from these activities and they are shown in column 3. They can be used for managing the design process using the stage-gate method. Specialist subject knowledge components needed at different stages are shown in the fourth column of Figure 2.6.

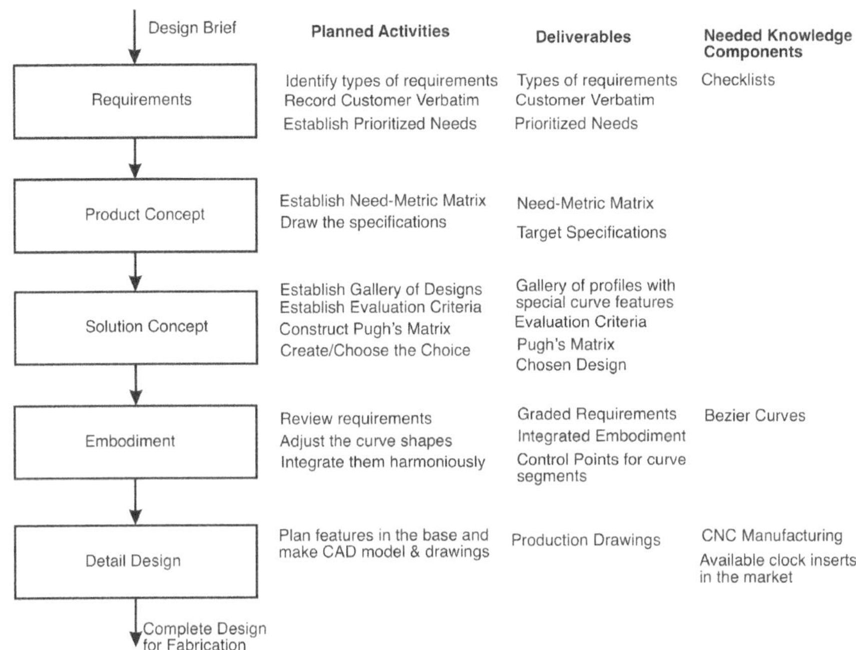

Figure 2.6 Design plan for designing the memorabilia clock.

REFERENCES

1. Haik Y., Sivaloganathan S. and Shahin T.M., *Engineering Design Process*, 3rd edition, Cengage, Noida, 2017.
2. Dixon J.R., *Design Engineering: Inventiveness, Analysis and* Decision *Making*, McGraw-Hill, New York, 1966.
3. Pugh S., *Total Design*, Addison-Wesley Publishers Ltd, Boston, MA, 1991.
4. Pahl G. and Beitz W., *Engineering Design*, *The Design Council*, London, 1984.
5. Cooper R.G., Stage-gate systems: A new tool for managing new products, *Business Horizons*, 33, 44–54, 1990
6. Ullman D.G., *The Mechanical Design Process*, 4th edition, McGraw Hill Higher Education, New York, 2010.

Chapter 3

Establishing design brief

3.1 INTRODUCTION

Accomplishments of technology has great influence in the standard of living, assisted and built by designed products. `Product Development' using the available technology is at the epitome of all engineering endeavours and technology is the aggregate of, techniques, skills, methods, systems, and manufacturing machines and processes, resulting from the application of scientific knowledge. A group of people, who collectively are equipped with the ability to increase their knowledge, technology and designed objects, are said to have a civilization. In this era of advanced civilization, product ideas are conceived by identifying societal needs or market opportunities. These opportunities are identified generally by marketing and sales divisions and, the senior management of a company analyses and studies these potential opportunities in several perspectives. They transform the opportunities into product ideas, add them to the company's portfolio and declare the company's expectations on a successful product. *The design brief is the company's description of the product idea derived from the market opportunities, and the company's expectations on a successful product.* It provides the essential information such as, what is its broad conceptual idea, what are the expectations from a successful product and who are the buyers and stakeholders. Design brief is the official document given to the design team and the richness of the resulting product depends on it. This Chapter describes design brief and its constituents and seven examples.

3.2 CONSTITUENTS OF DESIGN BRIEF

Design brief is the company's description of the product idea given by the senior management to the design team. It provides sufficient information for the design team to understand what the product is and what are the expectations of the company. The details that are included in a design brief should give sufficient information for the design team to start the design process. The design brief given and the information given can vary

DOI: 10.1201/9781003484950-4

significantly, and even their format can vary significantly. The elements the design brief should have are identified as the following in this book:

a. *Product description or what*: This is a broad description of 'What the product is' and can be helpful to identify the essential characteristics of the product.

   Example: The description of a luxury model of a motor car may read as '*A machine that would give the ultimate driving experience and noise, and packed with energy for your hands and legs*' [1]. On the other hand, a car model suitable for use as taxis may read as 'A car that has good fuel efficiency, large boot and leg space, comfortable seats and isolation of outside noise while the engine makes very little noise'. The decision to make which car the company is going to make is a decision for the company and not for the customers.
b. *Product concept or how*: This gives an outline description of how the product will look and perform the basic functions expected from it.

   Example: The product concept for a washing machine can read as follows. A power-driven machine fully contained within a volume of $1.0\,m^3$ that will perform the functions such as wetting, applying soap, agitating and dislodging dirt, rinsing off the dislodged dirt, and squeezing off the excess water, in an algorithmically controlled fashion, and suitable for permanent fixing in domestic houses in cities. Another one can be, the washing machine is compact in size, lightweight and portable with no washing programme or temperature control and requires manual squeezing. It does only mixing the soap with water, wetting the clothes, agitating and rinsing off the dislodged dirt. The decision on which kind of machine the company wants to make, is up to the company and not to the customers.
c. *Benefits to be delivered or why*: These are the differentiating characteristics that the company expects the product should have over other competing company's products. These are used as part of the specifications and criteria for concept selection.
d. *Positioning and target price (business goals)*: This is a decision for the company and its strategy. Some may be aiming at the top of the range while some may be aiming the middle of the range. Some may target the market share while some may target bigger margins.
e. *Target market or to whom to sell*: These are the paying customers for the intended product. It is from these customers the company achieves sales.
f. *Assumptions and constraints*: This section discusses the local conditions, standards to be met and any other constraints or favourable conditions.

   Note: In certain parts of the world, hydraulic lifting mechanisms are not in use and all mechanisms in use are cable-driven. In such

Table 3.1 Elements and format of a design brief

Project title	
Drafted by	*Simple Simon Company Ltd*
Product description	
Product concept	
Benefits to be delivered	
Positioning and target price	
Target market	
Assumptions and constraints	
Stakeholders	
Possible features and attributes	
Possible area for innovation	

areas, the expertise to use and maintain hydraulic mechanisms may be rare. Therefore, it may be prudent to use cable-driven mechanisms in the design. Local conditions such as these should be taken into consideration during the design phase.

g. *Stakeholders*: This is a list of paying and non-paying stakeholders. It is better to use the contacts and knowledge that the marketing and sales divisions have.
h. *Possible features and attributes*: This is an optional item that may be useful when a new member product is added to an existing family or when a new version of an existing product is under development.
i. *Possible areas for innovation*: This is also an optional item and new technologies can be considered for inclusion at this point.

A schema in the form of a table has been developed to facilitate the formulation of a design brief (Table 3.1). The left-hand side shows the element name and the right-hand side is left for filling by the user.

3.3 MODEL QUESTIONS AND ANSWERS

Example 3.1: Design brief for a step ladder with hand rails

A ladder can be described as a set of supported rungs that permits a person to climb up and reach heights to perform tasks at the elevated points and return back after completing the tasks. Ladders have several uses, and depending on the use or application and form, they have different names. A step ladder is a self-supporting portable ladder hinged in the middle to form an inverted 'V', with stays to keep the two arms of the 'V' at a fixed angle. Step ladders have flat steps and a hinged back rail. A sketch of a step ladder with hand rails is shown in Figure 3.1

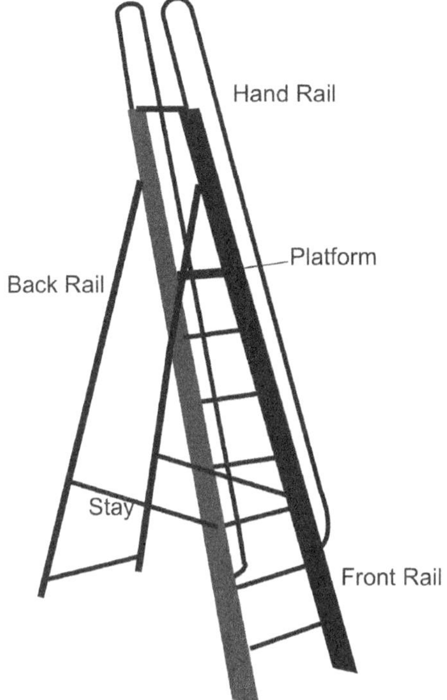

Figure 3.1 Step ladder with hand rails.

to give an idea of the product. Write a design brief with the elements outlined in Table 3.1 for a step ladder with hand rails.

Answer

The schema shown in Table 3.1 can be used to establish the design brief for the step ladder with the hand rail. Table 3.2 shows the design brief developed.

Table 3.2 Design brief for a step ladder with handrail

Step ladder with handrail	
Drafted by	*Simple Simon Company Ltd*
Product description	A lightweight step ladder that would help to access heights up to 3.5 metres with a conveniently placed handrail for easy and safe climbing, staying and working at heights and getting down.
Product concept	A collapsible step ladder with (a) front rail pair fitted with steps (b) back rail pair and stay providing stability and (c) hand rail to hold while climbing and standing on the platform.
Benefits to be delivered	a. Hand rail to hold when standing on the platform b. Broad rungs c. Rungs at close distances for easy climbing d. Bigger rungs to ensure stability (sidewise) e. Lightweight f. Compact package
Positioning & target price	Middle of the range with a price of $150–$250.
Target market	Domestic users, ladies, DIY enthusiasts.
Assumptions and constraints	Complies with ANSI ASC A14.2-2017 standard
Stakeholders	Housewives DIY users Professional builders Builders' merchants Hire shops
Possible features and attributes	Hand rail starts at the middle and extends above the platform.
Possible area for innovation	Robust and easy to setting up 'stay'. Location of the back rail hinge.

Example 3.2: Design brief for a ladder cart

A ladder cart is a combination of two basic machines, a class 2 lever and wheel and axle. This is very useful in replacing fused street lamps and other similar works where a person has to climb the ladder to perform light duties. The CAD model of a typical one is shown in Figure 3.2. Some people design and build one to suit their needs. Write a design brief with the elements as outlined in Table 3.1 for a ladder cart.

The design brief for the ladder cart using the schema given in Table 3.1 is shown in Table 3.3.

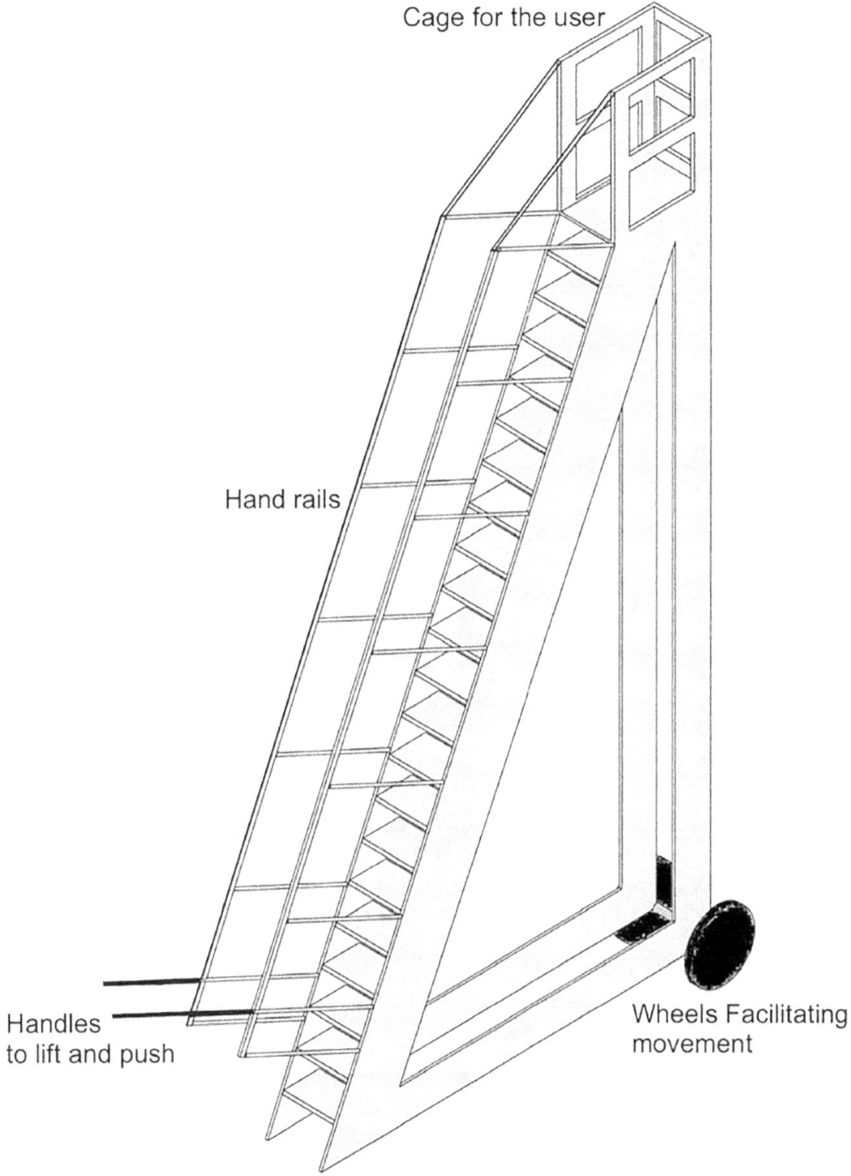

Figure 3.2 CAD model of a conceptual ladder cart.

Model Answer

Table 3.3 Design brief for a ladder cart

Ladder cart	
Drafted by	*Simple Simon Company Ltd*
Product description	A ladder fitted with wheels for easy movement and location at any required outside position for light work. A typical use is the changing of burnt street lamps.
Product concept	Combination of two basic machines (a) a class 2 lever and (b) an axle and wheel in pair arranged and fitted as a cart. One side of it is fitted with the wheels and the other side is having the handles. The ladder is located in the middle and thus brings the load in the middle.
Benefits to be delivered	a. Lightweight b. Easy manurable c. Facilitating hand rail d. Broad Rungs e. Extendable side riggers f. Easy moving around with minimum effort
Positioning & target price	Middle of the range with a price of about $500–$750.
Target market	Street light maintenance Decorating contractors Building contractors
Assumptions and constraints	
Stakeholders	a. Street light maintenance b. Exterior building painters c. Building contractors
Possible features and attributes	• Extendable side riggers
Possible area for innovation	• Resistance to abuse or wrong use.

Example 3.3: Design brief for a manually operated kitchen can crusher

Many food products are packaged in metal tins and cans that, once empty, take up valuable space in bins. Can crushers can compact the used cans, reducing the volume they occupy substantially. This will facilitate collecting materials for recycling. There are several can crushers available in the market and many of them are power-driven. The concept of a typical hand-driven can crusher is shown in Figure 3.3. Write a design brief with the elements as outlined in Table 3.1 for a manual kitchen can crusher (Table 3.4).

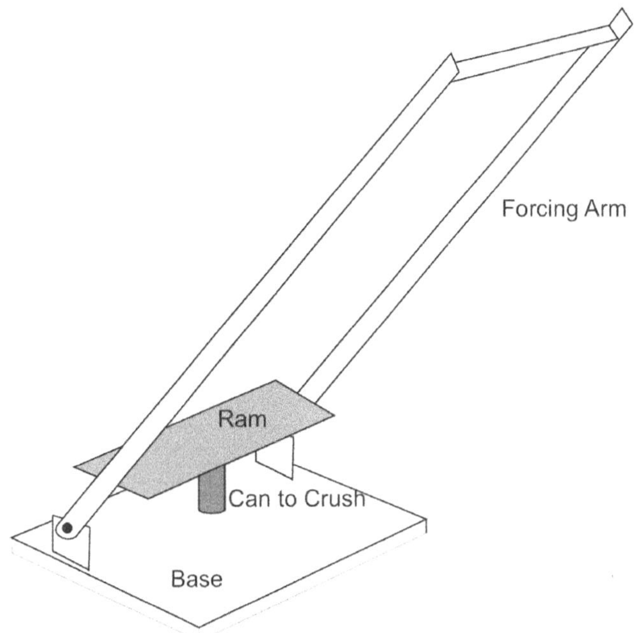

Figure 3.3 Concept of a kitchen can crusher.

Model Answer

Table 3.4 Design brief for a manual can crusher

Manual can crusher	
Drafted by	*Simple Simon Company Ltd*
Product description	A table-top, manual can crusher that can compact the used cans of up to 150 mm tall and operable by housewives.
Product concept	A second-class lever is a stick where the fulcrum is at one end of the stick, the effort is on the other end, and the load is in the middle of the stick. The load is the resistance offered by the can that is crushed. See Figure 3.3 to get an idea of the can crusher.
Benefits to be delivered	a. Crush cans up to 150 mm tall b. Easy effort up to 60–100 N force c. Better magnification of the effort d. Wall mountable option e. Settable on table tops
Positioning & target price	Middle of the range with a price of \$100–\$150.

(*Continued*)

Table 3.4 (Continued) Design brief for a manual can crusher

Manual can crusher	
Drafted by	*Simple Simon Company Ltd*
Target market	Housewives Elderly persons
Assumptions and constraints	–
Stakeholders	a. Housewives b. Elderly persons c. Children in early teens d. Retail hardware shops
Possible features and attributes	Firm fixtures and extendable lever
Possible area for innovation	• Positioning of the hinge close to the load (can) so that the lever arm for load becomes small when taking load about the hinge.

Example 3.4: Design brief for a combined shoes storage/stool in the porch

A storage for all pairs of shoes in use by a person and looks like an innocent stool when closed is needed at the entrance of houses. As an additional function, the person should be able to sit and wear or remove the shoes. Figure 3.4 shows a typical porch stool-storage. Write a design brief with the elements as outlined in Table 3.1 for a porch stool-storage (Table 3.5).

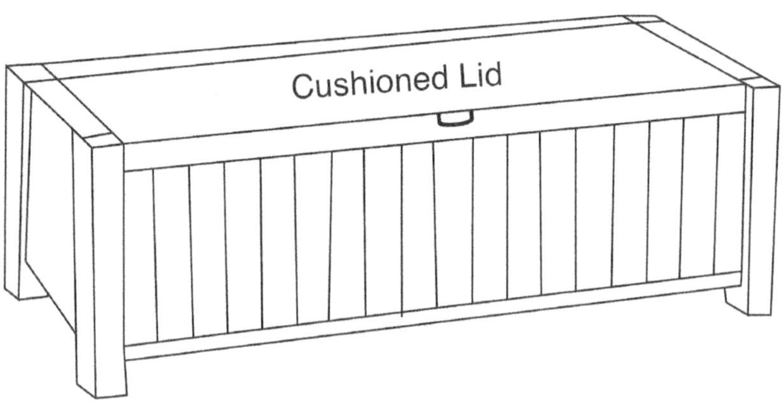

Figure 3.4 A typical porch stool-storage.

Model Answer

Table 3.5 Design brief for a porch stool-storage

Porch stool to sit and wear or store shoes	
Drafted by	*Simple Simon Company Ltd*
Product description	A storage to store, shoes and sundries like shoe polish, brush, towel and socks, in a cosy and easy way to access through a properly fitting door, and a top seat where the user can sit and wear shoes. This shoe storage stool combines the storage function and comfort of changing shoes in an attractive manner.
Product concept	A kind of shelf that permits orderly arranged shoes and sundries, covered in three sides and a door on the fourth side. The top is built to carry a sitting person and is covered by a luxury material. From the outside it will appear as a cosy stool.
Benefits to be delivered	a. Attractive and strong seat b. Proper arrangement for many pairs of shoes c. Proper storage of sundries d. Natural colours and finish e. Rich and high-class look
Positioning & target price	Middle of the range with a price of $60–$100.
Target market	Luxury houses Luxury hotel rooms
Assumptions and constraints	
Stakeholders	a. Luxury housewives b. Temples and shrines c. Hotel owners
Possible features and attributes	Multiple layers of shoe storage
Possible area for innovation	• Type of racks for shelves

Example 3.5: Design brief for a smart handbag

Sizes of the handbags available in the market are becoming bigger and bigger and often small things go to the bottom of the bag and cannot be located and picked up in a hurry. Often users have to go to a private corner to empty part of the personal items in the bag before locating and picking the wanted piece while the rest of her group waits for her. The major cause of this embarrassing problem is identified as the lack of light inside the bag. Write a design brief with the elements as outlined in Table 3.1 for a handbag that has lighting fitted inside it, which switches on when the bag is opened and shuts when it is closed. It also sounds like a left-behind-alarm when the user forgetfully leaves it behind (Table 3.6).

Model Answer

Table 3.6 Design brief for a smart handbag

Smart handbag	
Drafted by	*Simple Simon Company Ltd*
Product description	A luxury handbag that has lighting, fitted inside the bag, that switches on when the bag is opened and shuts when it is closed. It also sounds a left-behind-alarm when the user forgetfully leaves it behind. The small battery should last for long periods (before the handbag becomes out of fashion) and is made an integral part of a handbag.
Product concept	A fully integrated lighting inside the handbag that adds little weight to the handbag, lights up automatically when opened and shuts down once it is closed. It also provides an alarm when the user leaves it and goes away when activated as 'in use'.
Benefits to be delivered	a. Adequate lighting to all parts of the bag b. Shuts down the light when closed c. Long battery life d. Indicator bulb when activated as in use and e. Alarms when left behind by the user
Positioning & target price	Upper quartile of the market with a price tag of $250–$500
Target market	Professional girls and ladies
Assumptions and constraints	The bag is made to a high standard with genuine leather suitable for the upper quartile of users.
Stakeholders	Professional girls and ladies Handbag retailers
Possible features and attributes	Distributed lighting On-off alarm
Possible area for innovation	–

Example 3.6: Design brief for a micro fridge for storing insulin

Health facilities and developments of medicines have increased the life expectancy and ability of people to work, after the present retirement age, even if they have a medical condition. Diabetics have to have their regular supply of insulin to maintain a near normal life. The insulin pens have to be kept refrigerated. If a portable battery-operated micro fridge to store a few pens of insulin is available, diabetics can travel all over the world carrying the required insulin with them stored in the micro fridge. Write a **Design Brief** with the elements as outlined in Table 3.1 for a micro fridge that can store up to six pens, switches on whenever cooling is needed and switches off whenever there is excess cooling (Table 3.7, Figure 3.5).

Model Answer

Table 3.7 Design brief for micro fridge for storing insulin

Micro fridge	
Drafted by	*Simple Simon Company Ltd*
Product description	A small and portable fridge to store up to six insulin pens when travelling out of home. It is battery-driven and switches on when the temperature is above a defined value and switches off when it is below a certain value.
Product concept	A battery-driven fridge, with automatic control of the temperature inside the fridge between 5°C and 10°C . It is lightweight and robust so that it can be stored in a suitcase during flights.
Benefits to be delivered	i. Lightweight ii. Robust for air travel in suitcases iii. Consumes little power iv. Long battery life v. Rejects little heat
Positioning & target price	Upper quartile of the market with a price tag of \$150–\$300
Target market	Insulin-dependent diabetics with leisure use (visits, holidaying, etc.) Insulin-dependent diabetic professionals who travel
Assumptions and constraints	
Stakeholders	GP surgeries Diabetic clinics Pharmacies Diabetics
Possible features and attributes	
Possible area for innovation	Easy and slow heat dissipation

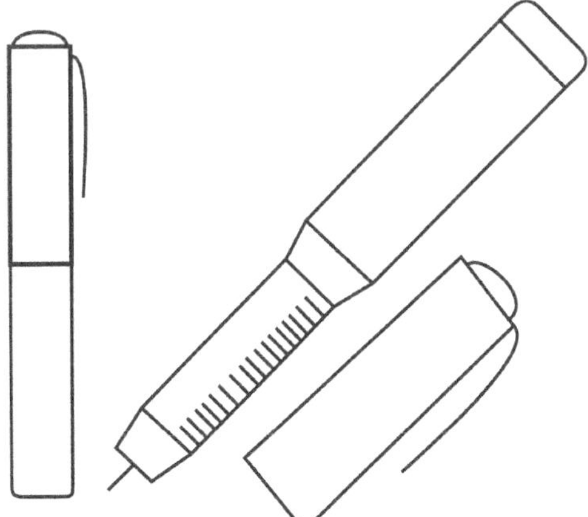

Figure 3.5 Outline of an insulin pen.

Example 3.7: Design brief for vertical access platform for use in buildings

Vertical access platforms carry a cage mounted on a telescopic mast which is mounted on a mobile base. The cage can carry a person and his accessories up where the user can do some work and return safely. A sketch showing the main components of an access platform is given in Figure 3.6. Write a design brief with the elements as outlined in Table 3.1 above for a vertical access platform (Table 3.8).

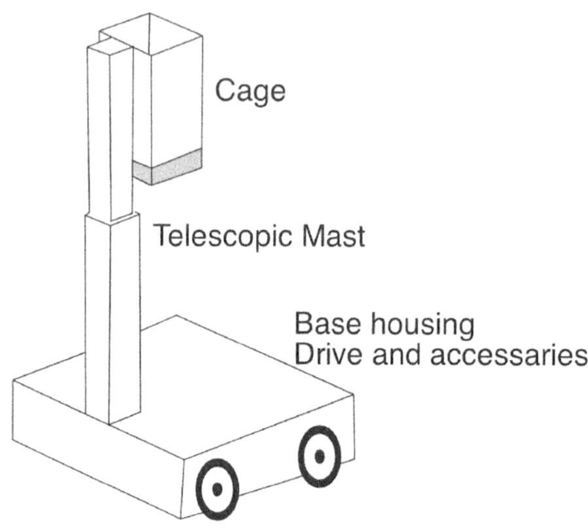

Figure 3.6 Concept of a vertical access platform.

Table 3.8 Design brief for a vertical access platform

Vertical access platform for use in buildings		
Drafted by	*Simple Simon Company Ltd*	
Product description	A low-level (3–4 m), power-driven lightweight access platform with a small footprint, vertical telescopic boom, outrigger legs, non-marking tyres and no hydraulic mechanisms.	
Product concept	A wheel-mounted platform small in size to go through domestic doors and lifts, with a telescopic boom carrying the cage, and electrically driven up and down by chain drives	
Benefits to be delivered	Access through narrow passage Low collapsed height Outrigger for additional stability	Mobile with hand-push Power available in cage Battery option

(*Continued*)

Table 3.8 (Continued) Design brief for a vertical access platform

Vertical access platform for use in buildings		
Drafted by	*Simple Simon Company Ltd*	
Positioning and target price	Middle of the range $10,000–$12,000	
Target market	Building maintenance companies Equipment hiring companies DIY users	
Assumptions and constraints	Design satisfies the required European Standards Operator completed IPAF 3A/3B Mobile Vertical/Mobile Boom Operator Course	
Stakeholders	Building maintenance companies Equipment hiring companies Qualified operators	DIY users
Possible features and attributes		
Possible area for innovation	Replacing chains with lightweight wire cables.	

REFERENCE

1. AMG Technology. https://www.mbmelbourne.com/mercedes-benz-amg-technology-melbourne-fl.html, accessed on 22nd April 2019.

Chapter 4

Establishing requirements

4.1 INTRODUCTION

The design brief, in addition to defining what the product is, identifies the stakeholders for the product. The stakeholders look at the functions or behaviour the product can perform, and their preference on the behaviour, greatly influence the success of the product. *A customer or stakeholder requirement is an expression of a behaviour of the product, as desired by the customer. The Design Brief describes `what the product is' while the customer requirements describe `what the product should do, from the users' point of view'*. Stakeholders can help by providing information in the areas desired by the design team, but not in every area. For instance, if `the car should reach from 0 to 60 mph in 15 seconds' is identified as a requirement from a stakeholder, it is a legitimate stakeholder requirement, while the size and layout of the cylinders in the engine are matters for the designer to deal with. To maximise the benefits, the design team should decide the areas where they need inputs from the stakeholders before they meet the stakeholders. Obtaining requirements is not a random activity; it is an activity that should be planned carefully. The design team should address two main items (a) deciding what inputs are needed from the stakeholders and (b) obtaining them from the stakeholders. In order to get a reliable set of requirements the stakeholders chosen for interviewing should be a true representative sample. A five-stepped methodology shown in Figure 4.1 is prescribed for obtaining the requirements in this book.

a. ***Planning the Type of Requirements - is the establishment*** of the types of requirements that need to be obtained from the stakeholders, by referring the checklists.
b. ***Planning and choosing Stakeholders*** - is the process of identifying different types of users and stakeholders so that the obtained requirements represent the requirements of the full spectrum of stakeholders.
c. ***Recording Stakeholder Verbatim*** - records the statements as made by the stakeholder. It may include the feeling the stakeholder attaches to the product or its behaviour.

DOI: 10.1201/9781003484950-5

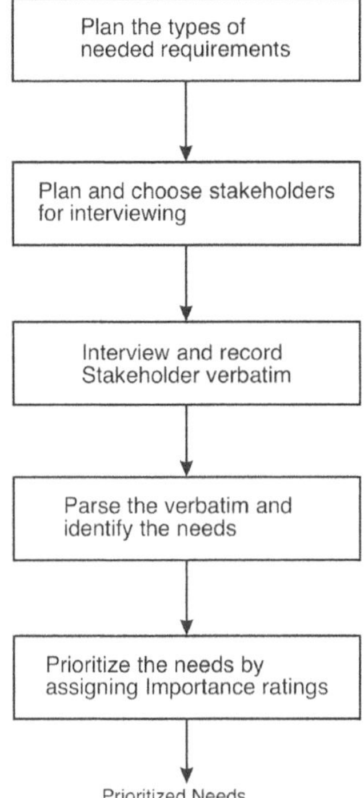

Figure 4.1 Methodology for obtaining requirements.

d. ***Parsing the verbatim and Identifying the Needs*** - is the process of discussing the recorded verbatim within the design team, and establishing the needs which represent them in a more technical manner and similar in content.
e. ***Obtaining the importance ratings*** - refers the needs back to the stakeholders to obtain the relative importance on a 1 to 10 scale. If it is difficult to contact stakeholders again, the interviewed designers can rate them based on their observations.

4.2 PLANNING THE TYPES OF REQUIREMENTS – WHAT INPUTS FROM CUSTOMERS

The inputs needed from stakeholders vary from sector to sector and product to product. Therefore, it is difficult to identify specific items that should form these inputs. The process is particularly difficult for students or beginners who have limited experience in design. In these situations, 'Checklists' produced by experienced designers and researchers, can be used as a guide.

A checklist is a type of informational 'job aid' used to reduce failure by compensating for potential limits of human ability at a given time. It helps to ensure consistency and completeness in carrying out a task. Experienced designers and researchers produce such lists describing the types of stakeholder requirements that should or could be elicited from the customers. In place of these generic lists, companies can prepare their own checklists for specialist products, they design and produce. A checklist for a birthday cake is given here as an example. The checklist can become very useful when a company involves in adding new members to an existing product family. They could tailor the checklist using their past experience.

Using customer requirements to assist the design team itself, was and is an expanding paradigm and as such, the items included in a checklist also were changing with time. Two such lists that can be used to plan the types are those produced by Pahl and Beitz [1] and David Ullman [2]. As can be seen, the former list is more biased towards technical details while the latter is more biased towards functional details, closer to the customer (Table 4.1).

Table 4.1 Types of customer requirements by Pahl and Beitz [1]

Main group	*Sub-groups*
Geometry	Size, height, breadth, length, diameter, space requirement, number, arrangement, connection and extension
Kinematics	Type of motion, direction of motion, velocity and acceleration
Forces	Direction of force, magnitude of force, frequency, weight, load, deformation, stiffness, elasticity, stability and resonance
Energy	Output, efficiency, loss, friction, ventilation, state, pressure, temperature, heating, cooling, supply, storage, capacity and conversion
Materials	Materials Physical and chemical properties of the initial and final product, auxiliary materials, prescribed materials (food regulations, etc.)
Signals	Inputs and outputs, form, display and control equipment
Safety	Direct safety principles, protective systems, operational, operator and environmental safety
Ergonomics	The man–machine relationship, type of operation, clearness of layout and lighting
Production	Factory limitations, maximum possible dimensions, preferred production methods, means of production, achievable quality and tolerances
Quality	Control possibilities of testing and measuring, application of special regulations and standards
Assembly	Assembly special regulations, installation, siting, foundation, transport limitations due to lifting gear, clearance, means of transport (height and weight), nature and conditions of dispatch

(Continued)

Table 4.1 (Continued) Types of customer requirements by Pahl and Beitz [1]

Main group	*Sub-groups*
Operation	Quietness, wear, special uses, marketing area, destination (e.g., sulfurous atmosphere, tropical conditions)
Maintenance	Servicing intervals (if any), inspection, exchange and repair, painting and cleaning
Recycling	Reuse, reprocessing, waste disposal and storage
Cost	Costs Maximum permissible manufacturing costs; cost of tooling, investment and depreciation.
Schedule	End date of development, project planning and control, delivery date

Ullman [2] identifies types of specifications under seven major types and provides details for each. This can be used as a checklist to identify the needed types for customer requirements. It is easy to consider and decide whether the particular sub-group is relevant to the specific project. The seven types identified by Ullman can be explained as follows:

1. *Functional performance*: This includes flow of energy, flow of information, flow of material, operational steps and operation sequence. This categorization is very useful in mechatronic product design.
2. *Human factors*: This includes Appearance, Force and motion control and Ease of controlling and sensing state. It can be seen that aesthetically pleasing factors and human inputs fall in this category.
3. *Physical Requirements*: In this category, Ullman identifies two items (i) available spatial envelope and (ii) physical properties.
4. *Reliability*: Ullman includes two aspects (i) mean time between failures and (ii) safety hazard assessment which refer to product's life span and its safety during use.
5. *Lifecycle concerns*: Many widely varying factors are grouped under this category: distribution (shipping), maintainability, disposability, repairability, cleanability, easiness for installation and retirement.
6. *Resource concerns*: This again is a broad collection of factors including, time, cost, capital, unit, equipment, standard and environment.
7. *Manufacturing requirements*: Materials, quantity and company capabilities are considered under this category.

Referring these two lists and choosing the items relevant to the project concerned is the method used in this book. Details can be added to the chosen list to make the inputs more relevant to the product.

4.3 PLANNING AND CHOOSING THE STAKEHOLDERS

The requirements obtained from the survey should represent the requirements of the full spectrum of stakeholders, and a representative group of stakeholders should be assembled, to get a representative set of requirements. Ulrich and Eppinger [3] identify some as lead users, extreme users and normal users. The lead users identify the needs and shortcomings in existing or adjacent products, extreme users use the product heavily on a regular basis while normal users use the product as and when they require it. Other categories include service stations for the maintenance of the product, distributors and hire shops. Depending on the type and nature of the product, the constitution of the group for surveying would vary. Inputs from sales and marketing can be very useful in choosing stakeholders.

4.4 HOW TO OBTAIN REQUIREMENTS – RECORDING THE VERBATIM

There are several methods that are being used to obtain stakeholder requirements. They include (i) questionnaire survey, (ii) individual customer interview, (iii) focus group interview and (iv) observing the customer using the product or an adjacent one. Out of these methods, the questionnaire survey is deemed as least effective and the others are seen as equally effective, though individual customer interview can be more informative and is the method adopted in this book. Planning the interview is the key to establish a comprehensive list of requirements. The following steps are recommended.

a. From the checklists and otherwise, establish the primary goals or specific questions that need answers.
b. Plan how the interview would be started and proceeded, and the elicited needs are recorded.
c. Plan the questions
d. Structure the interview with a time plan by allotting a certain amount of time for each interview question or section.
e. Plan the phrasing to be used, in the interview, precise and concise. It is important for the stakeholder and the interviewer to have the same understanding.
f. Record what the stakeholder statements are and check the final take of the interview with the interviewees.

The interviewers should be economical with the questions they ask, and offer follow-up questions at the correct intervals. They need to be active listeners and observers while both probing for follow-up questions and evaluating non-verbal cues or body language. Sometimes it is worth having a pilot run on interviewing and learning from it to fine-tune the main run. It is important to debrief with the Internal Team to consolidate thoughts and observations captured, and themes identified. This will become very useful if importance ratings have to be given by the interviewing team later on. Normally the verbatim is recorded as said by the stakeholder and, the verbatim is parsed by the design team when they translate them into needs for prioritizing.

4.5 PARSING THE VERBATIM AND IDENTIFYING THE NEEDS

Stakeholder requirements is a prime factor in determining the functions or the behaviour of the product. Every member in the design team should therefore have similar understanding of the requirements. In addition to this, each of the stakeholders verbatim may not have similar information content. Because of this, a single verbatim may have to be divided into multiple requirements. To eliminate these and other similar difficulties, the design team meets and analyses the verbatim and translates them into customer needs which are similar in information content and understood by all of them. In this book, they are identified as needs or stakeholder needs.

4.6 OBTAINING THE IMPORTANCE RATINGS

The final step is to prioritize the needs by assigning importance ratings on a 1–10 scale. Ideally, these ratings should be given by the stakeholders themselves. In this method, the needs are referred back to the stakeholders, after parsing and translating the verbatim into needs, and their ratings are obtained. In the alternative method, the ratings are given by the interviewing team members and agreed by the design team.

4.7 MODEL QUESTIONS AND MODEL SOLUTIONS

Two kinds of examples (i) Planning the types of the needed requirements and (ii) carrying out the plan to obtain the needed requirement are given below. To give a clear idea of the product, a sketch is given where appropriate.

Example 4.1: Planning to obtain customer requirements for a rolling shoe rack

A rough sketch of a rolling shoe rack to store a large number of shoes is shown in Figure 4.2. Assuming that you are a team of four product designers establish the types of requirements you would like to elicit from your stakeholders using Ullman's checklist.

Table 4.2 shows the types of requirements that can be useful to the design team that designs the rolling shoe rack. The first two columns give the list as given by Ullman [2] and the third column indicates why that sub-group may be relevant to the requirements of a shoe rack.

The chosen types would be useful in establishing questions that need answers from the stakeholders and to determine the logistics (path followed and recordings to be made) during the planning and structuring of the interview.

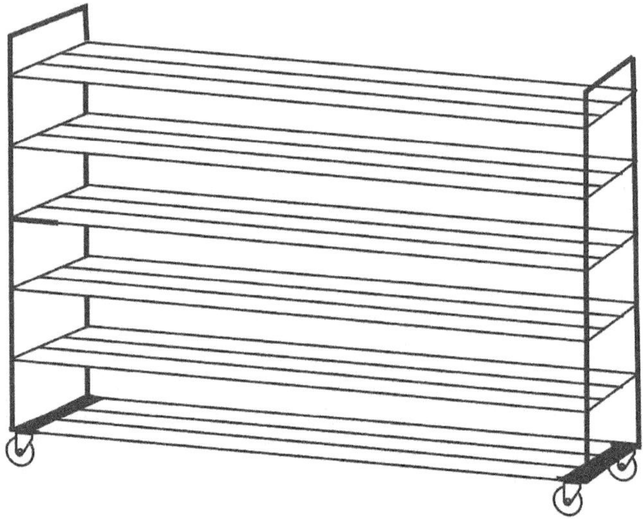

Figure 4.2 Rolling shoe rack.

Model Solution

Table 4.2 Choosing types

Main group	*Sub-group*	*Chosen types and reason*
Functional performance	Flow of energy	Easy to roll around
	Flow of information	
	Flow of material	Ability to store different shoe sizes
	Operational steps	Easy assembling
	Operation sequence	Height adjustment etc

(*Continued*)

Table 4.2 (Continued) Choosing types

Main group	*Sub-group*	*Chosen types and reason*
	Appearance	The look of the shoe rack
Human factors	Force and motion control	Less effort needed to move Lockable in position
	Ease of controlling and sensing	Easy to roll around
Physical requirements	Available spatial envelope	Number of layers Number of shoes in each layer
	Physical properties	Light in weight
Reliability	Mean time between failures	
	Safety hazard assessment	Safe at all different loadings
Lifecycle concerns	Distribution (shipping)	Collapsible for packaging
	Maintainability	No maintenance Should be stable once assembled
	Disposability	Recyclable or bio-degradable
	Repairability	
	Cleanability	Easy to clean
	Installability	
	Retirement	
Resource concerns	Time	
	Cost	Low cost
	Capital	
	Unit	
	Equipment	
	Standard	
	Environment	
Manufacturing concerns	Materials	Light materials Fashionable materials
	Quantity	
	Company capabilities	

Example 4.2: Planning to obtain customer requirements for an open book marker

Assuming that you are a team of four product designers and using the list by Pahl and Beitz [1] to establish the type of customer requirements for an open book marker to keep the book at the required page. A rough sketch of the product idea is shown in Figure 4.3 (Table 4.3).

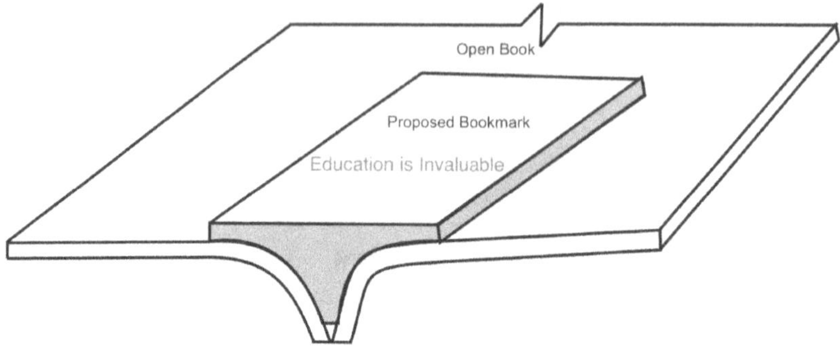

Figure 4.3 Open book marker.

Model Solution

Table 4.3 Types of customer requirements for a book marker

Main group	*Sub-groups*	*Chosen types and reasons*
Geometry	Size, height, breadth, length, diameter, space requirement, number, arrangement, connection and extension	Breadth of the marker Length of the marker Depth of the splitter
Kinematics	Type of motion, direction of motion, velocity and acceleration	N/A
Forces	Direction of force, magnitude of force, frequency, weight, load, deformation, stiffness, elasticity, stability and resonance	Weight of the book marker
Energy	Output, efficiency, loss, friction, ventilation, state, pressure, temperature, heating, cooling, supply, storage, capacity and conversion	N/A
Materials	Materials Physical and chemical properties of the initial and final product, auxiliary materials, prescribed materials (food regulations, etc.)	High density material is preferable. Use of material forms available in the market
Signals	Inputs and outputs, form, display, control equipment	N/A
Safety	Direct safety principles, protective systems, operational, operator and environmental safety	Elimination of accidental injuries
Ergonomics	The man–machine relationship, type of operation, clearness of layout and lighting	N/A
Production	Factory limitations, maximum possible dimensions, preferred production methods, means of production, achievable quality and tolerances	The largest book size Surface properties

(*Continued*)

Table 4.3 (Continued) Types of customer requirements for a book marker

Main group	*Sub-groups*	*Chosen types and reasons*
Quality	Control possibilities of testing and measuring, application of special regulations and standards	Fine surface finish
Assembly	Assembly Special regulations, installation, siting, foundation, transport limitations due to lifting gear, clearance, means of transport (height and weight), nature and conditions of dispatch	N/A
Operation	Quietness, wear, special uses, marketing area, destination (for example, sulfurous atmosphere, tropical conditions)	N/A
Maintenance	Servicing intervals (if any), inspection, exchange and repair, painting and cleaning	Polish to maintain the smoothness and glow of the surface
Recycling	Reuse, reprocessing, waste disposal and storage	
Cost	Costs Maximum permissible manufacturing costs; cost of tooling, investment and depreciation.	What kind of market is there? High end, middle, How much they can pay?
Schedule	End date of development, project planning and control, delivery date	

Example 4.3: Customer verbatim by Pahl and Beitz checklist - solar irrigation system

Access to water for agricultural purposes remains critical in some areas such as in arid regions of Africa and Southern Asia. Many Indian and African farmers fetch the water directly from the well or the rivers and irrigate their fields using buckets. If farmers of those regions could have access to a motorized pump, they would increase their yield by 300% [4]. Harvesting solar power and converting it to electricity is a matured technology and solar panels are available in commercial outlets. The development of battery systems for storing the harvested electric power also has reached or is reaching maturity. DC and AC motors and water pumps driven by them, belong to the well-established sectors of engineering and technology. A typical solar-powered system is schematically shown in Figure 4.4. But there are no solar-powered irrigation systems that one can buy from a sale counter and put to use. This entrepreneurial challenge remains unsolved, and the purpose of this project aims to address this challenge. The problem can be summarized as "To establish a commercially viable solar irrigation system for smallholder farmers, from areas with adequate solar light". Establish the customer verbatim for a solar irrigation system for a small farm (Table 4.4).

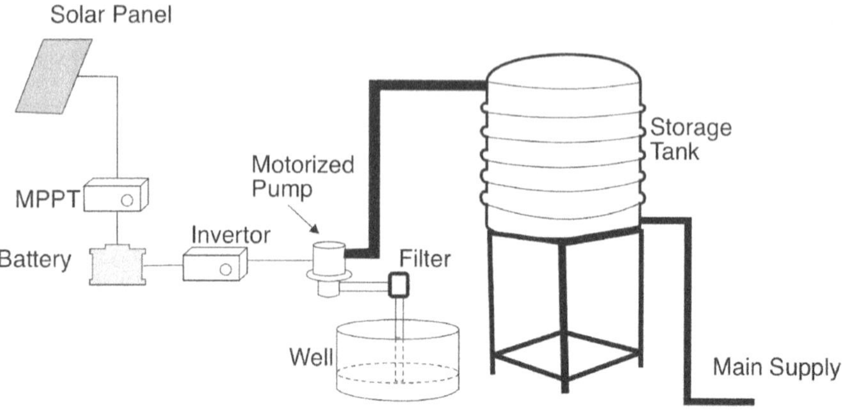

Figure 4.4 Schematic of a small-scale solar powered irrigation system.

Model Answer

Table 4.4 Establishing customer verbatim for solar irrigation

Type	*Prompting question*	*Verbatim*
Geometry	Distance between tank & distribution system	1. The maximum distance can be assumed as 100 m
	Available area to collect sunlight	2. The area for collection can be 3–3.5 m^2
	Area of land to be irrigated	3. The area of the farm-lands on average is 1 hectare
	Water depth	4. Maximum of 6 m
Kinematics	Type of irrigation system (drip line, channels..)	5. This varies in the village but assume open channels
	Manual or automatic	6. Manual irrigation where man/woman directs water
Force	Height of water tank	7. The height cannot be more than 3 m (bottom of tank) 8. Tank may be around 1.5 m tall (about 1,000 L)
Energy	Flow rate	9. A safe estimate is 6,000 L per hour
Signals	Battery level	10. I need an indicator showing battery charge in %
	Tank level	11. The water level in the tank visible from distance
	Switch	12. Simple on/off switch to start and stop the pump

(*Continued*)

Table 4.4 (Continued) Establishing customer verbatim

Type	*Prompting question*	*Verbatim*
Safety	Storage during long period of non-use	13. Monsoon season lasts from October to January
	Panel protection (rain, wind)	14. Protection from wind, sand and rain is needed
	Preventing overcharge of battery	15. Battery protection from over charge and depletion needed
	Isolate components from the rest	16. Isolate components from the rest
	Safe housing	17. Should have vandal proof enclosure
	Safe for different climates in countries	18. Operating Temperature varies from 15°C to 45°C
Ergonomics	Layout	19. The pump, batteries and other components should not take more than 1 m^2
		20. The system should be placed on a platform around 0.6 m height
	Easy to access	21. Should be accessible from all sides
Production	Units produced per week (production rate)	22. 50 units per week
Standards	Meet the standards	23. Use search engines 24. BS EN 16603-20-08:2014 and water quality standards
Assembly	Easy to assemble	25. Easy to assemble and disassemble
	Good packaging	26. Packaging should protect the panels
	Safe to transport	27. Safe to transport
		28. Weight of package should not exceed 20 kg
Operations	Clear instructions	29. Clear operating instructions
		30. Clear Labelling of wires
	Few Operation prior checks	31. Few checks prior to operation
	Alternate power sources	32. Stand alone with no grid connection
		33. Stand-by battery preferred
	Routine cleaning	34. Easy for routine cleaning of the panels
		35. Easy access to panels
	Accessories	36. Provide lighting for night operations
Maintenance	Corrective maintenance	37. Should be modular for easy corrective maintenance

Example 4.4: Customer verbatim by checklists - cycle for leg amputees

Natural disasters, civil wars and wars result in humans losing their legs. Often these places or countries would be poor and these leg amputees have to work and earn their living. They move around in crutches for short distances but going long distances becomes a major challenge. A cycle that can be driven by hand would provide redress from this difficulty. In this example, the checklists by both Pahl and Beitz [1] and Ullman [2] are used to establish the stakeholder verbatim. Table 4.5 shows the verbatim when the checklist by Pahl and Beitz [1] was used. Table 4.6 shows the verbatim obtained using the list outlined by Ullman [2]. Figure 4.5 shows a conceptual idea for the bicycle.

Table 4.5 Verbatims using checklist by Pahl and Beitz [1]

Main group	*Verbatim*
Geometry	1. *Height*: Height should be matching that of a normal chair 2. *Size*: Seat should be easily bridgeable with a normal chair for easy getting in and getting out 3. *Length and breadth*: The length and breadth of the seat should be big enough for a comfortable seating 4. *Arrangement*: The drive system should not foul with the leg of the rider. 5. *Connection and extension*: A foldable bridge should be available for moving to a chair from the bicycle and from the chair to the bicycle
Forces	6. *Magnitude of force*: The force to be exerted by the rider should be minimal 7. *Stability*: The bicycle and rider should be stable at all conditions. 8. *Stability*: Should be lockable with brakes at a given position during ascending and descending
Safety	9. *Safety*: Should be safe during ascending, during operation and during descending. 10. Safety – Should be safe in downslopes, level ground and upslopes, during ascending, descending and staying in position
Ergonomics	11. *Appearance*: Aesthetically pleasing appearance 12. *The man–machine relationship*: Type of operation, clearness of layout, lighting, 13. *Man–machine relationship*: Easy ascending, seating during operation and descending
Production	14. *Maximum possible dimensions*: Should be able to go through domestic elevators 15. *Preferred production methods*: Normal Cutting and welding, some machining in lathes and mills only
Quality	16. *Standards*: Should use standard bicycle parts as far as possible 17. Designed parts should match the standard bicycle parts
Assembly	18. Installation: Assembly should be easily made with few operations
Operation	19. Easy to start, ride and stop
Maintenance	20. Routine cleaning should be easy 21. Annual servicing to ensure lubrication should be sufficient
Cost	22. Preferable maximum cost is about \$300–\$400

Table 4.6 Requirements using the list by Ullman [2]

Main Group	*Requirements*
Functional performance	1. *Flow of energy*: Should be able to drive by one hand 2. *Flow of energy*: Should be able to control by one hand 3. *Flow of material*: Able to carry a person and his belongings (leg amputee) 4. *Operational steps*: Easy to Ascend 5. Easy to sit and operate 6. Easy to descend from the bicycle 7. *Operation sequence*: Should able to stay in upslopes, flat ground and downslope
Human factors	8. *Appearance*: Aesthetically pleasing appearance 9. *Force and motion control*: Should able to stay in upslopes, flat ground and downslope 10. *Ease of controlling and sensing state*: Lockable berakes for ascending and descending
Physical requirements	11. *Available spatial envelope*: Drive should not foul with the leg 12. Hands should reach the steering with ease 13. *Physical properties*: Should be able to climb from crutches or from a seated chair 14. Should be able to go into domestic lifts 15. Should be able to pass through domestic doors
Reliability	16. *Mean time between service*: One year 17. *Safety hazard assessment*: Warning for other vehicles should be installed
Lifecycle concerns	18. Distribution (shipping) 19. *Maintainability*: One annual service should be sufficient 20. *Cleanability*: Easy to clean
Resource Concerns	21. Time to assemble should not be more than 20 minutes 22. Cost should be within $300–$400 23. Standard bicycle components should be used as far as possible
Manufacturing requirements	24. *Company capabilities*: Parts that needs making should be makeable with normal lathes, mills and welding.

Figure 4.5 A concept of the bicycle for leg amputees.

Example 4.5: List of needs for 'event tables for renting'

It is a common practice for people to celebrate lunches and dinners in outside venues. To have such events they rent tables and chairs. A group of entrepreneurs are planning to open such a renting company. Establish the set of requirements for the table for renting. A typical arrangement to visualize the need is shown in Figure 4.6.

The methodology adopted in this book, as explained in Section 4.1, has the recorded verbatim as the output of contacts made with stakeholders. The design team parses the verbatim and prepares a list of requirements, called needs in this book, that are equal in technical content and more technical in their presentation. The needs obtained from the recorded verbatim for the event tables are shown in Table 4.7.

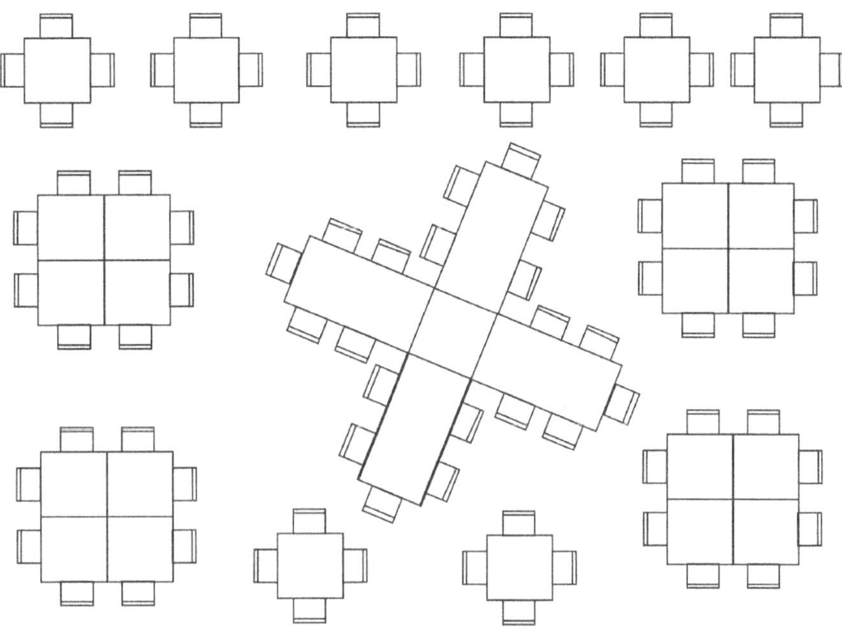

Figure 4.6 Typical arrangement of seats in a wedding reception.

Table 4.7 Verbatim need translation for 'event tables for renting'

Verbatim	*Needs*
Arrangeable into different combinations	Provide different patterns to customers
	Small sizes that can be arranged big
	Possible to mix different patterns
Easily dismantlable and assembleable	Require a minimum number of tools
	Require little time to assemble
Portable	Easy bundling of dismantled parts
	Lightweight
Suitable for multiples of a small number	Standard size of the small building unit
Arrangeable as dense or with gaps	Arrangeable as dense or with gaps
Stand repeated dismantling & assembling	Withstand repeated dismantling
	Withstand repeated assembling
Compact packing	Suitable for compact packing
Easy to maintain	Robust with minimum maintenance
Easily repairable	Modular design
	Replaceable modules
Low cost	Low cost
Use standard fixtures	Use standard fixtures

Example 4.6: Needs of a table-top reading assistant

A reading assistant helps a reader to refer and cross-refer books that are kept and presented to him in the open condition. This is kept on the table so that the reader can access the book without getting up from his seat. A typical one is shown in Figure 4.7. Establish a set of customer verbatim and translate them into needs (Table 4.8).

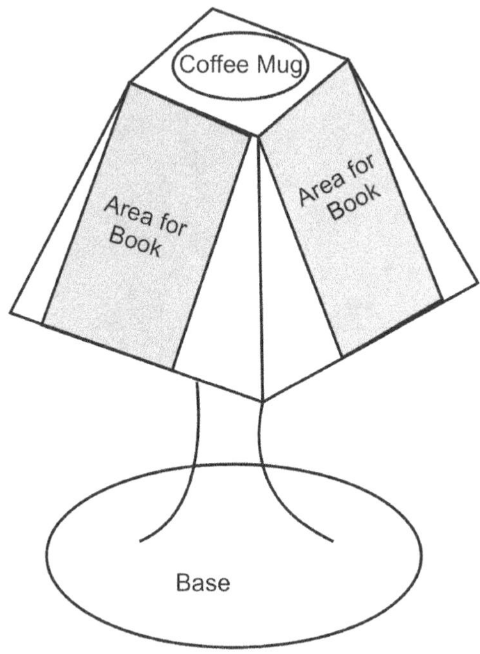

Figure 4.7 A table-top reading assistant.

Model Solution

Table 4.8 Customer verbatim needs translation for a table-top reading assistant

Verbatim	*Need*
Should be able to hold up to four A4 size books	Should hold up to four books The maximum size of the book is A4
Should be able to hold my coffee mug	Safe place for coffee mug
Should have four built-in book marks	Should have four built-in book marks
Easy access to books	Able to rotate for easy access
Should be looking elegant	Should be looking elegant
Should be lightweight Can be easily moved	Should be light in weight Can be easily moved

(*Continued*)

Table 4.8 (Continued) Customer verbatim needs translation for a table-top reading assistant

Verbatim	*Need*
Easy to control the book	Have clips to keep book open
Envelope should be minimum	Foot space around 600–650 mm^2 Overall height should be around 400 mm
Base should be heavy	Base should be heavy
Stable at all possibilities	Should be stable at 1, 2,3 or 4 books on the reading assistant
Should be collapsed and packed	Should be collapsible and packable
Easy to clean	It should be easily cleanable
Affordable	Low to medium cost

Example 4.7: Development of a checklist for a birthday cake designer

The lists by Pahl and Beitz [1] and Ullman [2] are generic and applicable to general products. But there are specialist products where information about specific characteristics of the product from the stakeholders are more relevant and needed for a successful design. This comes from the experience that designers had on designing the product. Consider the birthday cake as an example. While the expertise in getting the right balance between the four ingredients (flour, egg, sugar and fat), leavening agent, temperature time control etc and skill in decoration to get the right appearance are individual merits of the person making the cake, the inputs from the stakeholders are important to a grand birthday cake. Develop a checklist using the experience of bakers who design and bake birthday cakes (Table 4.9).

Model Solution

Table 4.9 Checklist for types of requirements for a birthday cake

Main group	*Description of important points*
Appearance	Appearance is one of the, if not the, most important features of a cake. The cake should be attractive to catch the very first glimpse of every individual present on the floor. It is difficult to describe and there can be several adjective phrases that make the baker understand what is needed. Some phrases are (i) inexplicable fairy, (ii) baked unleavened, (iii) miniature, (iv) big and magnificent, (v) gorgeous features, (vi) calm and cumulative, etc.
Taste	One of the very basic yet important features of any cake is its taste. A cake should be enriched in taste and should satisfy completely. Some phrases are (i) Mature and ripe honey, fruity, etc., (ii) grainy, (iii) fibrous, (iv) watery, (v) mild chocolate, (vi) strong strawberry, etc.

(Continued)

Table 4.9 (Continued) Checklist for types of requirements for a birthday cake

Main group	*Description of important points*
Frosting	Frosting of the cakes adds a new dimension to the enjoyment of cakes. Frosting is a sweet topping that accompanies the cake. Frosting adds to flavour, appearance and quality by providing a protective coating. Six broad categories are (i) buttercream frosting, (ii) cooked frosting, (iii) whipped cream frosting, (iv) royal icing, (v) ganaches and (vi) glazes.
Flavour	The most necessary feature of any cake is its flavour. Chocolate, vanilla, butterscotch, strawberry, and many more flavours are added to serve every taste and flavours are the centres of attraction of any cake.
Personalized characteristics	These make the cake more personalized and unique for the individual. Shapes of animals, birds, places, monuments, characters, messages like happy birthday, golden 50, relevant frostings, etc.
Layering	A layer cake is a cake consisting of multiple stacked sheets of cake, held together by frosting or another type of filling, such as jam or other preserves. Fillings make the cake unique.
Tiering	Tiered cakes consist of multiple layer cakes of different sizes stacked on top of each other. Structural supports are placed in all tiered cakes with a dowel rod through all tiers and into the baseboard.

Example 4.8: Prioritized needs for a portable mini refrigerator

Summer is hot, but it is the time for family outings as well. Often the children and elderly will ask for chilled drinks during such an outing. A portable mini fridge would be very useful in these situations. Establish a list of verbatim and their translation into prioritized needs for such a fridge.

In this example, the verbatim is collected from stakeholders, translated into needs by the design team and weighted by the stakeholders subsequently (Table 4.10).

Model Answer

Table 4.10 Verbatim need translation of a mini portable fridge

Verbatim	*Need*	*Importance*
Able to cool 3–4 bottles of water	Able to cool 3–4 bottles of water	9
Should indicate the temp in fridge	Should indicate the temp in fridge	9
Should show the remaining battery	Should show the remaining battery	9
Low power requirement	Low power requirement	7
Easy to move around	Lightweight	8
	Provide handle for lifting	6
	Have belts to put on the shoulder	7
	Detachable trolley	6
Manageable size	Compact construction	7
	shape to fit on back or shoulder	8
	Shape to fit on a trolley	7
Reliable	Reliable	8
Easy to clean	Easy to clean	6
Should be quiet	Less noise	8
Fast cooling	Cool in a short time	8
Reasonable price	Reasonable price	6

Example 4.9: Prioritized needs for a presentation table in exhibition halls

Open days in universities, exhibitions in conference halls and other similar shows in hotels have several company representatives presenting their products and speaking to customer groups as they come to their stalls. They need a compact presentation station and several of them are needed in each event. Figure 4.8 shows such a presentation station. Establish the prioritized needs for such a table (Table 4.11).

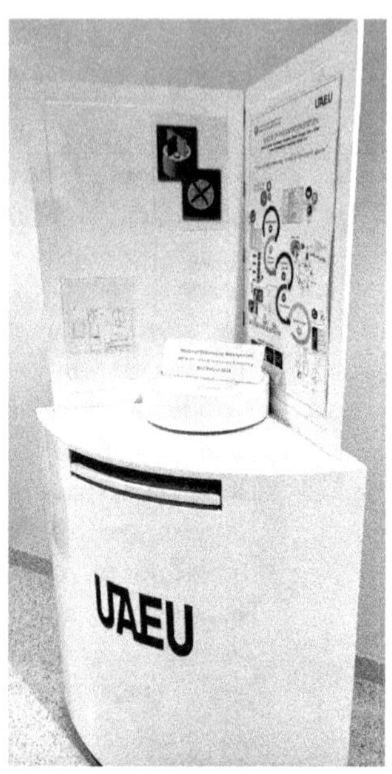

Figure 4.8 Presentation table in exhibition halls made by students.

Table 4.11 Customer verbatim, needs and importance ratings

No	*Verbatim*	*Needs*	*Imp*
1	Should be able to display prototype	Should contain a stand for the prototype	9
2	Must have a holder for the poster	Should have a dedicated poster space	9
3	Nice if it has a screen holder	Better to have a screen holder	8
4	Sometimes I display pictures	Should have place picture points	7
5	I prefer a laptop space	Space for laptop is desirable	7
6	Should look nice and attractive	Should look cosy	8
7	I need storage space	Provides drawers and shelves	6
8	Easy to assemble and dismantle	Easily installable & removed	7
9	Would be nice to have plug point	Provides additional plug point	6
10	Able to access all presentation tools	Provides various presentation tools.	5
11	Should be storable in a small space	Dismantled and packed as small packs	4
12	Should be easy to move around	Should have wheels to move around	7
13	Should be easily cleanable	Cleanable and accessible surfaces	7
14	Would be nice to have spot lights	Should contain spot lights	6
15	Prefer accessories like speakers	Can have optional speakers	4
16	Should occupy small area	Should occupy small exhibition area	8
18	Prefer to be easily maintainable	Made up of standard parts	6
29	Should be reasonably priced	Middle of the range price	8
20	Should have long life	Made up of durable material	7

Example 4.10: Prioritized needs for a building hopper

In campus universities, buildings are spread in a vast complex and it is difficult to go from one building to other. Often the universities operate a free bus service to go from one point to another. This however needs a driver for each bus and users have to wait at specified bus stops till its arrival. Users find this ineffective and often, some choose to walk. This project addresses this problem. Building Hopper is a project where battery-operated mini-vehicles are parked at various docking stations which users can borrow and use to their destination and leave it at next docking station near their destination. They can borrow another one to return after finishing their work (Table 4.12).

Model Solution

Table 4.12 Verbatim need translation of the building hopper

Verbatim	*Need*	*Importance*
Size enough for two passengers	The Hopper car has two seats	8
I need additional space for cargo	Cargo box included	7
Able to go to any destination in the university	Equipped with monitoring and controlling system.	9
Should have lights for night use	Should have lights for night use	8
I like medium speed	Should be able to go at 10–15 km/hr	8
Safe against collision	Should have protecting bumpers	7
Battery should last the whole day	A single charge should last 12 hours	9
Should be compact	Able to pass through narrow paths	7
Easy access to the car	Should have multiple doors	7
Should be safe to travel	Should have safety belt etc	9
Easy to operate	Few controls to operate	8
Aesthetically pleasing colour and finish	Aesthetically pleasing colour and finish	8

REFERENCES

1. Pahl G. and Beitz W. *Engineering Design*, The Design Council, London, 1984.
2. Ullman D.G., *The Mechanical Design Process*, 4th edition, McGraw Hill Higher Education, Boston, MA, 2010.
3. Karl U., Eppinger S. and Yang M. C., *Product Design and Development*, 7th edition, McGraw Hill, Boston, MA, 2019.
4. Stéphanie R., *Solar-Powered Irrigation: A Solution to Water Management in Agriculture*. https://www.renewableenergyfocus.com/view/44586/solar-powered-irrigation-a-solution-to-water-management-in-agriculture/, accessed on 8th April 2017.

Chapter 5

Functional modelling

5.1 INTRODUCTION

Function is a description of the action or effect performed by the product to meet the requirements and, functional representation or modelling allows one to describe design problems and solutions in terms of their functions which permits reasoning about them. Function can be seen as `the intended behaviour of the product or what the components of the product do'. There are two main approaches to functional modelling. In this book they are called function tree modelling, and function structure modelling. Function Tree provides an overview of the whole system in a hierarchical manner. A Function Structure on the other hand breaks down the overall function as given in the `Black Box Diagram' into sub-functions. These sub-functions are the operations that the product performs on a flow or a set of flows to transform them from its input state to its output state. In this context a Black Box model is an input-output model that assumes that the product's function is to change the input flows of material, energy and information (signals) into output flows of the same kind.

5.2 FUNCTION: DEFINITION

5.2.1 Key concepts

Vermaas [1] identifies five key concepts related to functions that completely describe a device. They are:

1. *Goals of device*: These define a state of affairs that a user wants to have or realize by using the device
2. *Action*: The behaviours of the users when they use the device
3. *Functions*: The desired roles played by the devices
4. *Behaviours*: The state changes the device exhibit
5. *Structure*: Materials and fields of devices, their configurations and their interactions

DOI: 10.1201/9781003484950-6

The design of a device starts with the goals or the purpose for the device. The next step is to ensure the right environment and the inputs to the device. These are the actions of the user. When the actions are performed by the user the product produces the functions, which are the desired roles. The functions are the results of the behaviour of the structures or structural components.

As an example, to understand these key concepts, consider the bench vice shown in Figure 5.1. The goal is to keep part A shown stationary, while forces and moments are applied on it to shape it. This means that equal and opposite force or moment should be generated as and when a force or moment is applied on part A. This opposing effort is generated by keeping the object between two rough surfaces, the surfaces of the jaws, to generate friction. To generate the frictional force a normal force has to be created between the surface of the jaw and the object, and this is generated by tightening the power screw. This tightening is the setting up of the right environment by the user. Now the user applies a shaping force on part A. An equal and opposite force is created by friction, the behaviour of the jaw surfaces. For the jaw surfaces to generate the frictional force or moment on the object, part A, the moving jaw should be strong and constrained. This defines the structure.

The functional hierarchy during design can be described in the order as shown in Figure 5.2. At the top, the goal or purpose of the product is placed. In the next position, the user's input in setting up the environment and providing the input, is placed. In the next position, the desired role, the function, is placed. But functions are generated by the behaviour of the structure and hence the desired behaviours come to the next position. In the last position comes the structure of the device. During use, however, the

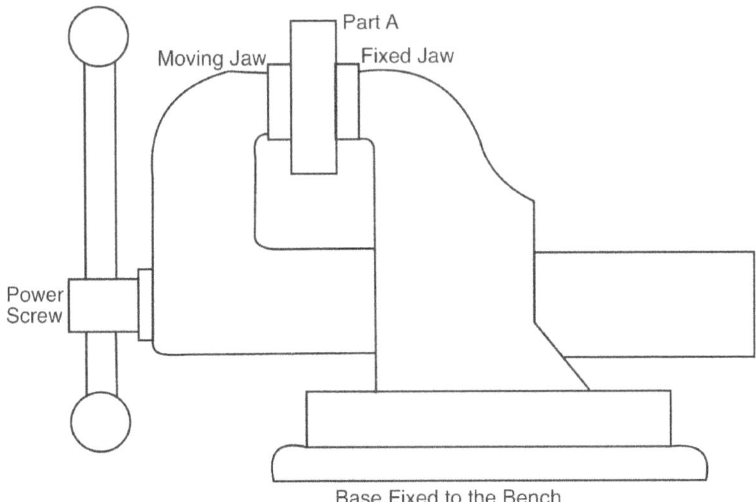

Figure 5.1 Bench vice holding Part A between jaws.

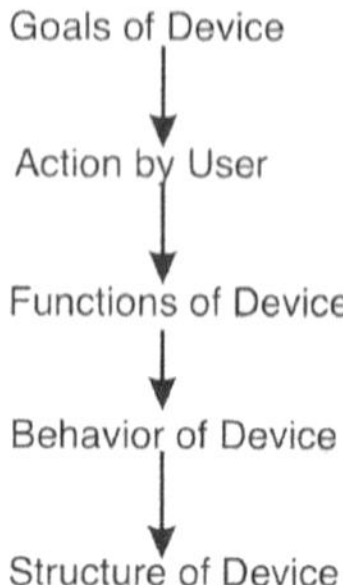

Figure 5.2 Hierarchy during the design process.

user sets the environment and gives the input to the structure and it creates the behaving of the structure. This achieves or generates the function meeting the goal of the device.

5.2.2 Definition and classification of functions

Function of a product can be described in the following ways:

1. *Function as the goal*: This is defined at a high level of abstraction and often is called the purpose function.
2. *Function as the desired behaviour*: since the device does something exhibiting behaviour, the desired behaviour from the device is defined as the function.
3. *Function as the desired effects of behaviour of the structure*: These are the actions that the structure performs to provide the expected behaviour.

Because of this possibility of describing function in different ways, there are many definitions for function. Some of them are given by Haik et al. [2].

1. Functions fulfil the expectations of the purposes of the resulting artefact.
2. Function is the intended purpose of a device.
3. Function is the logical flow of energy, material, or signal between objects or the change of states of an object due to flows.
4. Functions are the intended input/output relationships.
5. Function is the description of the behaviour abstracted by users in order to make use of it.
6. Function is the fulfilment of a goal or purpose.
7. Function is the intended behaviour.
8. Function is referred to at a higher level of abstraction and as the purpose of the product.

9. Functions convey what components do.
10. Function is the intended effect that a device has on its environment.
11. Function is an action of product on its inputs and outputs.

Constituents of function: A function is made up of the following two components:

1. An action verb
2. A noun representing the object on which the action described by the verb takes place

Any given function can be described with the combination of an action verb and a noun.
Example: Rotate (*Action Verb*) Arm B (*Object on which action takes place*)
Drive (*Action Verb*) *Main shaft* (*Object on which action takes place*)

Purpose and Action Functions: A product's purpose is to provide a certain set of high-level functions. Deng [3] classifies them as purpose functions. For example, consider a jack for lifting a car to change a flat tyre. The purpose functions of a Jack are: (i) Lift the car by receiving the human effort; (ii) hold the lifted car at the lifted position and (iii) release the lifted car on the ground safely. Each of these purpose functions is provided by a group of sub-functions which are called the action functions.

5.3 FUNCTION TREE

Function tree defines the functions of the product arranged in a hierarchical or top-down, inverted tree structure with the root at the top and the branches below. In it, the top function represents the goal or the overall activity that the product is intended to perform. This top function is then decomposed into other functions that must be provided for the top-level function to successfully occur or happen. This successive decomposition continues until the functions become basic in nature. The main outcome of the function tree is the identification of the basic functions through the decomposition of the higher-level functions. These ideas are illustrated by the schematic shown in Figure 5.3. The root or top of the tree shows the main goal in verb–noun phrase format. At the first hierarchical level, the goal is decomposed into first-level purpose functions. These can be used as the basis for a morphological analysis to generate conceptual designs. The purpose functions are further decomposed until the resulting functions are sufficiently simple.

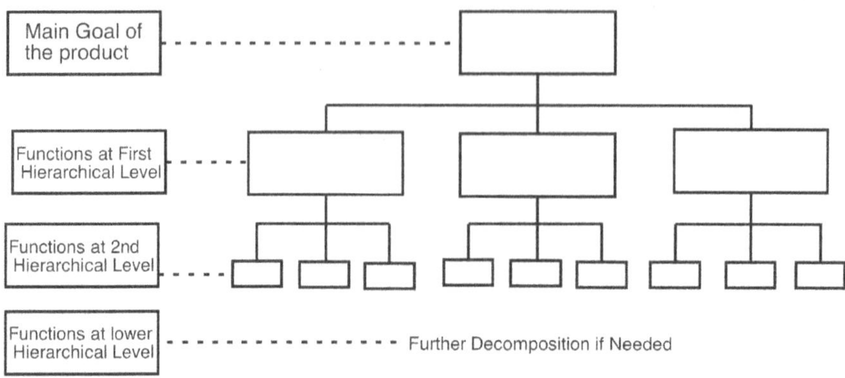

Figure 5.3 A typical function tree.

5.4 FUNCTION TREE MODELLING

Function tree modelling is carried out in three different ways depending on the individual situation. They are:

1. Function tree of an existing product
2. Function tree of a new product – Top-down approach
3. Function tree of a new product – Bottom-up approach

5.4.1 Steps for creating the function tree of an existing system

- *Step 1. Extract the functions performed by the product*: Take a picture or sketch of the product and label its parts at a sufficiently high level of abstraction to avoid cluttering with details. Consider each labelled item and establish the purpose for its inclusion in the product from a functional point of view. Then list down all the functions that the component performs, to achieve this purpose.
- *Step 2. Arrange the functions as function tree*: Establish a concise function for the product in the 'verb, noun and adjectives' format. This will form the root of the tree. Now consider the functions list and extract the purpose-functions describing the purpose for the inclusion of each unit. These form the 'next level purpose functions' for the product. The purpose functions and their providers are considered for harmonious integration during conceptual design. Now consider the remaining functions and arrange them under each group. These are the action functions meaning that they result from some behaviour of a part of the product. Thus, the leaves of the tree will be action functions and all higher-level nodes will be purpose functions. A function tree shows only the main philosophies of the system and focuses on the evolution of the system.

5.5 MODEL QUESTION AND ANSWERS

Example 5.1: Function tree of a winged corkscrew

A corkscrew is an implement specifically having included a sharp screw-shaped part for opening bottles that are sealed by a cork. To use the corkscrew, lift the wings or arms slightly so that the base can be grasped and, place the base or foot on the top of the bottle's neck. Hold it and the neck of the bottle, with one hand. With the other hand, centre the tip of the screw in the middle of the cork and twist the handle (the part consisting the eye) clockwise until the screw is almost completely embedded in the cork and the wings or arms are raised up in a V shape. Use both hands to press down on the wings, pulling the cork up and out of the bottle. Establish the function tree for the corkscrew.

Model solution

In the first step, take a photograph of the product and label its parts as shown in Figure 5.4. The product is divided into three sections namely (i) base assembly, (ii) screw assembly and (iii) the arms.

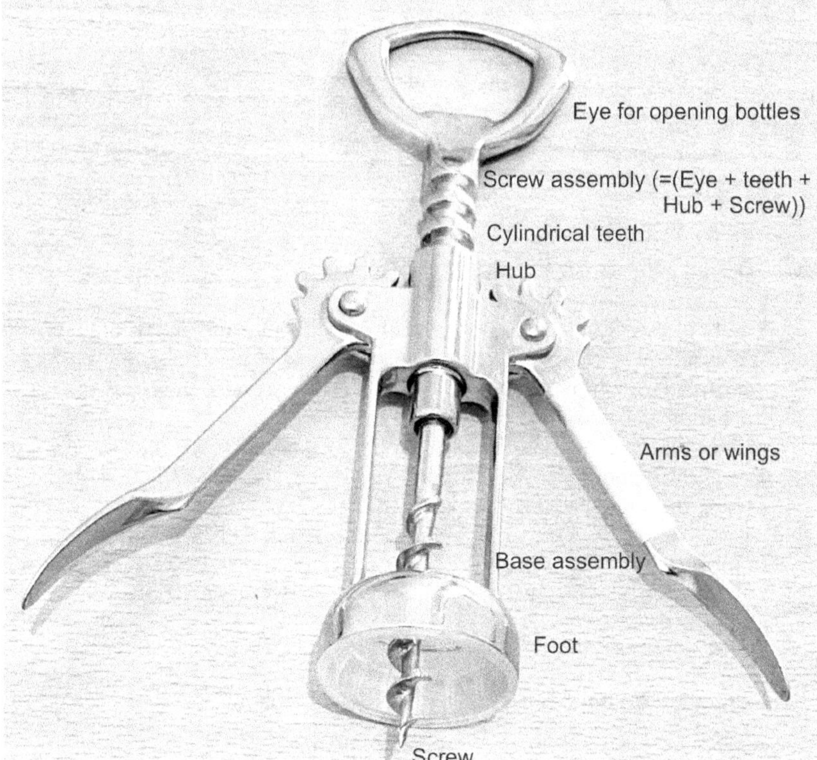

Figure 5.4 Winged corkscrew.

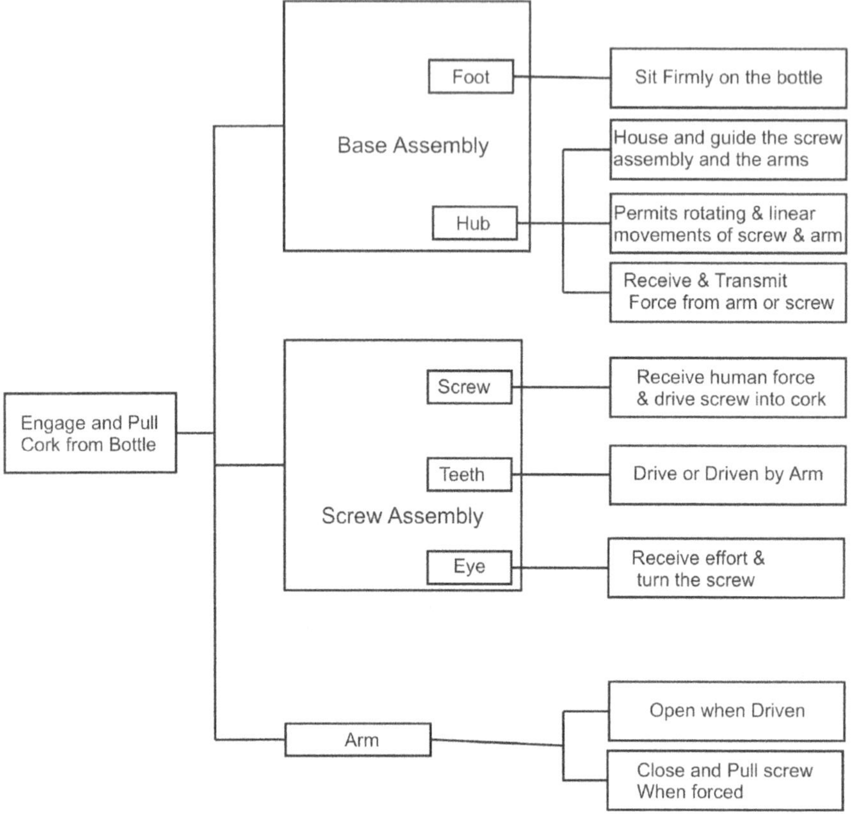

Figure 5.5 Function tree for the winged cork screw.

The base assembly consists of the hub, connecting rods and foot. The screw assembly consists of the screw, connector (a linear portion without threads), cylindrical teeth and eye. The arms receive the effort from the hands and pull the cork up.

Relating to functions the base assembly sits on bottle and acts as an interface and guide for efforts to drive the screw into the cork and pulling the cork. The screw assembly drives the screw into the cork and receives effort from the arms to pull the cork out. The arms got driven to form the 'V' shape when the screw is driven and receives effort and pull the cork out when used to pull the cork. The function tree is shown in Figure 5.5.

Example 5.2: Function tree of a supermarket trolley

A supermarket is a shop where many household items are on display and shoppers walk around and pick the items they need before going to the checkout where they pay for the items. As the sizes of the supermarkets

Figure 5.6 Labelled supermarket trolley.

and the variety of items on display increased, it became necessary to provide trolleys to shoppers to facilitate picking and bringing items to the checkouts. This trolley is a familiar product to everyone. The task here is to establish the function tree for a typical supermarket trolley shown in Figure 5.6.

Model solution

The product is divided into three sections namely (i) wheeled frame, (ii) handle and force application area and (iii) the basket as shown in Figure 5.6. The wheeled frame essentially carries all other units and facilitates movement with the wheels. The handle and force application area receives the human effort and facilitates manoeuvreability. The basket accommodates all the goods the customer chooses and provides a seating area for the child. The purpose functions for these sub-systems are: (i) house the entire system, (ii) manoeuvre the trolley and (iii) hold the payload. The function tree is shown in Figure 5.7.

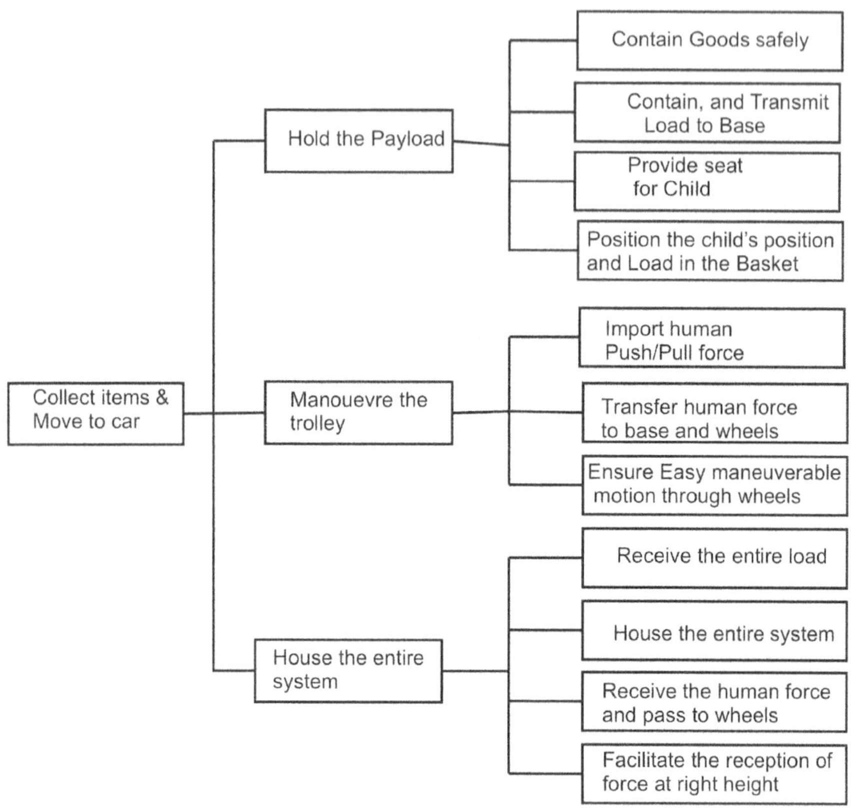

Figure 5.7 Function tree of a supermarket trolley.

Example 5.3: Function tree of an electric kettle

Electric kettle is a familiar household appliance to everyone. The kettle has a jug to contain or fill with water. Electricity is passed through a heating coil which converts the electricity into heat. The generated heat is used to heat and boil water. Thus, there are three main functional components for a kettle (i) filling system, (ii) electric heating system and (iii) the container housing everything or the housing system. As time passed, additional facilities like the viewing window and light indicator were introduced. A typical cordless kettle is shown in Figure 5.8. Figure 5.9 shows the function tree of the kettle. Table 5.1 shows the functions of every part in the kettle and how they join together to provide the purpose functions.

Examples 5.1–5.3 analyzed the function trees of existing and familiar products. Examples 5.4 and 5.5 show the function trees of industrial components and systems.

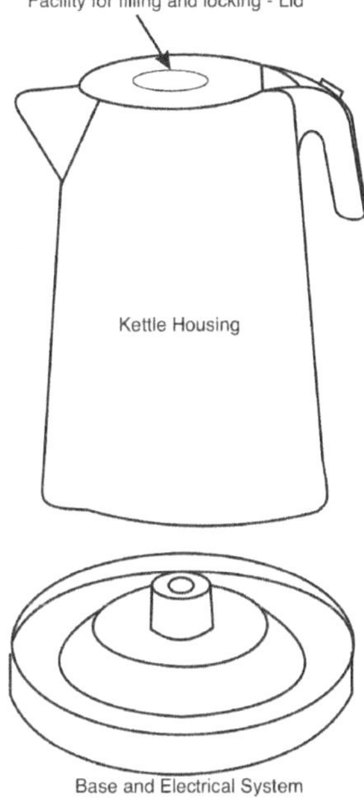

Figure 5.8 An electric kettle.

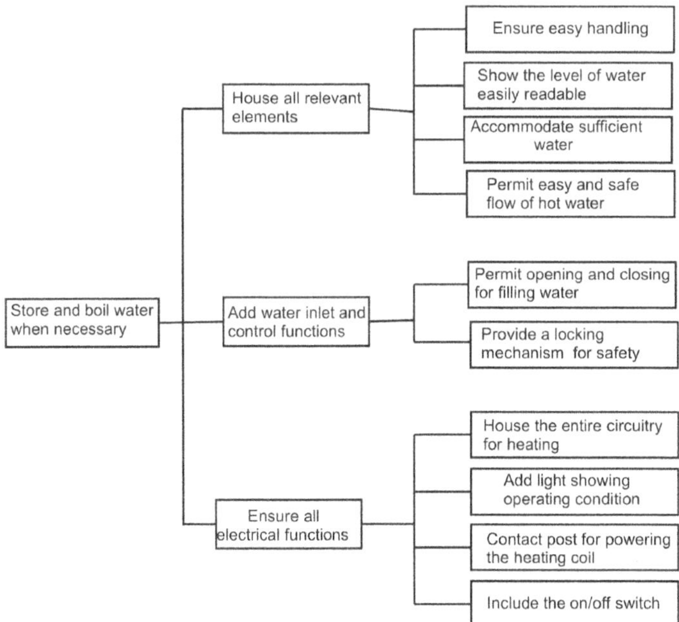

Figure 5.9 Function tree of an electric kettle.

Table 5.1 Function analysis of kettle

Part no	*Group function*	*Part*	*Functions performed*
1	Filling water	Lid lock release	Open and close in a controlled manner
2		Lid	Open the lid to allow filling and close
4	Supply electricity as and when needed	Electrical contact post	Allow electricity available in a safe way
5		Base stand	The base houses the circuitry and supports the kettle bowl
6		On/off button	Allows control of electricity flow
		Heating coil	Converts electrical to thermal energy
3	House everything compactly	Handle	Permits easy moving & handling of kettle
7		Indicator lamp	Shows the functioning mode
8		Water window	Shows the level of water in the kettle
9		Kettle body	Accommodates water
10		Water sprout	Permits the flow of hot water from kettle
		Heating coil	Use electricity to heat water

Example 5.4: Function tree of motor mounting frame

In industries, belt-driven systems are quite common. This is because the motors generate low torque at high speeds while the system requirements often are of the high-torque-low-speed type. To achieve this torque-speed requirement, the motors are fitted with small diameter pulleys, and the system components are fitted with larger diameter pulleys, and the pulleys are connected by belts as shown in Figure 5.10. The pulleys have to be aligned properly as misalignment will shorten the belt life. The problem is exacerbated by the elongating nature of the belts in operation because they are often made with rubber and canvas. This necessitates the motor to be moved, in the x and y directions as shown in Figure 5.11, by small distances every now and then. To meet this requirement, a motor mounting frame is designed and used. It is made up of two structures: structure A and structure B.

Now consider structure A in the frame shown in Figure 5.11. It has slotted holes through which bolts are inserted. The lugs 1 and 2 in the motor align with these holes as shown. Similarly, the lugs, on the other side, also align with the slotted holes 3 and 4. The slotted holes permit positioning the motor in the correct location in the x direction. Structure A is supported by horizontal rails. Structure A is positioned along the y-axis by locating screws in the rails or structure B. Thus, the motor can be positioned accurately for alignment in the xy plane.

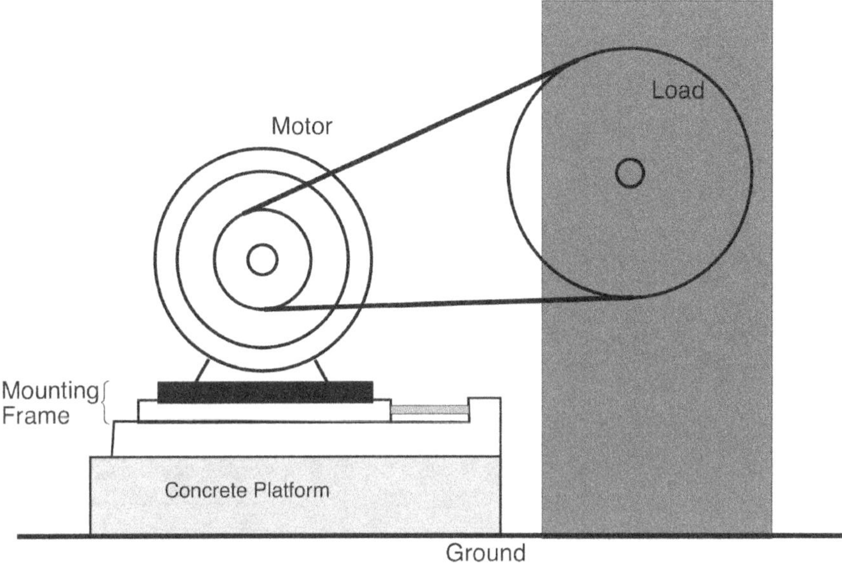

Figure 5.10 Frame mounted motor and load connected through belt.

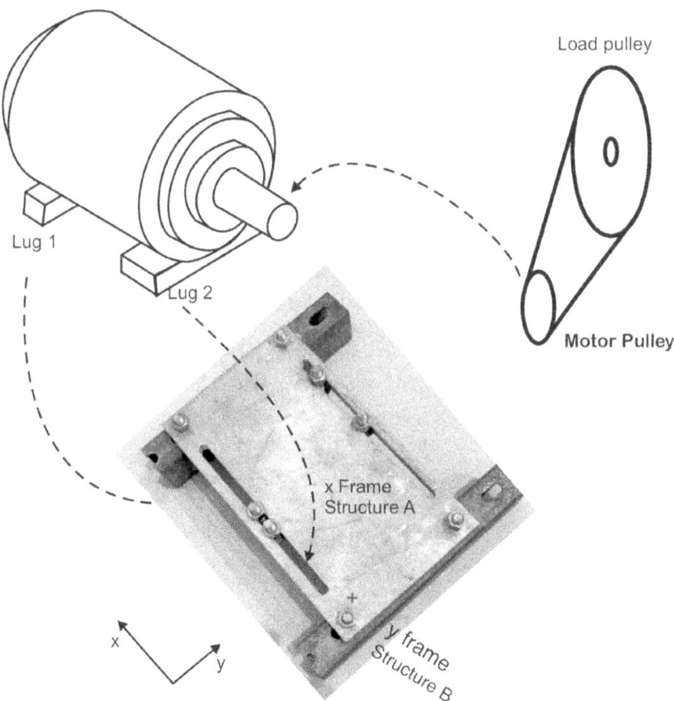

Figure 5.11 Layout of a motor mount frame in an industrial setup.

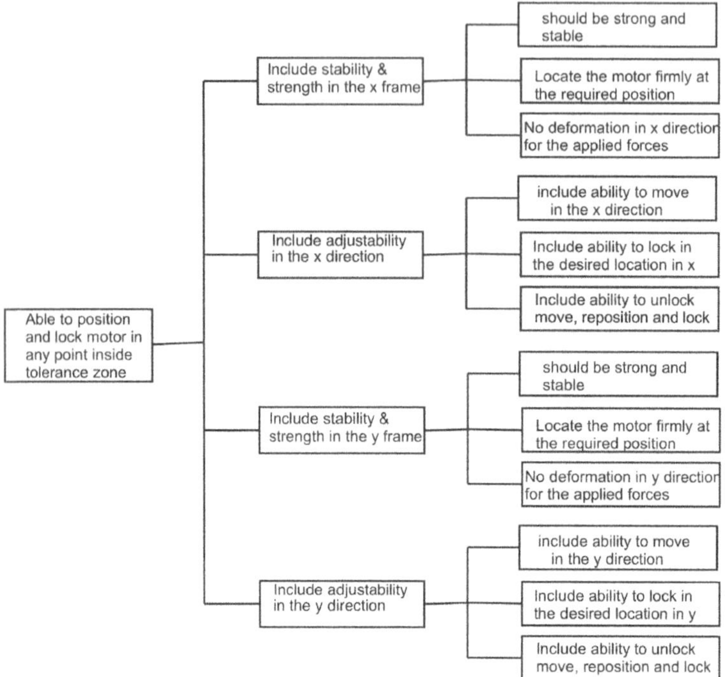

Figure 5.12 Function tree of a motor mounting frame.

Figure 5.12 shows the function tree established for such a motor mounting frame.

Example 5.5: Function tree of an industrial blower

Blowers or centrifugal fans are quite common in industrial applications. A typical blower is made up of four components: (i) motor, (ii) belt and pulley, (iii) impeller driven by a shaft and (iv) the volute casing. When the impeller is rotated the air inside the volute chamber is also rotated. This gives it a centrifugal motion and the air escapes through the outlet. Fresh air enters through the axial opening to fill the vacuum and the airflow continues this way. The big wheel of the belt and pulley drive is connected to the shaft in a typical system. The small pulley is connected to the motor. A typical blower layout is shown in Figure 5.13. The air inlet, outlet and the shaft of the impeller are marked in the figure.

The industrial blower system can be divided into five sub-systems: (i) casing, (ii) impeller, (iii) shaft carrying the impeller, (iv) bearing, housings and platform and (v) belt and pulleys. The function tree is shown in Figure 5.14.

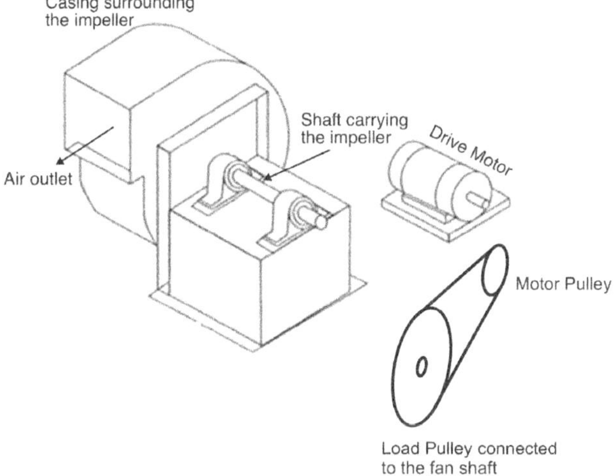

Figure 5.13 Exploded view of the layout of an industrial blower.

Model Solution

Figure 5.14 Function tree of an industrial blower.

5.6 FUNCTION TREE OF A NEW PRODUCT: TOP-DOWN APPROACH

Functional representation of a product is one of the first steps to be taken by the design team and therefore a systematic approach is necessary. This book follows the following procedure when the product's functions are fairly clear (top down). But as a prerequisite, brainstorm all the functions performed by the product in the eyes of the customer.

Procedure for establishing the function tree

i. Choose the prime function or goal of the product.
ii. Establish the purpose functions. These are high-level functions that the product should perform to achieve the goal of the product. In other words, they are essential to the achievement of the goal and they are direct causes of the occurrence of the prime function or goal of the product.
iii. The next level constitutes the 'Action Functions' for each of the purpose functions. In other words (i) they are essential to the performance of the relevant purpose function and (ii) they are direct causes for the occurrence of the relevant purpose function.
iv. In this way continue to arrange the functions such that each function is an Action Function for the function in the level above.
v. Formulate the tree by giving the topological connections.
vi. A function tree can be validated by asking 'How a particular High Order Function is performed'. The correct answer is by performing the connected functions one level below. The functions below should be necessary and sufficient.

5.6.1 Model questions and answers

Example 5.6: Function tree for a table-top display stand for A1 posters

Class room poster presentation is a common activity in design classes. An informative poster will normally be of A1 size 594 mm×841 mm. They are often used in tens in a class and therefore should not take much space. To keep the viewing and discussion easy, the poster should be at table level. Thus, the goal or prime function can be 'Displaying A1 poster in a stable manner' on a table-top or 'Provide a display area and support it on the table top'. This can be divided into three purpose functions (*i*) *Support the poster in open position* (*ii*) *present the poster support in a leaned position and* (*iii*) *house, the handle to carry leaned support, base and heels*. The function tree for the display stand is shown in Figure 5.15. The produced poster stands are shown in Figure 5.16.

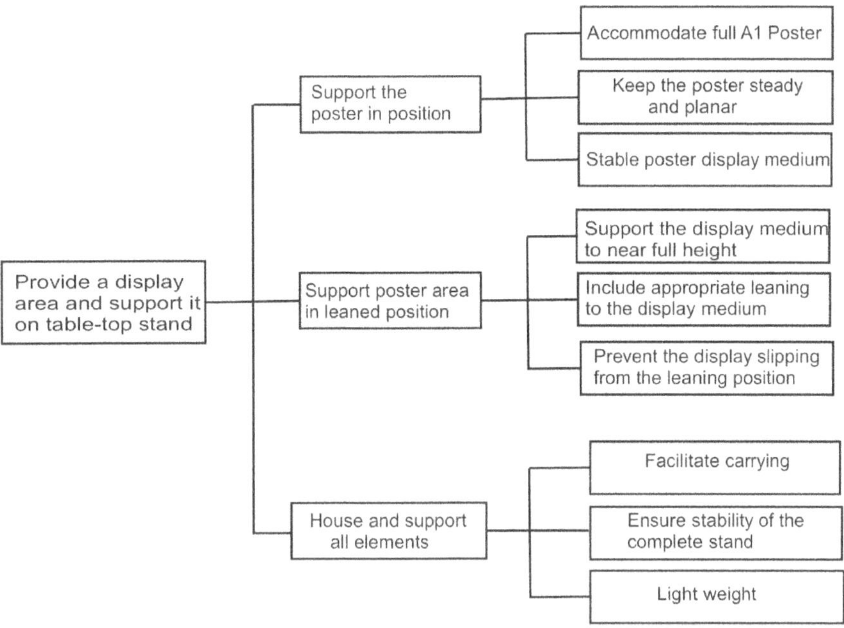

Figure 5.15 Function tree for a table-top A1 poster stand.

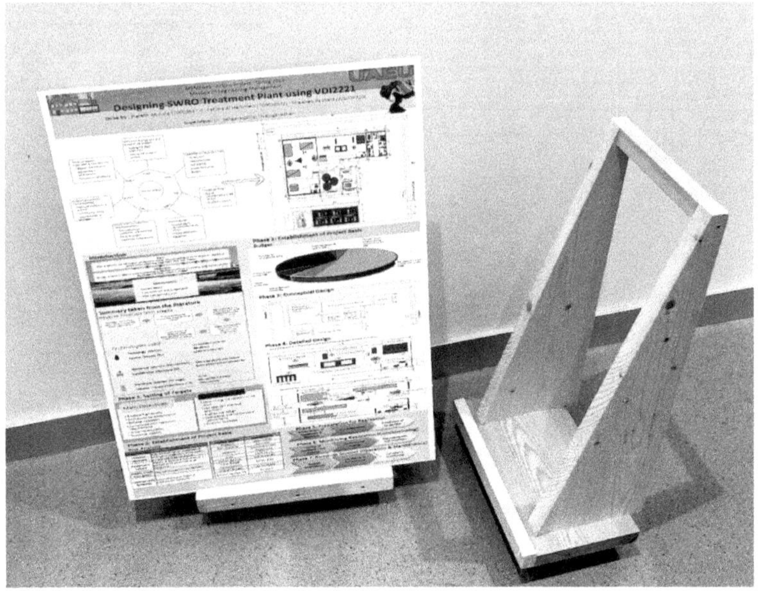

Figure 5.16 Table-top A1 poster stand.

Example 5.7: Function tree for a versatile and compact dessert display

A portable dessert display that can be, moved away when not needed, used to serve a single dessert item, or used with multiple and luxurious dessert items, is a need for the high-class society. The goal or prime function can be 'Displaying one or more dessert items in the dinner table in a cozy fashion'. This can be divided into three purpose functions (*i*) *presenting multiple luxury surfaces to place dessert items* (*ii*) *The surface holds and presents the desserts safely* (*iii*) *Attractive high-class housing of all components and* (*iv*) *facilitate choosing by the persons having the meal.*

Model Solution

The function tree for the display is shown in Figure 5.17. A conceptual design developed is shown in Figure 5.18.

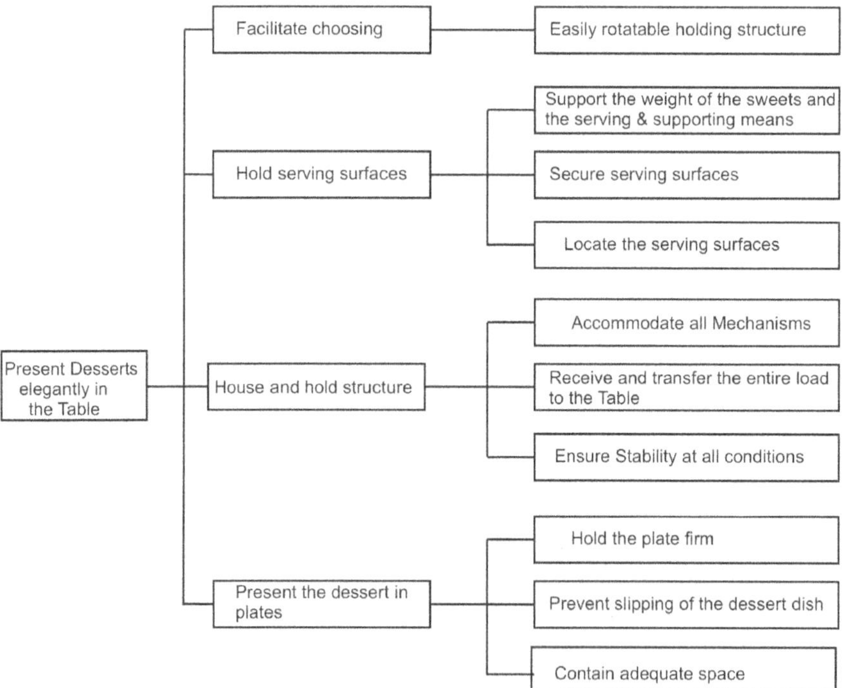

Figure 5.17 Function tree for a table-top dessert display.

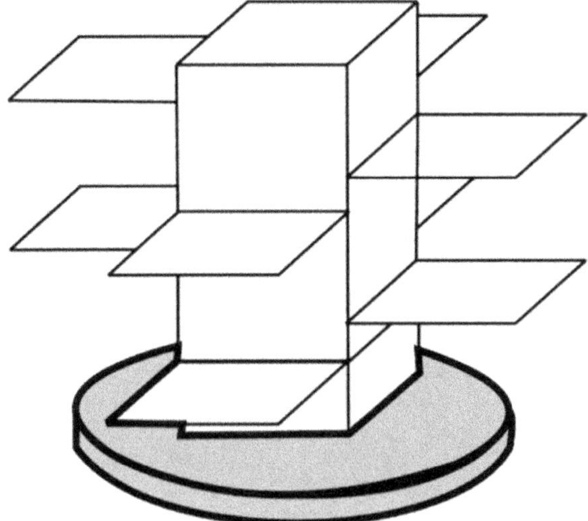

Figure 5.18 Conceptual design of a table-top dessert display.

Example 5.8: Function tree for a product stand for a precision component

Precision Engineering is the designing and making of machines, fixtures, and other structures that have exceptionally low tolerances, are repeatable and are stable over time. There are several thousands of companies that are engaged in precision engineering and they pride themselves as leaders in varied aspects or values. They normally display complex artefacts or components they have designed and/or manufactured so that the visiting customers can be impressed with the capabilities of the company. These components are often placed on individual stands. The stand should be within a foot area of 150 *mm* × 150 *mm* and should be of 50 mm high.

The goal or prime function can be 'Display and highlight the core characteristics of the company in an attractively complex geometry'. This can be divided into three purpose functions *(i) display chosen core characteristics (ii) present in an attractively complex geometry and (iii) house and support the precision item.*

The function tree for the display is shown in Figure 5.19. A conceptual design developed is shown in Figure 5.20.

In this example, the principal characteristic of the company is identified as its leadership in several other desirable characteristics. Six of them were identified as the desirables for reflecting in the stand.

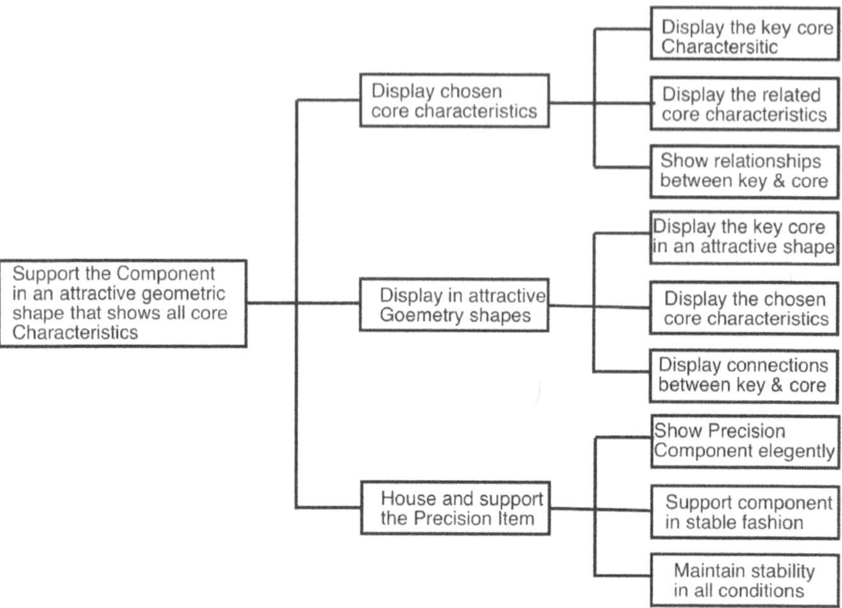

Figure 5.19 Function tree for a stand for a precision component.

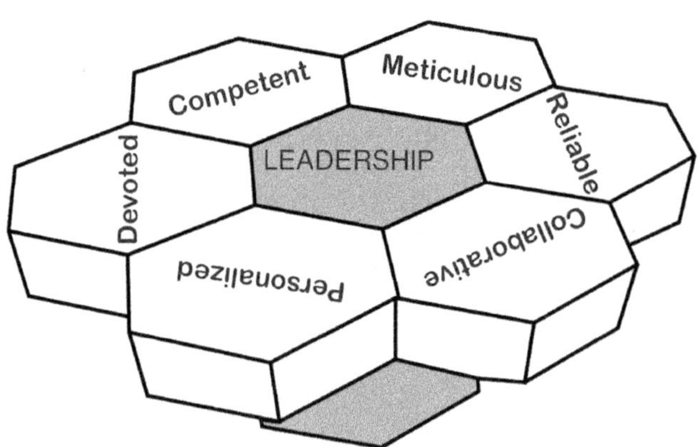

Figure 5.20 Concept for a stand for a precision component.

Example 5.9: Function tree for a graduation gift

Graduation is not only a major milestone for students, it is also a milestone for the family and friends who have supported them. They all feel proud for all the things the graduate has gone through, learnt and entrenched. A graduation gift should reflect all these. A paperweight to sit on the graduate's table is thought of as a graduation gift.

A pearl declares the greatness of the oyster and the graduate declares the greatness of the university. The oyster takes a tiny sand grain and nurtures it to become a pearl. A university takes a student and nurtures the student to become a valuable graduate with so much of ability to serve the community. The goal of making or becoming a graduate is *to nourish the student with so many direct and indirect learning experiences* so that they entrench in the student as specialist characters. The purpose functions can be: (*i*) *Personal experiences,* (*ii*) *integrating with the environment and* (*iii*) *educational learning.* Figure 5.21 shows the function tree of the graduation gift, and Figure 5.22 shows a pearl getting nurtured inside an oyster, as the conceptual design of a graduation project.

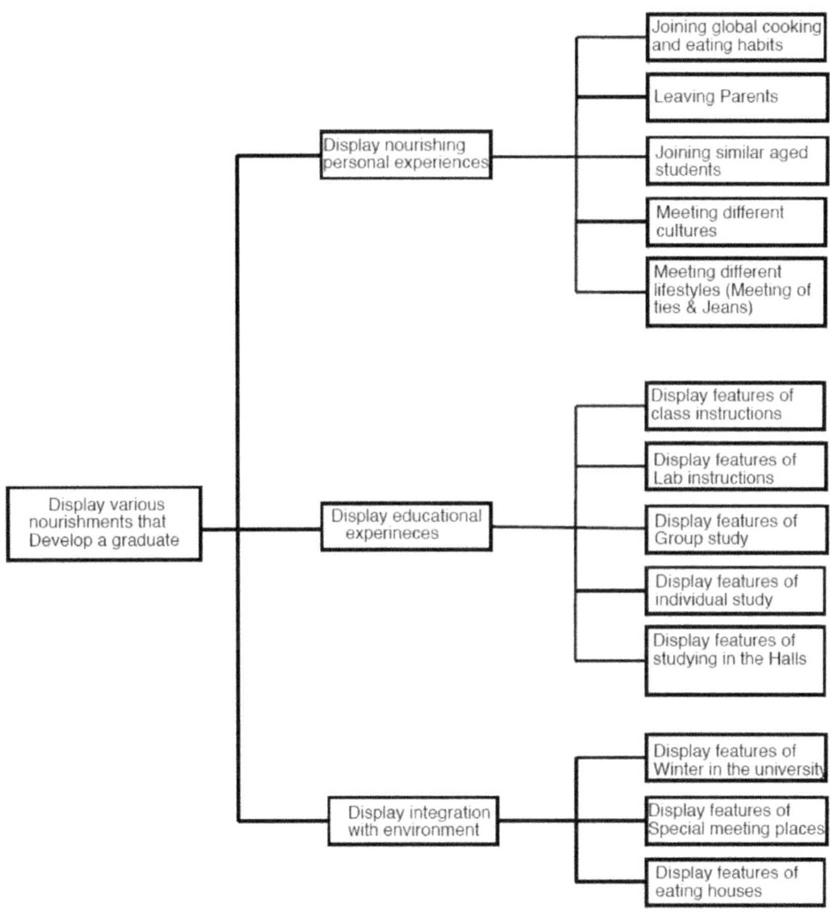

Figure 5.21 Function tree for a graduation gift.

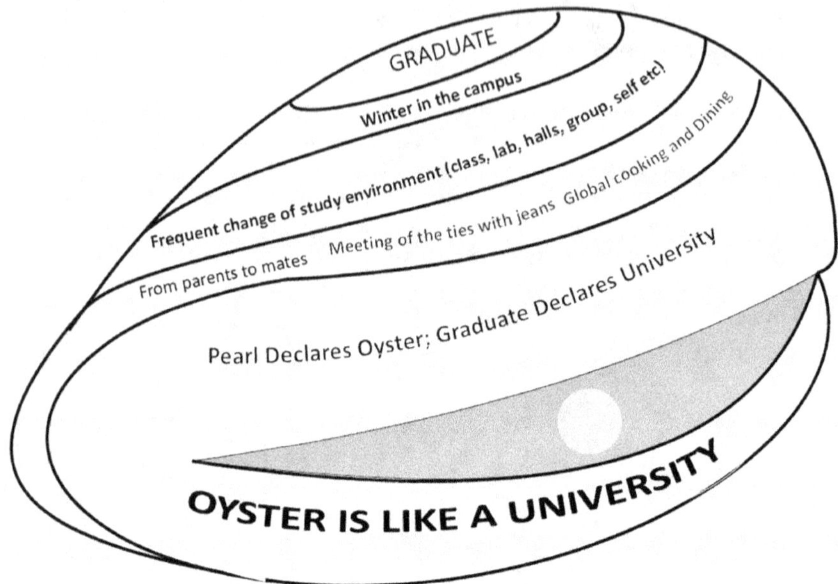

Figure 5.22 Concept for a graduation gift.

5.7 FUNCTION TREE OF A NEW PRODUCT: BOTTOM-UP APPROACH

Procedure for establishing the function tree:

i. Brainstorm all the functions performed by the product in the eyes of the customer. It can be easily done by writing down each function in a card. The functions should be in verb-noun format and in solution neutral form. Do not take any functions for granted. Get as many functions as you can.
ii. Now group these cards into sub-groups and give a name card that represents the sub-group in a verb-noun format.
iii. Choose the prime function of the product. This is the ultimate goal of the project.
iv. Arrange the group cards under the goal function. These group cards are the purpose functions to achieve the goal.
v. The functions in each sub-group become the action functions for the purpose function represented by the name card.
vi. Sometimes the groups may be grouped together to form a higher level of group. In that case, there will be more than one level for purpose functions.
vii. In this way continue to arrange the functions such that each function is a supporting function for the function in the level above.

5.7.1 Model questions and answers

Example 5.10: Function tree of a pith press for a laboratory

Coconut from the palm tree has the nut and husk as shown in Figure 5.23. The top row shows the nut from the tree, the appearance when the nut is half peeled, the nut taken out from the husk and the husk itself. The husk is further processed and divided into pith and coir as shown in the bottom row. This pith is a valuable material and is widely used in the agriculture sector. It is very loose and can be compacted to almost one-fifth of its original volume. It is said to have heat capacity similar to coal and can become a useful substitute fuel for bakeries.

A laboratory researching on this coconut pith needs a manual pith press. It is assumed that a compression chamber filled with the pith and pressed by a ram, is the mechanism employed here. With this approach establishing the function tree is the task here.

Answer

The first task in establishing the function tree from the bottom-up approach is to list down the functions needed in a brainstorming or other similar method.

The collected functions, in the verb-noun-phrase format, from a brainstorming are as follows:

1. Allow easy access to remove chamber for filling
2. Facilitate easy positioning of the chamber
3. Permit built-in alignment for the ram and chamber

Figure 5.23 Coconut and its constituents.

4. Ensure easy filling of the required quantity
5. Facilitate easy removal of block
6. Permit gradual loading
7. Permit intermittent stops for loading
8. Permit high pressure
9. Permit maintenance of high pressure
10. Permit partial release of load
11. Ensure minimum requirement of effort
12. Integrate all components in fixed places
13. Anchor firmly to the ground
14. Protect from accidental overload

These are the action functions and they are grouped to form the purpose functions and the prime function or goal.

The established function tree is shown in Figure 5.24.

A possible conceptual design is shown in Figure 5.25.

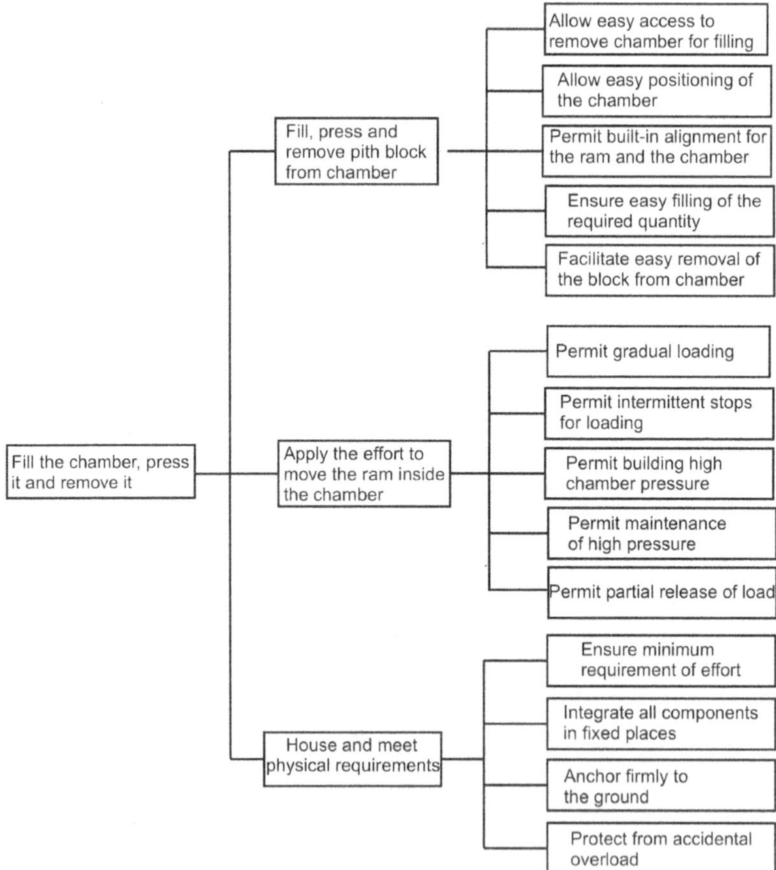

Figure 5.24 Function tree of a manual pith press.

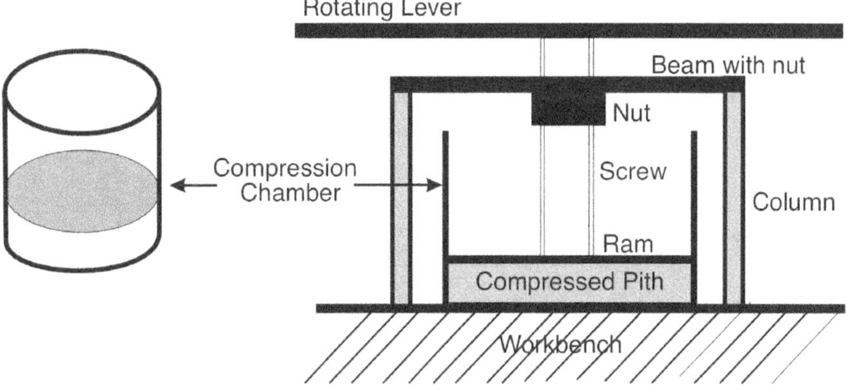

Figure 5.25 Possible conceptual design of a pith press.

Example 5.11: Function tree of a mechanical energy sink

Mechanical energy sinks are needed in development laboratories to test the power generated by trial designs and to brake rotating shafts in educational laboratories. A mechanical energy sink using band brakes, that can safely absorb 2 kW of power needs to be designed and fabricated. Establish the function tree using the bottom-up approach, for such a system.

The collected functions from a brainstorming are as follows:

1. Connect to the driving shaft smoothly
2. Allow friction-free support to the receiving shaft
3. Permit built-in alignment for the drive and driven shafts
4. Handle varied magnitudes of loads
5. Dissipate the received energy as heat
6. Control temperature difference as 30°C at most
7. Allow measurable temperature difference for a range of inputs
8. Measure the input and output band tensions
9. Measure the temperature of the wheel, band and atmosphere
10. Record the measurements continuously
11. House the entire system firmly
12. Allow convenient working height
13. Allow built-in safety in operation
14. Allow proportionate dimensions (Figures 5.26 and 5.27)

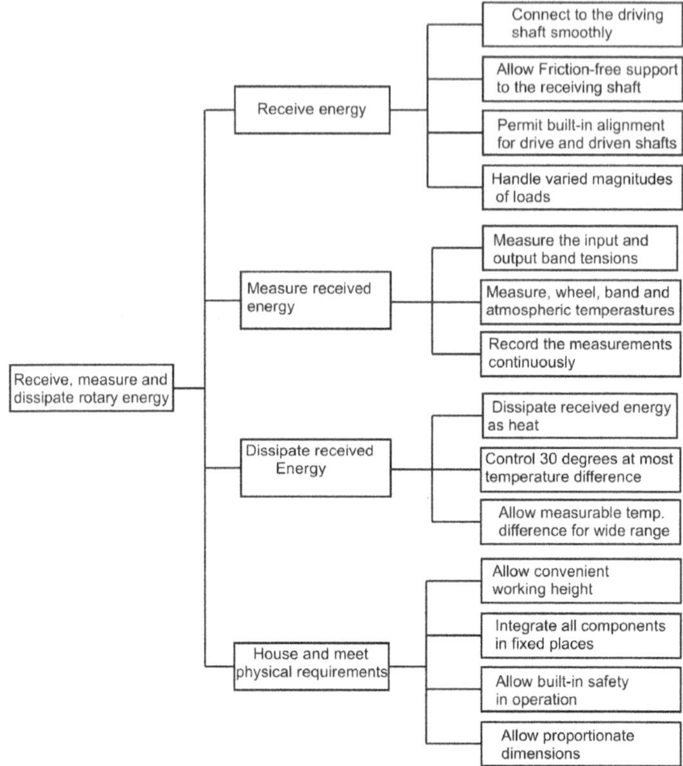

Figure 5.26 Function tree of a mechanical energy sink.

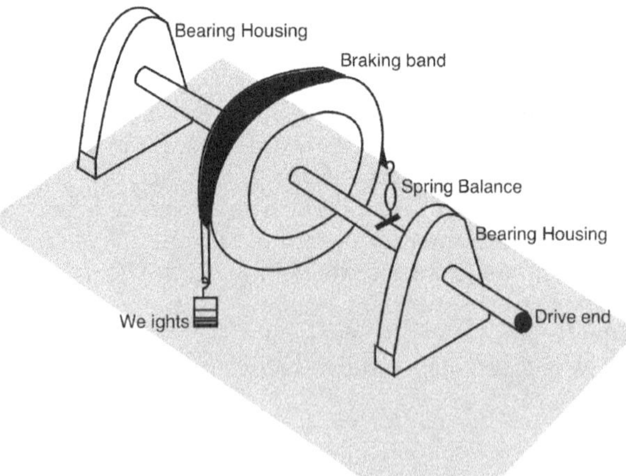

Figure 5.27 Typical conceptual design of a mechanical energy sink.

Example 5.12: A trolley to gather light material in farms

In rice fields, during harvesting time, people gather the harvested plants as shown in Figure 5.28, bundle them, and carry them to a common point where the grains are separated. An engineer, son of such a farmer, thinks that a lightweight trolley that could be dismantled and assembled easily can be made at very low cost, to carry the harvested plants. Establish the function tree of that trolley from a bottom-up approach.

Functions that would be Performed by the Cart are as follows:

1. Provide large area for stacking as the material is light
2. Facilitate easy bundling for loading and unloading
3. Allow easy access to harvested material
4. Allow easily removable side restraints if installed
5. Use lightweight components
6. Use modular components
7. Permit built-in alignment
8. Use simple activities to assemble the product
9. Eliminate carrying as far as possible
10. Provide easy rolling
11. Allow easy working height
12. Integrate all components in fixed positions
13. Ensure safety in all conditions
14. Let the dimensions be proportionate for operation by a single person

The function tree of the trolley is shown in Figure 5.29. A typical conceptual design is shown in Figure 5.30.

Figure 5.28 Harvesting rice.

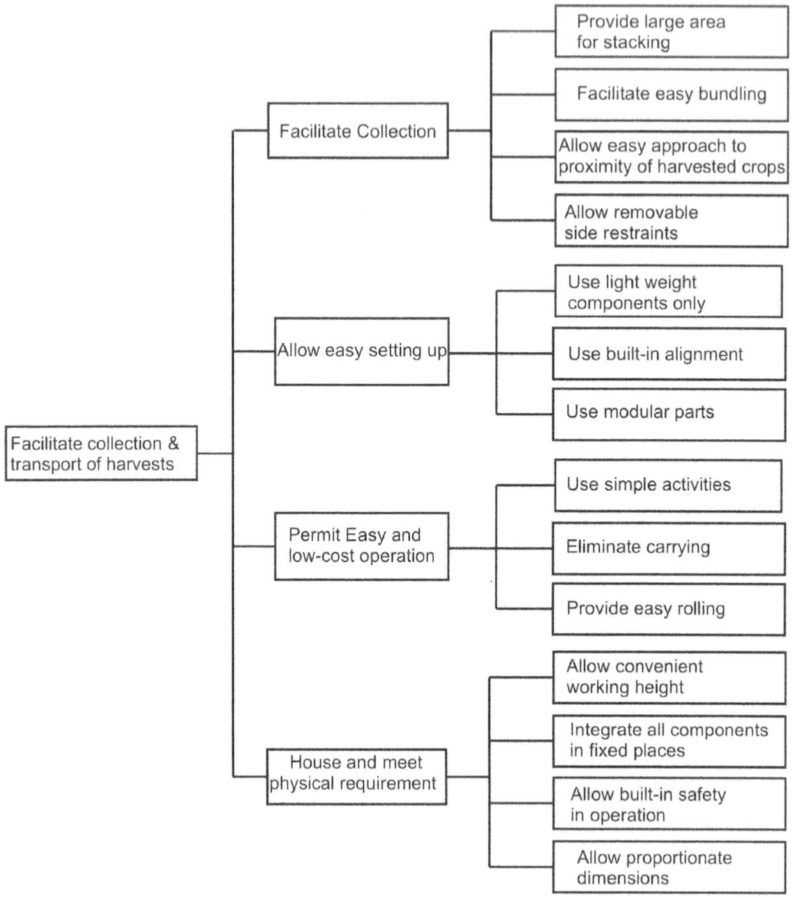

- Facilitate collection & transport of harvests
  - Facilitate Collection
    - Provide large area for stacking
    - Facilitate easy bundling
    - Allow easy approach to proximity of harvested crops
    - Allow removable side restraints
  - Allow easy setting up
    - Use light weight components only
    - Use built-in alignment
    - Use modular parts
  - Permit Easy and low-cost operation
    - Use simple activities
    - Eliminate carrying
    - Provide easy rolling
  - House and meet physical requirement
    - Allow convenient working height
    - Integrate all components in fixed places
    - Allow built-in safety in operation
    - Allow proportionate dimensions

Figure 5.29 Function tree of the cart for harvesting rice.

Figure 5.30 A conceptual idea of the trolley.

5.8 FUNCTION STRUCTURE AS INPUTS AND OUTPUTS OF FLOWS

A black-box model is an input/output model that is useful for figuring out the functions of engineering systems. In this model, a rectangle called a black box represents the system under consideration while arrowheads, on the left and right side of the rectangle mark, inputs and outputs respectively. The inputs and outputs can be the three types of flows: (i) material, (ii) energy and (iii) signals or information as shown in Figure 5.31. All inputs and outputs are entered to obtain a clear visualization of the system being considered. The *thick line* represents material, thin continuous line represents energy and *dashed line* represents signals or information in this diagram and in this type of diagrams in this book.

A function structure breaks down the overall function as given in the "black-box diagram" into a sequence of sub-functions, represented by the function blocks and, connected by the flows inside the black-box.

Functional description: A functional description of an activity or a product gives a function (verb) acting on a flow (object). Functional descriptions are solution neutral and do not refer to any particular solution. A function structure diagram is a graphical representation of the functions a product performs on its inputs and outputs. The sub-functions are connected by 'flows' on which they operate. Flows are materials, energy or information that is used by or affects the product.

Steps to construct a function structure:

1. *Construct the black-box diagram*: Understand and plan the functions the product performs. Identify the inputs as flows of material, energy and signals. In the same way, identify the outputs from the product. Using these, construct the black-box functional representation.
2. Pencil the function structure by carrying out the following steps:
   a. Normally a signal initiates the functioning of the product. This signal triggers the energy to do some activity (work) and the energy works on the material flow. Therefore, the first step is to plan the signals that generate actions inside the black box.
   b. The signals normally trigger power or energy. Therefore, the next step is to identify the activities where energy or power is involved.

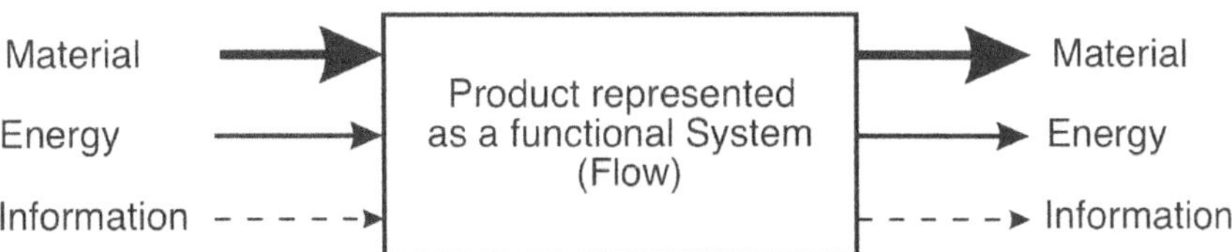

Figure 5.31 Black-box functional representation.

c. The power normally acts on the material. Now the material undergoes some change and in its final stages becomes the output.
d. The product normally effects changes in material or energy as the desired output. Check with the black-box diagram for consistency.
e. Carry out internal checks for the continuity of all flows.

3. Firm up the function structure by asking the following questions:
   a. What are the inputs and outputs?
   b. What happens to the input and the outputs inside the box?
   c. Where does the output come from?

5.8.1 Model question and solutions

Example 5.13: Function structure of an electric kettle

An electric kettle receives water (material), electricity from the mains (energy) and a switching on/off (signal) from the user. It produces hot water and LED/audio signals when switching on and switching off as outputs. Some heat is lost and it is called waste energy. The black-box diagram for the kettle is shown in Figure 5.32.

The process starts with the user filling the kettle jug, and placing it in the base. Then the user switches the kettle on. The switching passes electricity to heat the water and triggers power to the LED light. The power continues to heat water and the LED light continues to burn. Once the water is boiled it triggers a signal that turns the switch off. It cuts the power flow to the water and turns off the LED light. The outputs therefore are (i) boiled water (ii) wasted energy as heat and (iii) LED signal when switching off. These are illustrated in the function structure shown in Figure 5.33. In the figure thick lines show material flow, thin lines show signal flow and dashed lines show energy flow.

Figure 5.32 Black-box representation of functions (electric kettle).

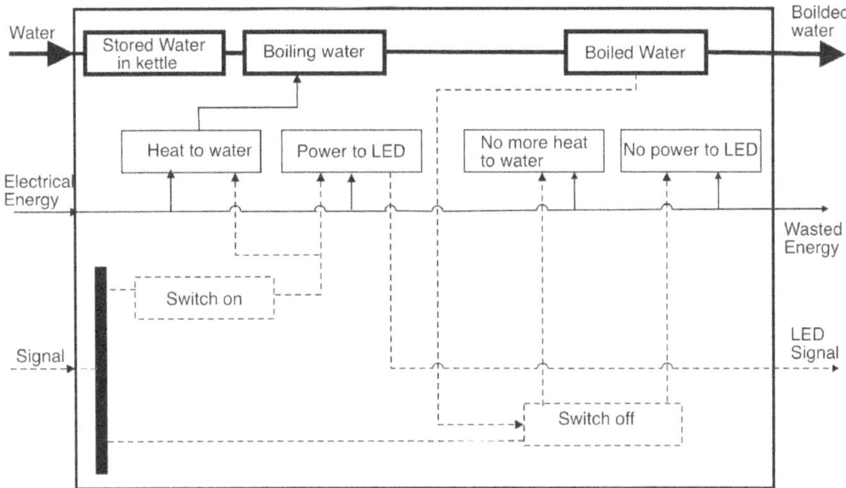

Figure 5.33 Function structure for an electric kettle.

Example 5.14: Function structure of a vacuum cleaner

A vacuum cleaner passes air through the loose debris to lift the debris. Then the laden air is subjected to a separation effort to collect the debris. Functioning of a vacuum cleaner enjoys two kinds of energy input namely electrical energy and human energy. The human energy is used to loosen the debris that is stuck to the floor. As shown in Figure 5.34, the inputs are (i) debris and (ii) air as material, (iii) electrical, and (iv) human energy in the energy category and (v) switching on and switching off as signals. The outputs are (i) waste air, (ii) collected debris, (iii) waste energy and (iv) debris level indication. Figure 5.35 shows the function structure. In the figure thick lines show material flow, thin lines show signal flow and dashed lines show energy flow.

Figure 5.34 Black box diagram of the functions of a vacuum cleaner.

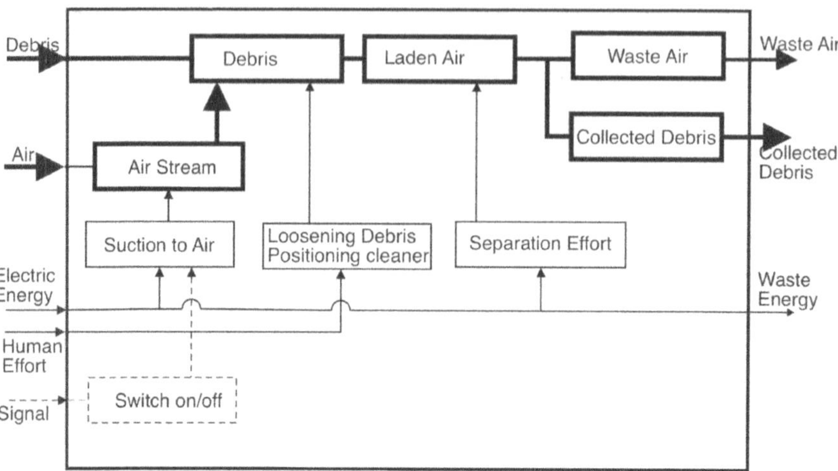

Figure 5.35 Function structure for a vacuum cleaner.

Example 5.15: Function structure of a microwave oven

A microwave oven generates microwave which changes polarity and increases the energy levels of the water molecules in food by causing them to vibrate. The vibration produces heat that cooks the food. The microwave is contained within the oven container by reflection. The black-box diagram shown in Figure 5.36 indicates the inputs and the outputs.

The process starts with the user filling the food to be cooked in a tray, and placing it on the turn-table. He chooses the program and sets the time. Then the user switches the oven on. Two signals are generated to start the turning of the table and generate the microwave. The process continues to generate microwave and turn the table to heat the food and cook it. Once the set time passes the power is cut off and the turn-table stops and generation of microwave stops. Cooked food, waste energy and audio signals are the outputs. Figure 5.37 shows the function structure.

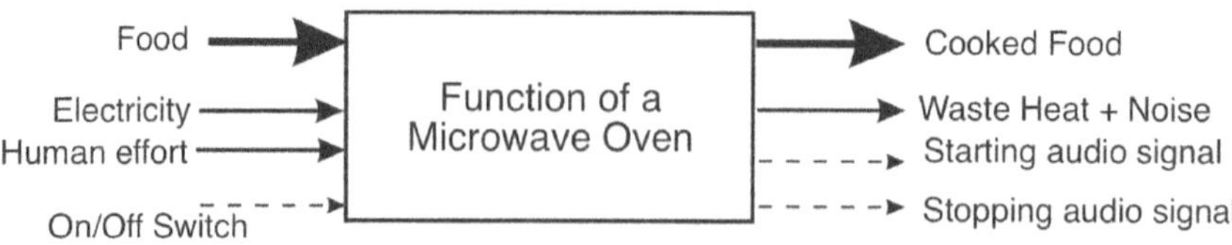

Figure 5.36 Black box diagram of the functions of a microwave oven.

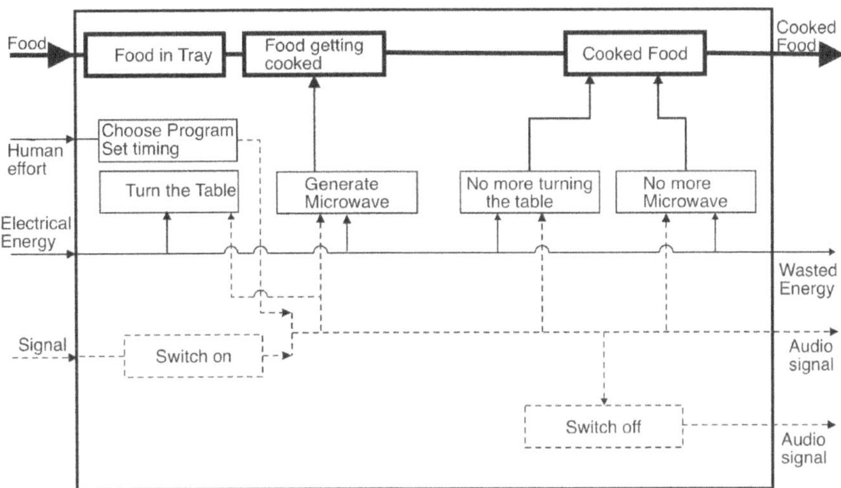

Figure 5.37 Function structure of a microwave oven.

Example 5.16: Function structure of a supermarket trolley

A supermarket trolley is used by customers, to optionally sit their child if they have one, to collect and temporarily store items till they collect all items they want and pay, and move the items and child to the car.

For the overall black box, the inputs are, two types of material, the child and items, manual effort as energy and human cognition as signal. This black-box notation is illustrated in Figure 5.38.

The flow pattern is as follows:

1. Human cognition decides or triggers the energy to pick the child and sit it in the trolley. Accordingly, this triggers the energy and the child is picked and seated on the trolley.
2. Then the signal from the cognitive system decides to pick items and store so that child and items are in trolley.
3. Then the cognition decides to finish. Accordingly, it goes to the counter and pays electronically.
4. The cognition then decides to go to the car park and the energy moves the trolley to the car.

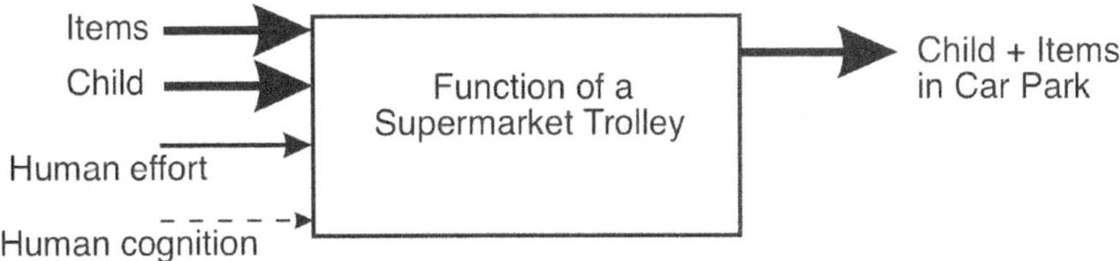

Figure 5.38 Black-box diagram of the functions of a supermarket trolley.

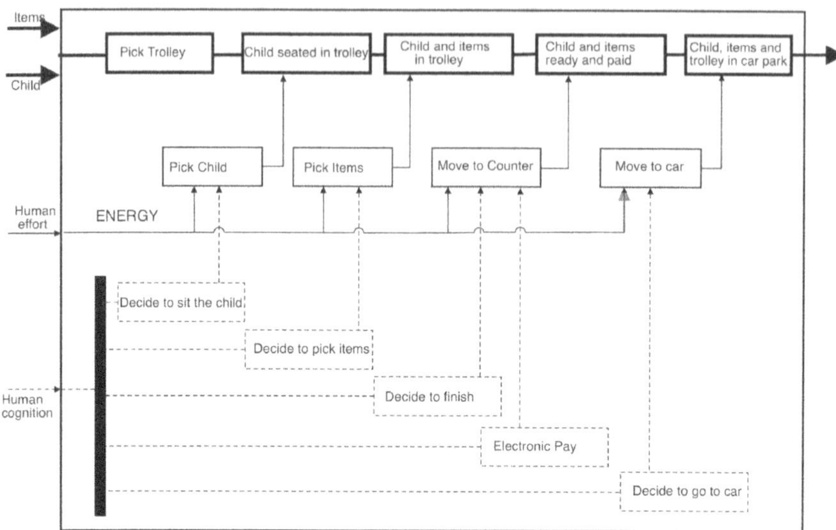

Figure 5.39 Function structure of a supermarket trolley.

These are illustrated in the function structure diagram shown in Figure 5.39.

Example 5.17: Function structure of a coffee roaster

Green coffee beans are heated to get the darker roasted coffee beans. The green beans have a humidity of around 10% and they need drying before the actual roasting process can start. The browning stage follows the drying stage where the aromas of the beans start to develop due to the conversion of the aroma precursors to aroma compounds. This process should have carefully controlled time and temperature. This follows the roasting process where roaster can fully shape the taste profile. Generally, there are three kinds of roast: light, medium and hard roasts. The waste that are produced during roasting are chaff and smoke. Figure 5.40 shows the black-box representation of the functioning of a coffee roaster (Figure 5.41).

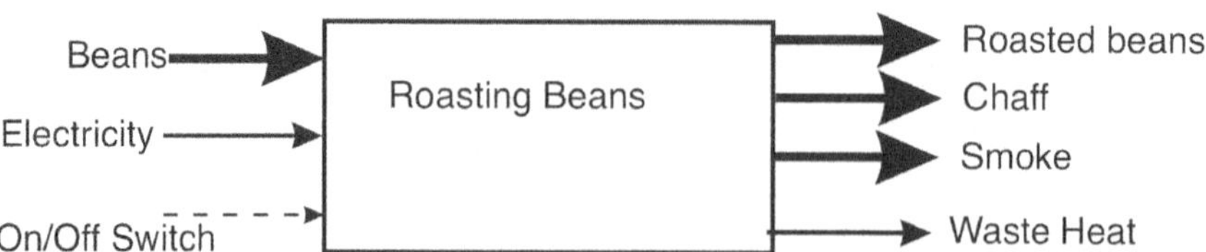

Figure 5.40 Black-box diagram of the functions of a coffee roaster.

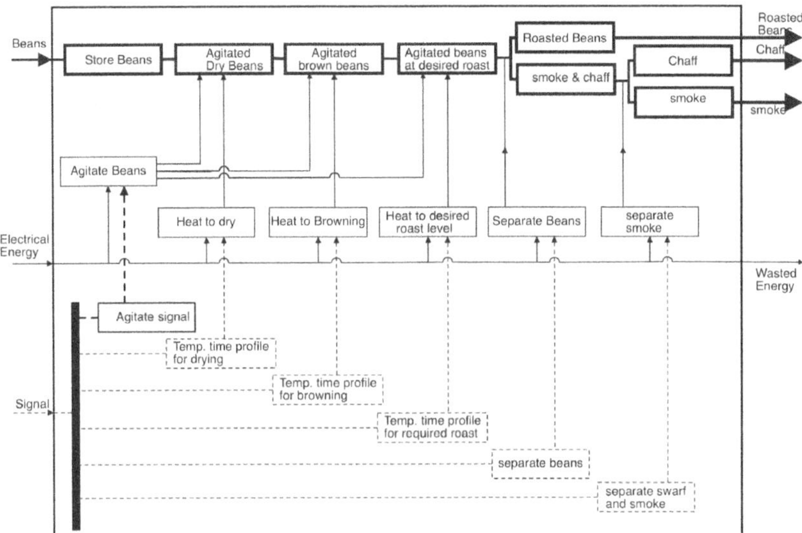

Figure 5.41 Function structure of a coffee bean roaster.

Example 5.18: Function structure of a washing machine

Function structure is established to structure the flow of signals which activates the energy to perform some action on the material. As the number of functions increases, some lower-level functions are omitted to keep the function structure simple and explain the important high-level functions. This process is called controlling the level of abstraction. In the development of the function structure of a washing machine, loading the laundry and finishing audio signal are some examples of this abstraction.

The black box representation of the functions has (i) laundry, (ii) detergent, and (iii) water as material inputs, (iv) electrical energy and (v) human energy as energy inputs and (vi) time-programmed control signals as information or signal inputs. Similarly, (i) clean clothes, (ii) two flows of wastewater as material outputs, (iii) waste energy and noise as output of the energy type and (iv) audio and LED signals as the signal type outputs. Figure 5.42 shows the black-box diagram.

The process starts with the user choosing the program and starting the machine. The action starts with the signal to open the tap to let water inlet. The water goes through the detergent and dissolves it and creates soap water. To move the soap water and mix it with the laundry and soak it, another control signal triggers the power. In this manner, the timed control signals trigger power to carry out various functions in a timed sequence to wash the laundry, into cleaned clothes. The function structure is shown in Figure 5.43.

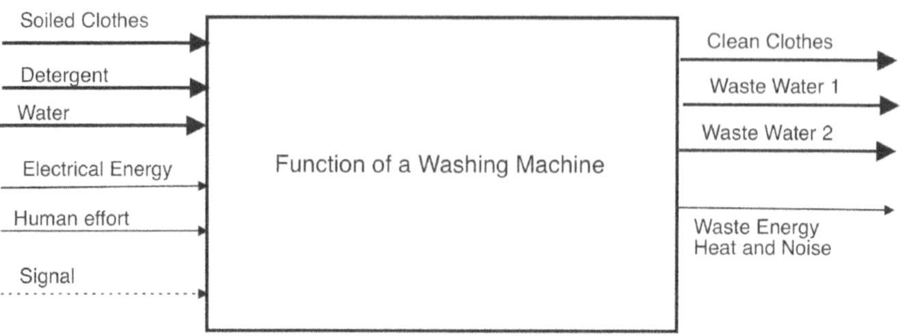

Figure 5.42 Black-box diagram of the functions of a washing machine.

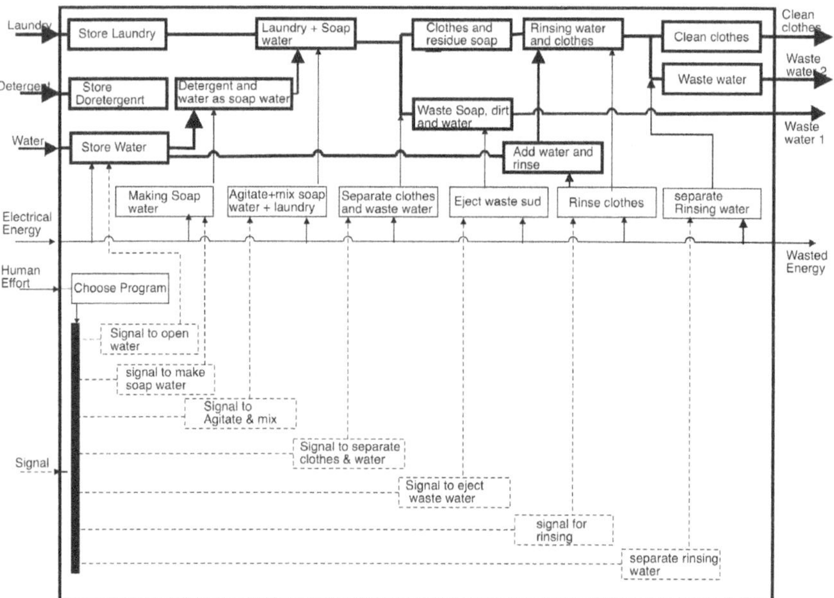

Figure 5.43 Function structure of a washing machine.

Example 5.19: Function structure of a picking and placing system

Figure 5.44 shows a vision-controlled system to pick the pastry coming out of the oven in a conveyor and place them in packs of four. The vision system advises the robot about the coordinate information of the positions of the incoming pastry and the locations of the positions 1, 2, 3 and 4 in the box. In the system under consideration here the pastry (material) is picked (energy) by the delta robot and placed in the position tray as instructed by the vision system (signal or information).

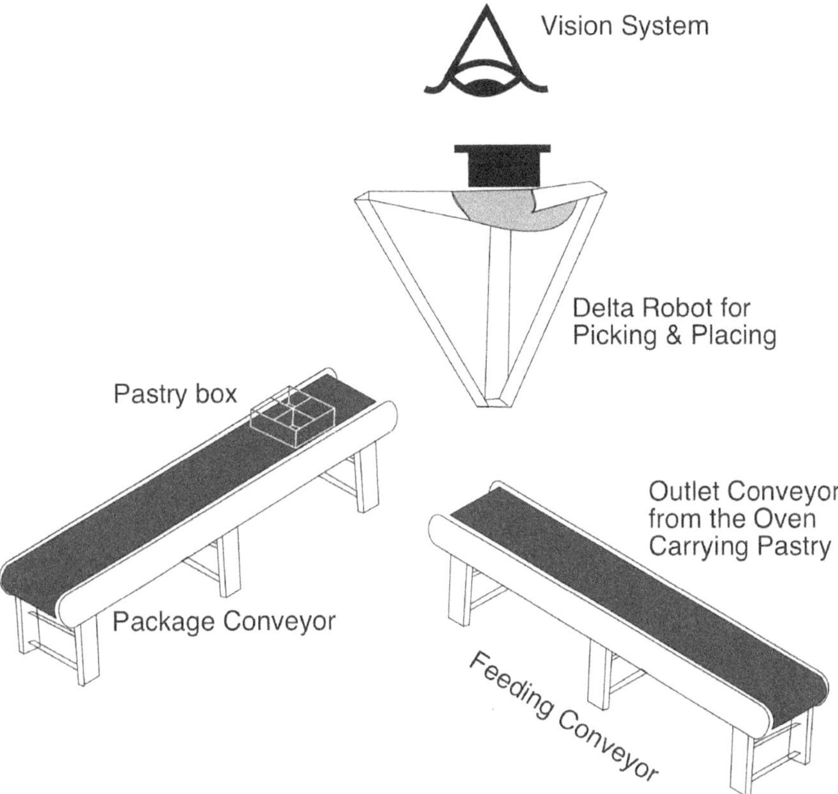

Figure 5.44 System for transferring pastry from feeding to package conveyor.

Each box is considered a cycle. The black-box diagram of the functional model is shown in Figure 5.45.

The control signals trigger the energy in a timed sequence to execute these functions. The function structure is shown in Figure 5.46.

Figure 5.45 Black-box diagram of moving pastry from feeding to package conveyor.

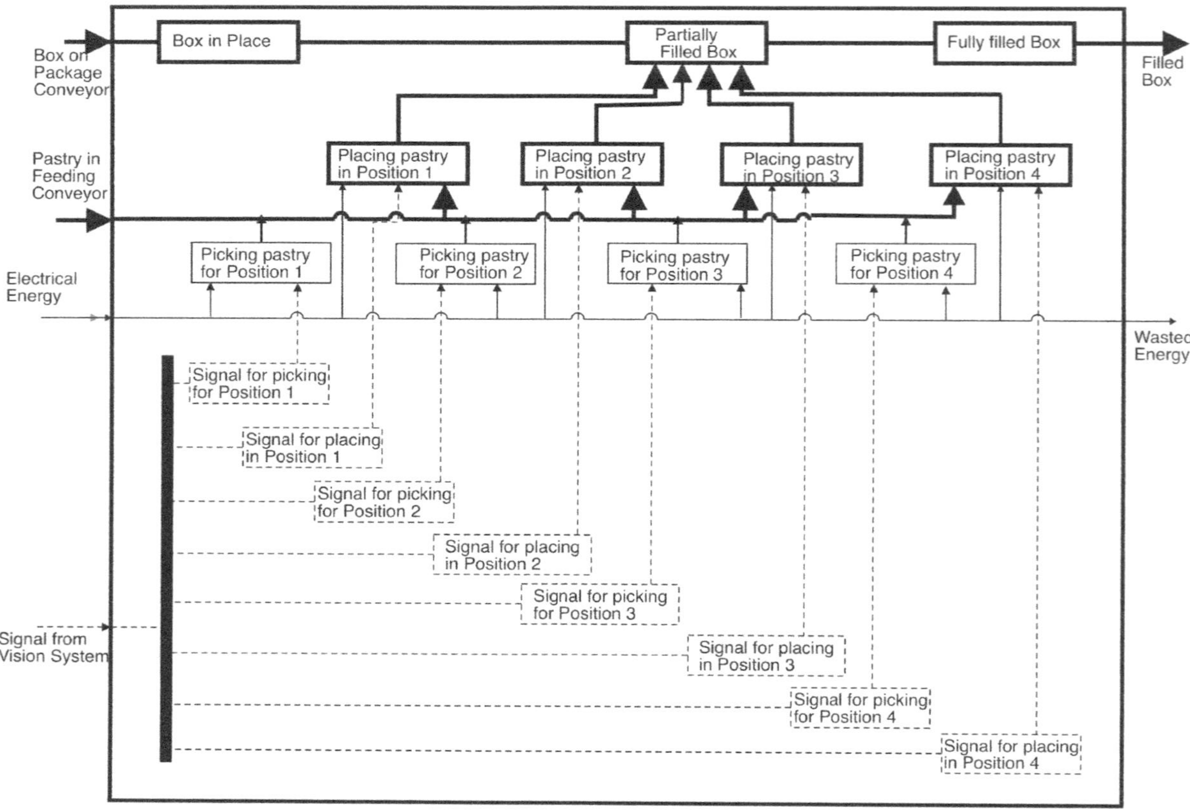

Figure 5.46 Function structure of pastry transfer from feeding to package conveyor.

REFERENCES

1. Vermaas P.E., Technical functions: Towards accepting different engineering meanings with one overall account. In: *Proceedings of the 2010 TMCE Symposium*, I. Horvath, F. Mandorli, and Z. Rusak (Eds.), Ancona, Italy, 2010.
2. Haik Y., Sivaloganathan S. and Shahin T.M., *Engineering Design Process*, 3rd edition, Cengage, Noida, 2017.
3. Deng Y.M., Function and behavior representation in conceptual mechanical design, *Artificial Intelligence for Engineering Design, Analysis and Manufacturing*, 16, 343–362, 2002.

Chapter 6

Need-metric matrix

6.1 INTRODUCTION

Specifications of a product can be seen as the product's performance in measurable quantities, and in the present customer-based design process the stakeholders provide their expectations on the performance of the product. The information content of their verbatim would normally vary and to make them similar in content, the design team translates them into needs. The needs are prioritized to facilitate meeting them by the product. '*If the implementation's effect cannot be measured, it cannot be improved*' is a statement attributed to Lord Kelvin. For this reason, each need has to have a representing characteristic that can be measured easily. A precise, measurable characteristic of the product, whose control will reflect the degree of satisfaction of a corresponding specific need, is called a metric of that need. This identification of metrics facilitates the definition of the product's performance in measurable quantities. The metrics involve measurements and the process of measurement has four levels namely, nominal, ordinal, interval and ratio and the needs may require measurement at any of these four levels. Need-metric matrix is a tool to define the metrics in these levels, for each need individually and an appropriate unit. The levels of measurement and the need-metric matrix are explained in this chapter. Examples are given to provide experience to the reader in the development of need-metric matrix.

6.2 LEVELS OF MEASUREMENT: THEORETICAL CONSIDERATIONS

Consider measurements as a topic for consideration. It can be defined as a 'means of assigning numbers or other symbols to characteristics of objects according to certain pre-specified rules.' The variables used in measurements can be classified into two different types: qualitative or quantitative variables. Qualitative variables, also referred to as categorical variables,

 DOI: 10.1201/9781003484950-7

describe data that are not numerical and fit into specific categories. Qualitative variables have no natural order and they cannot be added, subtracted, multiplied, or divided. Quantitative data, on the other hand, has numerical value and is something that can be counted or measured by a tool or scale. Both qualitative and quantitative variables are classified into four different levels, according to how they can be treated mathematically, in the following way:

1. *Nominal*: Nominal scales are at the lowest level of measurement, and they are simply names based on characteristics such as gender, ethnicity, place of birth with no specific order. Typical coding used are numbers, letters, colours, labels or any other symbol that distinguishes between the categories.
2. *Ordinal*: Ordinal scale names and then places variables in specific order, based on their attributes. It is a scale used to simply depict the order of variables and not the difference between each of the variables. It typically uses non-numeric categories for example, low, medium, and high. The ordinal scale is best used when rank is important, but when intervals between data points are not the same length. A typical example is a students' rank in class.
3. *Interval*: Interval scales goes beyond categorizing and ordering by establishing consistent distances known as intervals between categories or data points. Interval scale is a numerical scale where the order of the variables is known, as well as the difference between these variables.
4. *Ratio*: Ratio scale, at the highest level of measurement, is a measurement scale that not only produces the order of variables but also makes the difference between variables known, along with information on the value of true zero. This is considered the highest level of measurement because it satisfies all four levels of measurement, contains the most information about the data values, and includes the presence of zero.

6.2.1 Measurement of water in a tank: an illustrative example

In Northern Sri Lanka, there are many tanks, and a typical one is shown in Figure 6.1. The tanks get filled during the months of September to January due to the monsoon rain and will start losing water during the summer months of February to August. During these months, some of these tanks may get completely dry. Consider the measurement of the water in the tank.

Figure 6.1 Measurement of water in a tank.

1. The first and easiest measurement is to see whether there is water or not. The tanks will fall into two categories: those having water and those that are dry. This is a nominal category of measurement.
2. During the monsoon months, the water level increases. If water level during the monsoon months is arranged it can be said that the water level in October is more than that in September, water level in November is more than the waer level in October and so on. But by how much the level is more is unknown or not important. This measurement will answer 'which month will have more water? This is the second ordinal level measurement.
3. When quantity of water increases, several 'steps' go under water. The quantity or height of water can be measured by the number of steps under water. Thus, for example, it can be said the water level was at the ninth step by the end of October and at the fifth step in by the end of November. The levels can be arranged at regular intervals in terms of time in weeks or months, to see how many steps have gone under during that interval. The level from the top is measured by the number of steps that are clear. This is the third, interval level measurement.

4. If a pole point is fixed at the bottom of the tank, the water level at that point is zero. Keeping this as the zero point, the depth of water can be measured as a linear quantity. The water level can be measured in metres from this point. This is the fourth ratio-level measurement.

As can be seen, the above measurements quantify the water in the tank in different levels, and each one of them can be adequate for certain uses. The measurement has to be planned for each of the use. When considering specifications, some of them can be in the nominal level, some of them can be in the ordinal level, and some can be in the interval level or ratio level.

6.3 METRICS AND THEIR CHARACTERISTICS

A metric is a measurable quantity representing a requirement, which is an element of a societal need. Consider an oil lamp shown in Figure 6.2 as an example. The requirement is to provide light to the surrounding area. Within limits the lamp will provide more light when more oil is burnt by exposing more thread. Thus, the resulting light is controllable by the quantity of oil burnt. The change in light will be visually observable whenever the quantity of oil is changed. There is a one-to-one relationship between the light and the quantity of oil. A metric for the light therefore can be the amount of oil burnt, or alternately the level of thread exposed. In general, when choosing a metric, it is preferable to have controllability, observability and a one-to-one relationship. Also, they should be orthogonal to other metrics meaning they should not overlap.

The purpose of defining and using a metric in product development is to precisely specify the needed metric and use it, to impress new customers and

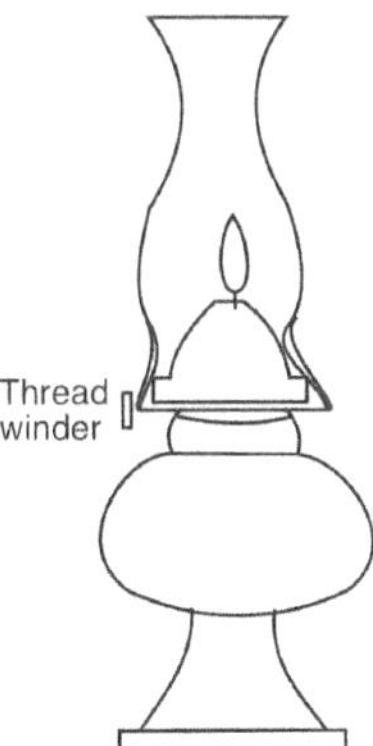

Figure 6.2 Oil lamp for the table.

recruit them, and to keep the existing customers happy. They can be at any level of measurement as described in Section 6.2. But a metric is needed to measure and ensure that originating need is fulfilled. Metrics to represent stakeholder needs can therefore take four kinds of units in the following way:

1. *Physical measurement units*: In this ratio category, physical quantities like millimetres and litres can be described precisely. This is at the highest level of measurement.
2. *Yes/no units expressing the presence or absence of a feature*: In this nominal category, a Yes/No answer is adequate to make a rational decision. For example, consider the 'Provision of a mechanism showing the water level in a tank' as a metric. It can take two values 'yes' it is present or 'no' it is not present. As seen in Section 6.2, this is the nominal category of measurement.
3. *Number or number of occurrences*: In some situations, the metric may be quantified by a number. For example, consider the need 'Easy for setting up' and the metric for this need as 'Number of steps for setting up'. Then the higher the number of steps, more difficult the setting up will be. This is of the ordinal category.
4. *Membership or not, in a list*: In many situations, recognized groups are available, and the need may be 'Belonging to these elite groups.' This is of the nominal category. For example, consider the need as 'Attractive colour', and the attractive colour group may consist of four colours 'Black, Silver, Red and Yellow'. Then, the need 'attractive colour' can be measured by the yes/no answer to the consideration 'whether the chosen colour is from the Elite list'. 'Membership in the list' is the metric.

6.4 NEED-METRIC MATRIX

Need-metric matrix is a mechanism developed to establish the metrics for the needs of a product. It is developed in the following way:

1. Create a matrix with the number of rows and columns equal to $(n+1)$ where n is the number of needs already established for the product.
2. Enter the needs in the first column starting in the second row.
3. Consider the first need which is in the second row and identify its metric. Enter it in the second column in the first row.
4. Now darken the cell in the second row and second column, indicating that the metric for the need in second row is in the second column.
5. Now consider the second need in the third row and identify or create a metric and enter in the third column. Darken the cell in the third row and third column.

Table 6.1 Generic need-metric matrix

	Unit for metric 1	Unit for metric 2	Unit for metric3				Unit for metric (n-1)	Unit for metric n
	Metric for Need 1	Metric for Need 2	Metric for Need 3				Metric for Need (n-1)	Metric for Need n
Need 1								
Need 2								
Need 3								
Need 4				-				
-------					-			
-------						-		

Need (n-1)								
Need n								

6. Continue in this fashion until metrics for all needs are identified and defined. The leading diagonal of the matrix will have the cells darkened.

A generic need-metric matrix is shown in Table 6.1. Need 1 is entered in the second row. Coming along the second row, the cell in the second column is darkened indicating that the metric for Need 1 is in that column. In the same way, if Need 2 in the third row is considered, the cell in the third column is darkened indicating that the metric for Need 2 is in that column.

Some needs may, in their present form, cannot be measured, and new metrics may have to be defined to accommodate these needs. Ulrich and Eppinger [1] give a good example. The requirement for a mountain bike is 'mud should not enter the axle during a ride through mud'. They suggest a mud chamber test, where the bike is put in a chamber and sprayed with mud for half an hour. The bike passes the mud chamber test if no mud is found in the axle bearings after the mud chamber test. The defined metric is pass or fail of the mud chamber test.

6.5 MODEL QUESTIONS AND ANSWERS

Example 6.1: Need-metric matrix for a 'lie to sit' device

A patient who cannot lie down or sit up without assistance is in the hospital. An attachment to the bed is planned to assist the care workers and patients. A typical bed with the attachment is schematically shown in Figure 6.3. The needs are given as follows. Establish the need-metric matrix.

1. Lie to sitting device is easy to use
2. One caregiver should be able to operate
3. Give the feeling of safe
4. Should be robust needing no maintenance
5. Should be able to setup in short time
6. Should not harm the operator or patient
7. Should have protection to the patient
8. Should be comfortable to patient
9. Able to support overweight person
10. Should not disturb other fitted devices
11. Should not be noisy during operation

Consider the first requirement – *Lie to sit device is easy to use*. In its present form satisfaction of this requirement is not measurable. A metric representing its fulfilment is needed. It is known that the device needs setting up. To set up several steps may have to be taken. They are more or less equal in terms of, time to do and difficulty to do. Then *number of these steps* can be a metric representing the easiness to use.

Consider the second requirement – 'One caregiver should be able to operate'. It is known that installing a mechanism to lift the weight of the patient would make the lifting operation easy. It is easy to check whether the mechanism is included or not. Thus, a metric '*Fitting of a mechanism to assist lifting*' with a 'yes/no' measurement is a suitable metric.

In the same way consider the third requirement – *Give the feeling of safe*. In its present form, satisfaction of this requirement is not measurable. A metric is needed. It is known that 'fitting the sideguards' makes the patients feel safe. It is easy to check whether the side guards are fitted or not. Thus, a metric 'Fitting of sideguards' with a 'yes/no' measurement is a suitable metric.

The requirement and corresponding metric normally should have a one-to-one relationship. When arranged in a matrix, the relationships will have a diagonal relationship as shown by the darkened cells in Table 6.2, which is the need-metric matrix for the lie-to-sit device.

Table 6.2 Need-metric matrix for a lie-to-sit device in a hospital bed

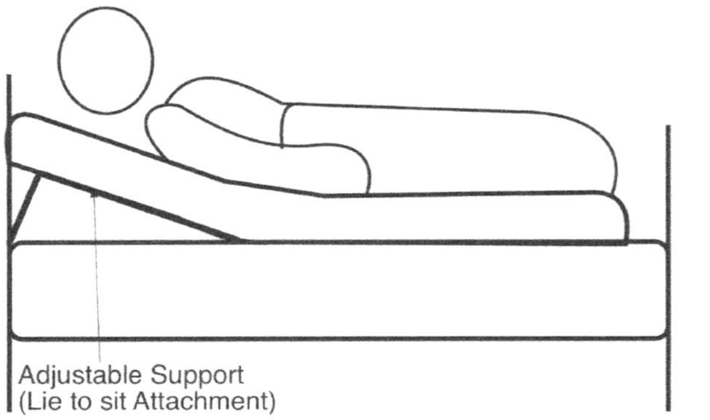

Adjustable Support
(Lie to sit Attachment)

Figure 6.3 Concept of the lie-to-sit mechanism.

Metric	Number of steps	Mechanised lifting	Fitted side guards	MTBF	Time to assemble	No manual lifting	Safety harness	No physical handling	Strong attachment	No disruption	No creaking noises
Units	Number	Yes/No	Yes/No	Hours	Minutes	Yes/No	Yes/No	Yes/No	Yes/No	Yes/No	Yes/No
Lie to sitting device is easy to use											
One caregiver should be able to operate											
Give the feeling of safe											
Should be robust needing no maintenance											
Should be able to setup in short time											
Should not harm the operator or patient											
Should have protection to the patient											
Should be comfortable to patient											
Able to support overweight person											
Should not disturb other fitted devices											
Should not be noisy during operation											

Example 6.2: Need-metric matrix for a roof top wind turbine

In a move to be environmentally friendly, an entrepreneur intends to manufacture and sell a particular type of 'Roof Top Wind Turbine'. The schematic of a house with the Roof Top Wind Turbine is shown in Figure 6.4. The entrepreneur has identified ten requirements from the stakeholders as given below:

1. Generates about 2 kw
2. Durable
3. Easy to Install
4. Minimum Noise
5. Compact size
6. Easy to maintain
7. Affordable
8. Not too tall
9. Not too big
10. Minimum maintenance

Each of identified the needs is taken and a suitable metric is identified as illustrated in Table 6.1. The need-metric matrix is given in Table 6.3. As can be seen there are three types of units (i) physical measurement units (ii) number or count of certain items and (iii) presence of a phenomenon with 'yes/no' unit. Another type of unit of measurement (that is not present here) is the membership in a list with a 'yes/no' answer.

Table 6.3 Need-metric matrix for a roof top wind turbine

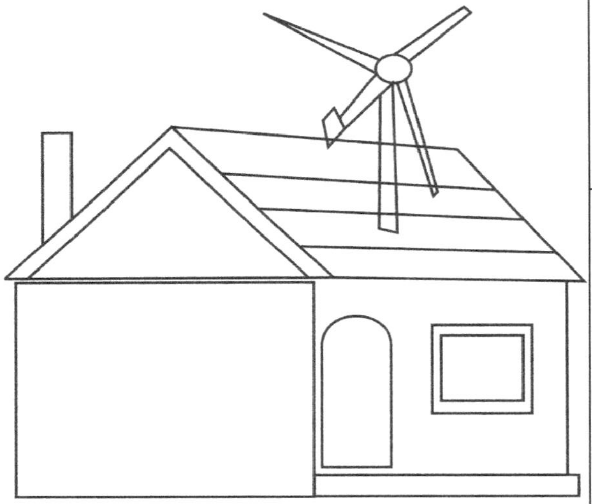

Figure 6.4 Roof top wind turbine.	*Units*	*Kilowatts*	*Operating hours*	*Number*	*db*	*metre*	*Number*	*$*	*metre*	*metre*	*Hours*	*Yes/No*
	Metric	*Intended power*	*Life expectancy*	*Number of activities*	*Noise measurement*	*Diameter of impeller*	*Number of steps*	*Cost in dollars*	*Height of the post*	*Diameter of the post*	*MTBF*	*Modular design*
Generates about 2 kw												
Durable												
Easy to install												
Minimum noise												
Compact size												
Easy to maintain												
Affordable												
Not too tall												
Not too big												
Minimum maintenance												

Example 6.3: Need-metric matrix for a folder trolley for medical records

A new hospital unit has been opened and several clinics were opened up for the consultants to see their patients. Their records are stored in the archive room. A trolley is needed to transport them between the archive room and the clinics. Ten requirements were identified for this trolley. The developed need-metric matrix is given in Table 6.4.

Table 6.4 Metric matrix for a folder trolley for medical records

Metric	*Volume per file*	*Number of files*	*Fitted with wheels*	*Enclosed drawers*	*Top space on trolley.*	*Position of C.G*	*Cost of the unit*	*Fitted lock*	*Extra drawer space*	*Time to failure*
Units	*Cubic millimetre*	*number*	*Yes/No*	*Yes/No*	*Sq. millimetre*	*Millimetre*	*Dollar*	*Yes/No*	*Percentage*	*hours*
Should have sufficient space for big files	■									
Should be sufficient for a full clinic		■								
Easy to move			■							
Should be closed for confidentiality during transit				■						
Should have a space to keep and check contents					■					
Stable at all conditions						■				
Affordable							■			
Should be lockable								■		
Should have extra space for temporary files									■	
Minimum maintenance										■

Figure 6.5 A typical folder trolley needs.

Example 6.4: Need-metric matrix for a bicycle repair stand

The need-metric matrix for the bicycle repair stand is shown in Table 6.5 (Figures 6.4–6.6)

Table 6.5 Need-metric matrix for a bicycle repair stand

	Handleable weight	Soft jaws in clamp	Actions to fix	Height adjustment	Fitted tool holder	Fitted wheels	Weight	Safe handling force	Operations needed	Cost	Pack size
Units	kg	Yes/No	Number	Yes/No	Yes/No	Yes/No	kg	Newtons	Number	Dollars	mm × mm × mm
Hold the fully and partially assembled bike											
Protect (should not damage) the paint											
Easy to fix bicycle to the stand											
Should have adjustable height											
Should have a tool holder											
Should be easily moveable											
Should be light in weight											
Should be stable at all conditions (handling force)											
Stand should be easy to assemble											
Should be affordable											
Easy to pack and post for mail orders											

Figure 6.6 Concept of a bicycle repair stand needs.

REFERENCE

1. Ulrich K., Eppinger S. and Yang M.C., *Product Design and Development*, 7th edition, McGraw Hill, Boston, MA, 2019.

Chapter 7

Drawing specifications

7.1 INTRODUCTION

Drawing specifications is the act of identifying important characteristics of a product and using them to define the product precisely. Specifications describe the product in precise detail with respect to functions, dimensions, materials and quality of work for the product to be manufactured. A single specification is made up of three elements namely (a) name of the characteristic (b) acceptable magnitude and (c) the units. The characteristics that form the specifications are drawn from the design brief, stakeholder needs and corresponding metrics, and the functional representation. Drawing specifications can be done in two stages namely (a) establishing the characteristics and (b) defining the preferred values, range of values or other properties for these characteristics. This book draws from the needs, need-metric matrix, design brief and the functional representation to identify the constituting characteristics. Then it identifies the permissible values and units for each characteristic. This chapter explains how specifications are drawn in a systematic design process by following the above two steps and provides example problems and solutions to gain experience in drawing specifications.

7.2 CONCEPTUAL FOUNDATION

The dictionary meaning of specification is the 'identification of something precisely or statement of a precise requirement'. It is a detailed description of (i) the criteria for the constituents, construction, appearance, performance, etc of a product, (ii) a material, apparatus, etc for its constitution or (iii) the standard of workmanship required in its manufacture. Consider a single rectangular block. It can be represented by three numbers representing its length, width and height in some units. Therefore, any rectangular block can be represented by a three-tuple. The same argument can be extended to products where all characteristics of a single product can be arranged in an n-vector where n is the number of characteristics. All similar products can be represented by n-tuples. In general, a product can be described by a vector like

DOI: 10.1201/9781003484950-8

$$\begin{bmatrix} \text{Characteristic 1} \\ \text{--} \\ \text{--} \\ \text{Characteristic } n \end{bmatrix}.$$

Having thus represented a product as a vector of its characteristics, a product with necessary condition, restrictions and values also can be represented using such a vector.

Specification is a description of the permissible area for a parameter or characteristic while imposing necessary restrictions. A specification is made up of three elements namely (a) name of the parameter or characteristic (b) acceptable magnitude and (c) the units.

As an example, consider the rectangular block with its dimensional specification along the x-axis to be more than 50 mm, along the y-axis to be between 55 and 65 mm and along the z-axis to be less than or equal to 75 mm. The block can be described as

$$\begin{bmatrix} \text{Dimension of } x & \geq 50 \text{ mm} \\ \text{Dimension of } y & 55\text{mm} \leq 65 \text{ mm} \\ \text{Dimension of } z & \leq 75\text{mm} \end{bmatrix}$$

Drawing specifications thus has two activities:

a. Establish the characteristic vector
b. Add conditions, restrictions or values to the elements of the characteristic vector

Consider the table-top poster display stand shown in Figure 7.1. It has seven pieces, two legs (heels), a base board, two back boards, a handle and a rib. There are six key parameters that decide the shape and size of the stand and the specifications define them. Though these are all geometric parameters, the argument can be extended to any performance characteristics.

The specifications for the stand can be $\begin{bmatrix} \text{Length of handle} \\ \text{Rib length} \\ \text{Base width} \\ \text{Backboard width} \\ \text{height of backboard} \\ \text{Height of the legs} \end{bmatrix}$

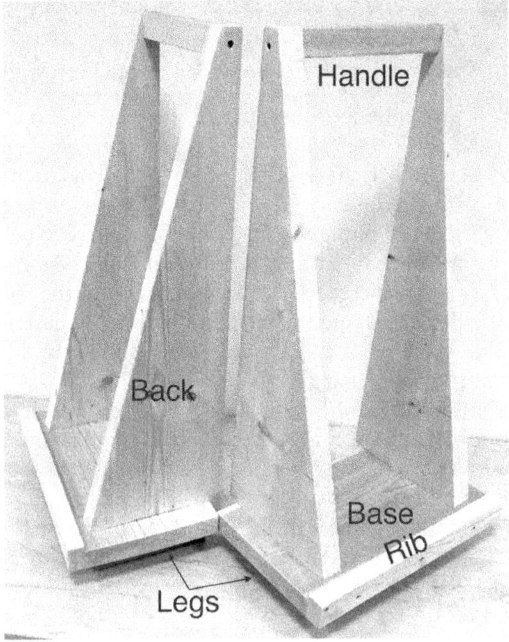

Figure 7.1 Table-top display stand.

7.3 DRAWING SPECIFICATIONS

A product is defined by the specifications and these specifications are based on the customer needs, design brief and the design solution. The specifications are drawn at the beginning based mainly on customer needs and later improved with the conceptual design.

Steps for drawing specifications:

1. Draw the characteristic Vector
   a. Obtain prioritized stakeholder needs
   b. Draw the need-metric matrix
   c. Establish the characteristic vector having the merged metrics and need descriptions as its elements
   d. Check the design brief and functional representation for any additional elements for the characteristic vector
2. Establish the conditions, restrictions or values
   a. Draw a metric-based competitive benchmarking if possible
   b. Choose the necessary condition, restrictions (for each metric) and values

7.4 MODEL QUESTIONS AND SOLUTIONS

Example 7.1: Specifications for the wooden frame for a wall clock

A wall clock is an appliance that indicates the time of a day. Normally it has a few standard shapes and colours for its frame. But they may not reflect the cultural richness of a society or the magnanimity of an organization or even the background of the family using it. In these circumstances, people buy a standard clock and fix it on a wooden frame that boast other factors with pride. In this assignment, such a wooden frame is considered. The task is to develop the specifications from the stakeholders' requirements. In order to support visualizing without the design brief a conceptual sketch shown by Figure 7.2 is given in the need-metric matrix.

The needs and corresponding metrics are given in the need-metric matrix shown in Table 7.1. It is worth noting that the units for a few metrics are shown as 'list membership'. This means that the parameter should be a member of a list. The need 'Reflect natural beauties' is measured by plantation items in the list. The list can be [Date palms, desert grass, cactus]. Then the measure is whether it is from the list or not. In the same way list for people characters can be [kindness, gentle, supportive]. In a similar fashion, a yes/no answer could be a unit confirming or denying a specification.

The characteristics vector representing the frame thus has the metrics as its elements. No additional elements came from the design brief or functional representation. These metrics are defined from the customer needs. The characteristic vector for the clock frame is

$$\begin{bmatrix} \text{Height of frame for the wall clock} \\ \text{Width of frame for the wall clock} \\ \text{Plantation (date palms, desert grass, etc)} \\ \text{Landscape (rock, sand dunes, etc)} \\ \text{People characters (kindness, simplicity)} \\ \text{Dimensions of the clock} \\ \text{Surface treatment (paint, varnish etc)} \\ \text{Weight of the frame} \\ \text{Type of clock (Analog / Digital)} \\ \text{General properties (easy to use, clean, maintain)} \end{bmatrix}$$

Using the elements of the characteristic vector as specifications, values and units are assigned to them. The importance ratings came from the need to originate this metric (Table 7.2).

Table 7.1 Need-metric matrix for a wall clock frame

12:00 Kindness and Love are Characters of the Soul *Figure 7.2* Clock frame.	Height of the frame	Width of the frame	Plantation	Landscape	People characters	Diameter	Surface treatment	Weight of frame	Type of clock	Characteristics
Metric / Units	mm	mm	List membership	List membership	List membership	mm	List membership	kg	Analog/Digital	List membership
Fit on reasonable wall space										
Reflect natural beauties										
Reflect the regional characters										
Proportional clock size										
Aesthetically pleasing										
Not very heavy										
I prefer digital clock										
Easy to use										

Table 7.2 Specifications of the wall clock frame

Specification	*Importance*	*Values*
Height of frame for the wall clock	9	500–550 mm
Width of frame for the wall clock	9	400–500 mm
Plantation shown (date palms, desert grass, etc)	7	Date palms
Landscape shown (rock, sand dunes, etc)	7	Sand dunes
People characters (kindness, simplicity)	8	Kindness
Dimensions of the clock	7	200–250 mm
Surface treatment (paint, varnish etc)	7	Varnish
Weight of the frame	6	2–3 kg
Type of clock (analog/digital)	6	Digital
General properties (easy to use, clean, maintain)	7	

Example 7.2: Specifications for the test rig measuring wear in half journal bearings

A test rig is needed to measure the wear of a half journal bearing. A camera takes pictures at specified time intervals and these pictures are analyzed using computer vision techniques and decisions are made based on them. The needs and metrics are given in Table 7.3. Establish the specifications.

The characteristics vector representing the test rig takes the metrics as its elements. These metrics are derived from the customer needs. The characteristic vector for the test rig is

$$\begin{bmatrix} \text{Uninterrupted light} \\ \text{Load on belt} \\ \text{Isolation of Other loads} \\ \text{Variable speed motor} \\ \text{Operating Temperature range} \\ \text{MTBF} \\ \text{Safety ensured by guard installation} \\ \text{Parts life} \\ \text{Easy operation with single}\dfrac{\text{on}}{\text{off}}\text{switch} \\ \text{Lubrication option available} \end{bmatrix}$$

The general rule for specifications is, If you need a particular feature in the product, include it in the specifications. Table 7.3 shows the specifications for the test rig.

The needs in Table 7.3 are taken one by one and their metrics are determined as shown in Table 7.4.

Table 7.3 Specifications for the test rig

Specification	*Importance*	*Values*
Top half opened	9	yes
Load on belt	8	10–15 kg
Universal joint (isolator)	7	yes
Variable speed motor	6	300–1440 rpm
Steady temperature	9	50°C–60°C
MTBF (mean time between failure)	8	5000 hoursC
Guarded moving Parts	9	yes
Reliable parts – Life of parts	8	>5000 hours
On/off switch	9	Yes
Optional lubrication	6	Yes

Example 7.3: Specifications for the convertible flower vase stand/display stand

Halls in big cities and towns host conferences and other smaller leisure activities and events. Normally they have several rooms and during a conference, parallel sessions of presentations take place in these rooms. The participants may cross between them to attend specific papers from more than one session. In order to facilitate their quick access a display board in front of each room is needed. When small functions take place in these centres only a small number of rooms will be in use. The display stands should be convertible as a flower vase stand in these circumstances. This project is concerned with the design of a convertible flower vase/display stand. The need-metric matrix is given in Table 7.5. Establish the specifications.

Here again the metrics are taken as the characteristic vector to establish the specifications. Table 7.6 shows the specifications for the convertible flower vase/display stand

Table 7.4 Need-metric matrix for a test rig measuring wear in half journal bearings

Figure 7.3 Test rig measuring wear in half journal bearings.

Metric	*Top half opened*	*Load on belt*	*Presence of Isolator*	*Variable speed moto*	*Temperature range*	*MTBF*	*Guarded parts*	*Life of parts*	*On/Off switch*	*Lubrication*
Units	*Yes/No*	*kg*	*Yes/No*	*Yes/No*	*0°C*	*Hours*	*Yes/No*	*>5000 hours*	*Yes/No*	*Yes/No*
Uninterrupted optical path to the bearing										
Adjustable Vertical loading on the shaft										
No additional load other than the vertical load										
Variable speed										
Suitable for continuous operation										
Suitable unsupervised operation										
Reliable operation – bought parts life										
Easy to operate										
Optional lubrication										

Table 7.5 Need-metric matrix for convertible flower vase stand/display stand

Figure 7.4 Convertible flower vase stand/poster display.

	Metric	*Length and width*	*Vase height*	*Storage chamber*	*Retract/Remove able*	*Portable*	*Side of square foot*	*Assembled height*	*High class finish*	*Culturally rich object*	*Cleaning time*	*Lifetime*
	Units	*mm*	*mm*	*Mm³*	*Yes/No*	*Yes/No*	*metre*	*metres*	*List membership*	*List membership*	*minutes*	*years*
Should be able to display A2 size poster												
Should be able to display about 400 mm tall Vase												
Should have storage space for the Vase												
Easily retractable/removable display board												
Should be portable												
Foot Area is less than 3600 mm^2												
Poster top around average person's height												
Aesthetically pleasing												
Easy to clean												
Durable												

Table 7.6 Specifications for the convertible flower vase/display stand

Specification	*Importance*	*Values*
Length and width	9	A2+50 mm
Height of vase	8	650–750
Storage chamber	7	Store Vase &board
Retract/remove ability for display board	6	Yes
Portable – has wheels	9	Yes
Square foot side	8	400–450 mm
Assembled height	9	1500–1700 mm
High class finish	8	Varnished
Culturally Rich resembling historic things	9	yes
Cleaning time	6	15–25 minutes
Lifetime	6	10 years

Example 7.4: Specifications for an inspirational paperweight

A paperweight is a small solid object heavy enough (usually a glass marble), when placed on top of papers, to keep them from blowing away in a breeze. While the main purpose of a paperweight is to keep papers safe, it contributes towards the atmosphere in the office. It can deliver inspirational, creative, unexpected, clever, delightful, special and thoughtful messages at times when decision-making becomes complex. Specifying such a paperweight is the task in this project.

In nature small creatures create great things like bird's nest, Beehive, and Termite hill and seeing them or getting reminded of them can inspire when attempting a significantly big task. In the same way, a gear wheel is a man-made thing, 'designed to be strong' to complete a given task, and can be inspirational when attempting a new venture. A gear wheel is taken as the base for the paperweight.

Attractiveness test: This is a test designed to measure the attractiveness of a product. Have the product with several (say ten) other similar products on a table. Bring ten individuals and ask them to pick three attractive pieces in order. If the product is chosen at least by eight persons as within the attractive three, it passes the test. Otherwise, it fails.

List membership: There can be two categories, inclusive and exclusive, of membership. Inclusive membership is the choice from a given list like attractive material from the list (stainless steel, silver, teak). Exclusive list seeks something outside the list like rare in the class should take anything other than from the list (flowers, sunset, sunrise, mountains) (Tables 7.7 and 7.8).

Table 7.7 Need-metric matrix for a paperweight

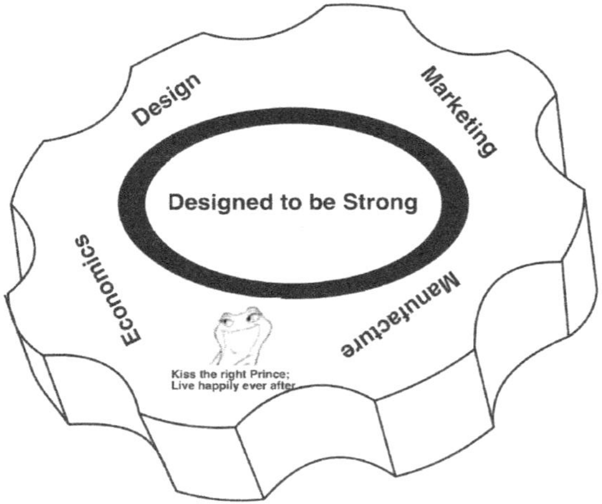

Figure 7.5 Inspirational paperweight.

Metric	Weight	Attractive test	Stated characters	Object category	Statements	Attractive material	Funny characters	Area of contact	Reminder present	Importance of the need
Units	kg	Pass/Fail	List membership	List membership	Yes/No	List Membership	Yes/No	Square millimetre	Yes/No	
Reasonable weight										8
Should be bright and attractive										9
Reflect character of the company										9
Should be rare in the class										7
Should have inspiring illustrations										8
Attractive material										9
Should be funny										8
Hold paper through a wide area										7
Remind well-thought-out decisions										8

Table 7.8 Specifications of a paperweight

Specification	*Importance*	*Values*
Weight	8	1 kg
Attractiveness of appearance	9	Pass
Display stated characteristics	9	Inclusive List
Object is outside the common paperweights	7	Exclusive list
Inspiring statements	8	Yes
Attractive material	9	Inclusive list
Contain funny characters and statements	8	Yes
Area of contact	7	>20 cm^2
Remind the necessity of thoughtful decisions	8	Yes

Example 7.5: Specifications for a portable refrigerator

A portable refrigerator is a handy equipment to cool and preserve coolness of drinking water during a travel in a hot country or hot season. A battery-operated small refrigerator can meet this requirement very well. Establish the specification for such a refrigerator using the need-metric matrix given in Tables 7.9 and 7.10.

Table 7.9 Need-metric matrix for portable refrigerator

	Metric	Charging time	Power consumption	Temperature display	Available power display	Noise level	Setting up time	Weight of the fridge	Number of bottles	MTBF	Cooling time	Cost of the refrigerator	Dimensions (L x W x H)	Importance
	Unit	hours	watts	Yes/No	Yes/No	db	Seconds	kg	Number	hours	minutes	Dollars	millimetre	
	Needs													
1	Require low time to charge													9
2	Require low operating power													8
3	Display temperature													7
4	Display Available power													7
5	Must produce less noise													6
6	Easy operation													7
7	Light weight													8
8	Abe to cool few bottles													9
9	Easy maintenance													8
10	Fast cooling													9
11	Reasonable cost													7

Table 7.10 Specifications for the portable refrigerator

No	*Specification*	*Value*	*Importance*
1	Charging time	<3 hours	9
2	Power consumption	<100 watts	8
3	Temperature display	Yes available	7
4	Available power display	Yes available	7
5	Noise level	<8 db	6
6	Setting time	<5 minutes	7
7	Weight of the fridge	<2.5 kg (all)	8
8	Number of bottles	4, 30 ml bottles	9
9	MTBF	>1000 hours	8
10	Cooling time	<30 minutes	9
11	Cost of the fridge	$80–$100	7
12	Dimensions	<400 mm	9

Example 7.6: Specifications for a paper cup for hot and cold drinks

Disposable cups have become part of modern life. The needs or requirements are identified as follows:

1. Comfortable for holding
2. Good quantity of liquid
3. Lightweight
4. Suit hot and cold drinks
5. Minimise heat flow/loss
6. Thick material
7. Cup sleeves etc for hand
8. Fire safety
9. High class
10. Recyclable
11. Low cost (Tables 7.11 and 7.12)

The benchmarking gives a snapshot of the market situation in terms of the metric which can permit the fixing of the target values for the product with confidence. The specifications for the product are shown in Table 7.13.

Table 7.11 Need-metric matrix for paper cups

Figure 7.6 Sketch of a paper cup.

	Metric	Dimensions of the cup	Volume of the cup	Paper density	Material used	Presence of ripples	Cup thickness	Provision of cup sleeves	Fire rating	Rating by a panel	% of recyclability	Cost per 1000	Importance rating
	Units	mm	ml	Gm/Square metre	List	Yes/No	mm	Yes/No	Yes/No	(1–5)	%	$	
1	Comfortable for holding												
2	Good quantity of liquid												
3	Light weight												
4	Suit hot and cold drinks												
5	Minimise heat flow/loss												
6	Thick material												
7	Cup sleeves etc for hand												
8	Fire safety												
9	High class												
10	Recyclable												
11	Low cost												

Table 7.12 Competitive benchmarking for paper cups

Metric	*Company A*	*Company B*	*Company C*	*Company D*
Dimensions	80 × 56 ** × 92	83 × 60 × 89	72 × 52 × 75	73 × 53 × 80
Volume	330	360	228	314
Paper weight	0.15	0.14	0.17	0.25
Material used	PE coated	SFI coated	Natural paper	Natural paper
Ripples present	Yes	No	No	No
Paper density	350 gsm		260 gsm	230 gsm
Cup sleeves	Yes	Yes	No	No
Fire resistance	Flammable	Flammable	Flammable	Flammable
High class	4 (good)	4 (good)	4 (good)	4 (good)
Recyclable	Part	Part	100%	100%
Cost	\$0.01–0.1		\$0.01–0.02	\$0.01–0.15

Table 7.13 Specifications for paper cups

Specification	*Imp.*	*Specification*	*Value*
Comfortable for holding		Dimensions	83 × 60 × 89
Good quantity of liquid		Volume	360
Light weight		Paper weight	0.15 g
Suit hot and cold drinks		Material used	SFI Coated
Minimise heat flow/loss		Ripples present	Yes
Thick material		Paper density	230 gsm
Cup sleeves etc for hand		Cup sleeves	Yes
Fire safety		Fire resistance	Flammable
High class		High class	4 (good)
Recyclable		Recyclable	Part
Low cost		Cost	\$0.01–0.1

CONCLUDING NOTE FOR PART 1

Part 1 of this book has explained how the design problem can be defined well following a systematic design process. If only a single output is expected from Part 1, it is the set of specifications. Specifications are characteristics of the product with acceptable measurements and quantities. A product can be adequately described in terms of its specifications.

The previous steps like identifying the types of needed stakeholder requirements, list of prioritized needs, need-metric matrix, and functional representation are all steps that contribute in the identification of characteristics in drawing the specifications. But the outputs from the interim steps are important in gathering a complete insight on the requirement. They will all be useful in the next challenge of 'Identifying a near optimal or optimal solution from the unknown solution space and developing it into the definition of a physically realizable product'. With the collection of the detailed information or tools gathered move to Part 2.

Part 2

Designing concepts, evaluating and choosing, and developing the final design

Chapter 8

Conceptual design

8.1 INTRODUCTION

Conceptual design is the act of formulating a scheme or a clever mental plan to bring together compatible function providers to achieve the functional goal of a product. A designer would have come across, function providers and theories leading to function providers, throughout his life and these would be in the repository in his memory. Conceptual design is the bringing together of a group of compatible function providers, which as a collection delivers the functions expected from the product. Creativity in this situation can be described as the activity of searching and matching suitable function providers discovering new combinations, whether it comes from existing established function providers, or from new ones resulting from advancements in science and technology.

French [1] identifies conceptual design as a solution to the design problem in the form of a scheme and specifies scheme as an outline solution carried to a point where the means of performing each major function has been fixed, as have the spatial and structural relationships of the principal components. Haik et al [2] describe it as a transition between (a) a set of required functions, (b) a set of behaviours of the chosen components that fulfil the functions, and (c) a preliminary arrangement of components or structures that make all required behaviours consistently. Thus, *A scheme with respect to conceptual design is a clever mental plan to employ various function* providers working together in harmony to achieve the required integrated functions.

There are several Design Methods for conceptual design and this chapter describes several methods to generate conceptual designs and, provides examples in each of them.

A typical conceptual design will normally have the following documentation:

1. A sketch (hand or computer-made) of the overall system showing the scheme, with necessary notes.
2. Important subsystems with sketches as necessary.
3. Rough dimensions of the proposed product.

DOI: 10.1201/9781003484950-10

8.2 UNDERSTAND THE SCHEME

Scheme in general is a large-scale systematic plan, or an organized configuration or arrangement, for putting a particular idea into effect. With respect to conceptual design scheme is the organized arrangement of function providers. To gain an understanding of what is a scheme and how a scheme is made, consider Example 8.1.

Example 8.1: Scheme by the farmer to send the horse to the stables

A farmer has a horse. It carries him inside the farm during the day like a caring friend. But it poses a difficulty in the evenings. It does not like to go inside the stables and the farmer had to struggle every day. Additionally, he has a cow, a dog and a cat in the farm. He has to formulate a 'Scheme' and execute it to send the horse to the stables in an easy way every day. He has to use his knowledge about his farm and the behaviour of his animals.

BACKGROUND THEORY OR KNOWLEDGE

Theory is fundamental to devise schemes. Normally better theoretical knowledge results in better solutions. The farmer's background knowledge about the horse, dog, cat and cow can be listed as follows:

1. The horse does not like the dog barking. It runs to the stables, the moment it hears the dog barking.
2. The dog normally does not bark. But it will bark when it is tickled by the cat when it comes to play with it.
3. The cat normally does not go to play with the dog unless its stomach is full. The farmer also knows that the cat loves fresh milk, as soon as it is milked from the cow.
4. The farmer can milk the cow twice, once in the early morning and then, late in the evening, or only once, and late in the morning.

Using the above knowledge, he formulated a five-stepped scheme.

1. He milks the cow twice a day, once in the morning and once in the evening.
2. He feeds the cat with some fresh milk in the evening. The cat drinks the milk to its heart's content.
3. Now the cat is in a good mood and goes to the dog for a play. It tickles the dog.
4. The moment the cat tickles, the dog starts barking.
5. The barking sound terrifies the horse, and it runs to its stables.

The main technique in formulating the scheme is to connect the behaviour of the various elements or function providers (in this case the animals) and trigger their behaviour as a sequence, as and when needed.

This formulation of the scheme is the essence of conceptual design. The scheme in conceptual design connects the behaviour of function providers in an orderly manner.

8.3 DESIGN METHODS

Conceptual design is the hard and important part in the design process. Therefore, several tools and techniques have been developed by designers and design researchers. These tools and techniques are called design methods. Several design methods or techniques have been reported with evidences showing their usefulness. In this context, a technique is a procedure, skill, or art used to complete a task. The following sections explain some of the methods followed by examples where the method is used.

8.4 DESIGN METHOD 1: STARTING WITH BASIC MACHINES

Simple or basic machines are mechanisms that are used to modify motion and increase magnitude and change direction of forces, in order to make work easier. The list includes lever, wheel and axle, pulley, inclined plane, gears and screw (Figure 8.1). At the beginning, it was believed that all the machines are composed of them. This method uses individual or combined basic machines as the scheme of the product to be designed.

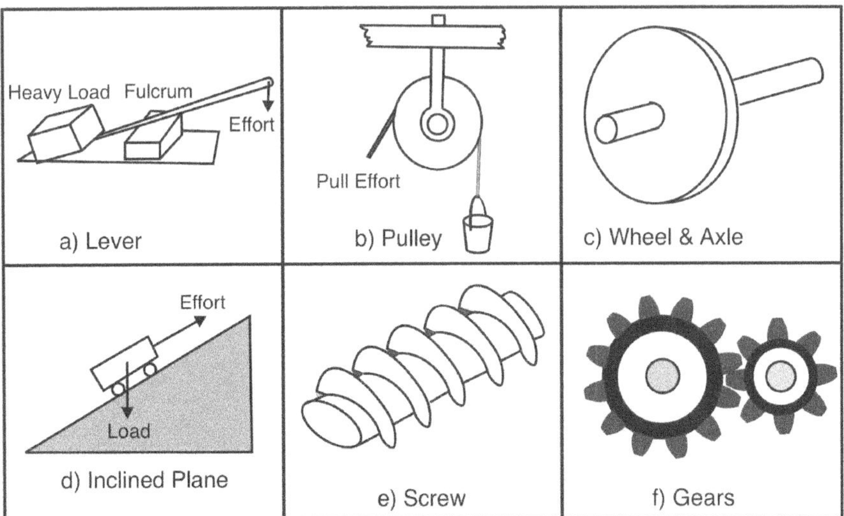

Figure 8.1 Basic machines.

Understanding and using basic machines is a good first step to learn conceptual design. Basic machines are devices with no, or very few, moving parts that make the work easier. Many of today's complex designs are just combinations or more complicated derivations of the six basic machines. They are used as function providers in simple and complex products. This set of examples under design method 1 is aimed to guide the reader to understand that *conceptual design is about identifying suitable 'Function Providers' and creating a harmonious integration of them.* These examples use the basic machines as the function providers. Their shapes, structural forms and dimensions have to be decided for each of the individual applications. The reader is advised to focus on the schemes or the formulation of schemes, the very act of conceptual design.

Example 8.2: Design of a well sweep

During the times when there were no water pumps, water had to be fetched manually in buckets from deep wells for irrigation purposes. Since large quantities of water had to be drawn from the well, an assisting device was needed. A well sweep illustrated in Figure 8.2 is such a device where the device is a class 1 lever rotating in the vertical plane

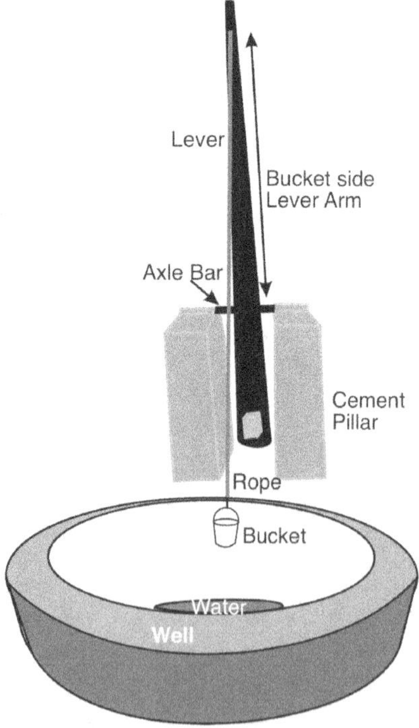

Figure 8.2 Illustration of a well sweep.

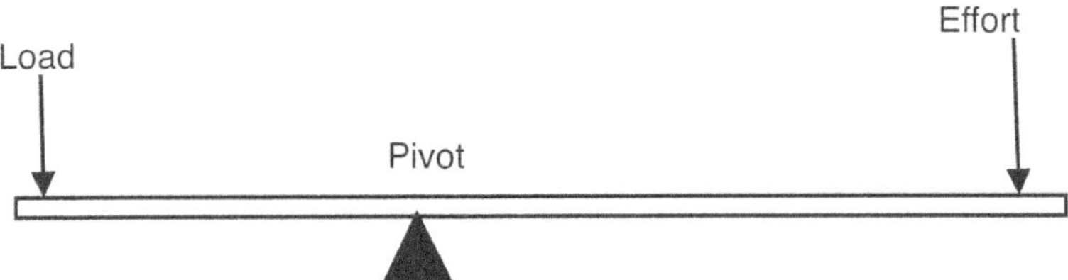

Figure 8.3 Class I lever.

about the axle bar. The aim in general is to increase the amount of water in the bucket as much as possible, without having to apply excessive effort. A class 1 lever is a device that balances the effort and load about a pivot point as shown in Figure 8.3. *In this example, the task is to develop a conceptual design for a well sweep that reduces the effort required to lift a bucket of weight W by half.*

The long lever ACB is the starting point of the design. The pivoting at the fulcrum is achieved by the axle bar supported by two similar vertical posts, often with a Y notch at the top. This device is mounted by the side of the well hole. The fulcrum is the axle bar attached at right angle to the sweep, which is the lever ACB. The lever ACB rotates in the vertical plane about the fulcrum at C.

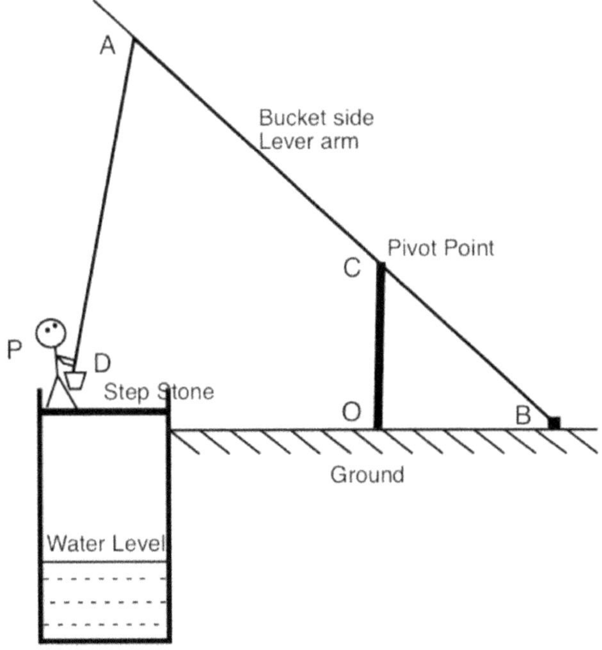

Figure 8.4 Conceptual design of a well sweep.

The scheme can be explained in the following way. To fetch a bucket of water, the user pulls the rope AD down so that the bucket reaches the water level, and water gets filled into the bucket. The user now pulls the rope up and the bucket of water comes to the user's hand. He empties the water and repeats the operations again to get the next bucket of water.

The sweep is heavy at the end B, which rests on the ground when not in use. The bucket side lever arm is longer so that the bucket can reach the bottom of deep wells with the longer rope. Also, from the principle of moments, it reduces the effort needed, to turn lever ACB. In Figure 8.4, OC represents the poles supporting the sweep. *P* is the point where the human user is standing. The full length of the rope is labelled AD. Normally, the effort required is *W*/2 where *W* is the weight of the full bucket. The dead weight will then be *W* and AC=2 CB. When sending the empty bucket into the well, the human has to impart a downward force of W/2 to balance the deadweight. When the bucket is full and needs to be pulled up, the human should again impart an upward force of W/2. Thus, the sweep helps to lift a weight of *W* by imparting a force of W/2.

Example 8.3: Design of a manual pith press for coconut pith

Coconut pith (also known as coconut coir dust, coir waste or fibre dust) is made up of short spongy fibres and is a by-product in the processing of husk to extract coir fibre. It accounts for about 70% of the weight of the husk. A coconut farm producing 10,000 nuts/year has the potential of producing about 3,000 kg of husks containing 2,100 kg of coir dust. Figure 8.5a shows raw coir pith and Figure 8.5b shows a pith block. Coir pith block is specially designed for commercial nurseries and greenhouses. A five-kilogram block has its dimensions as 30 × 30 × 13 cm and a compression ratio of 5:1 by volume. Develop a conceptual design for a machine to produce these blocks using a class 2 lever.

Figure 8.6 shows the free-body diagram of a class 2 lever. The resistance offered by the pith is the load acting upwards on the lever. The reaction at the hinge will be acting downwards to balance the lever.

Figure 8.5 Coconut pith before and after compression.

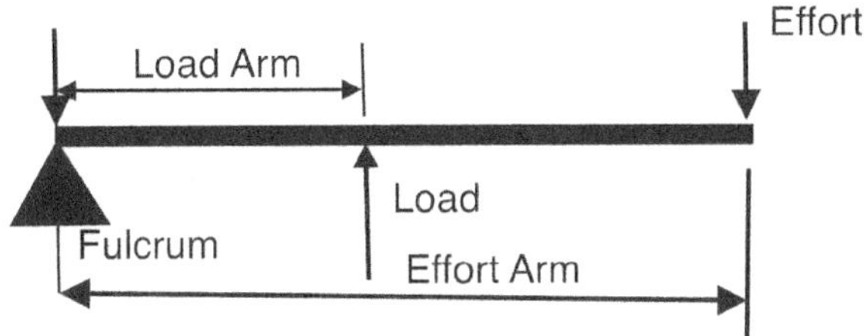

Figure 8.6 Free-body diagram of a class 2 lever.

If moment is taken about the hinge or fulcrum

$$\text{Load} \times \text{load arm} = \text{Effort } \times (\text{Effort arm})$$

From this, it can be said that the effort applied is magnified by an amount of the ratio effort arm/Load arm to act on the load (pith). Equating the forces gives

$$\text{Reaction} = \text{Load } - \text{ Effort}$$

The lever creates a force at the hinge to assist the effort and its magnitude can be manipulated by the ratio of the effort arm to load arm.

The scheme therefore is to create a class 2 lever with the hinge as close to load and the length of the lever as long as possible to get the maximum magnification of the effort.

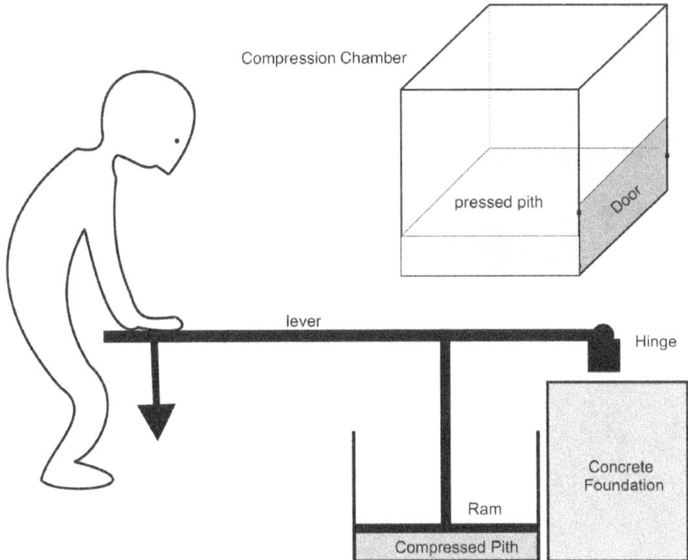

Figure 8.7 Coconut pith compressing machine.

The conceptual design of the manual press is shown in Figure 8.7. A compression chamber with a foot size of 30 cm × 30 cm and a height slightly more than 65 cm is the starting point. On one side of the chamber, a concrete platform to firmly install the hinge is added to the design. A lever arm of length 1.8–m (6–7 times the load arm), that can rotate about the hinge is added to the design. A ram that can travel a distance of 52 cm is attached to the lever. A door is attached to the compression chamber for removing the compressed block. From this example it can be seen that a smaller 'load arm' will give a bigger reaction from the concrete block. The lever, a basic machine, is used as the single function provider, in this design.

Example 8.4: Design of a wheelbarrow

The scheme for a garden wheelbarrow is formed by combining two function providers (basic machines) the wheel and axle, and a class 2 lever. This however requires more effort as the load needs lifting. It therefore is most suitable for bulky load with low density like garden waste while not suitable to transport heavy materials like steel balls or stones.

The load in a wheelbarrow is in the bucket. The distance from the axis of the wheel to the centre of gravity of the load is the load arm. Small load arm creates a bigger force on the axis of the wheel and this makes the needed effort to become small. Because of this reason, the front of the bucket is broader. Typical dimensions of a wheelbarrow are shown in Figure 8.8.

The scheme for the wheelbarrow is shown in Figure 8.9. It shows how the two function providers, the basic machine 'lever', and the basic machine 'wheel and axle', are combined to formulate the scheme. The scheme integrates the two basic machines by placing the fulcrum of the class 2 lever on the axle and exerting the driving force through the lever on the axle. The load in the bucket is balanced by the reaction by the axle on the lever, and effort. The reaction in the axle is transferred to the ground by the wheel. This shows the technique of combining function providers to work in harmony to make products. This is the essence of formulating schemes in conceptual design.

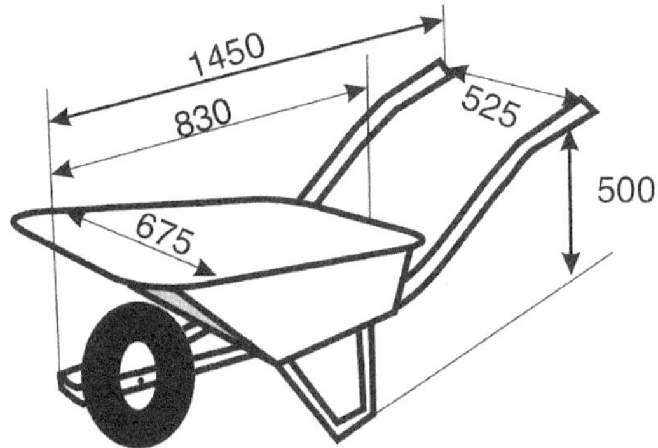

Figure 8.8 Typical dimensions of a wheelbarrow.

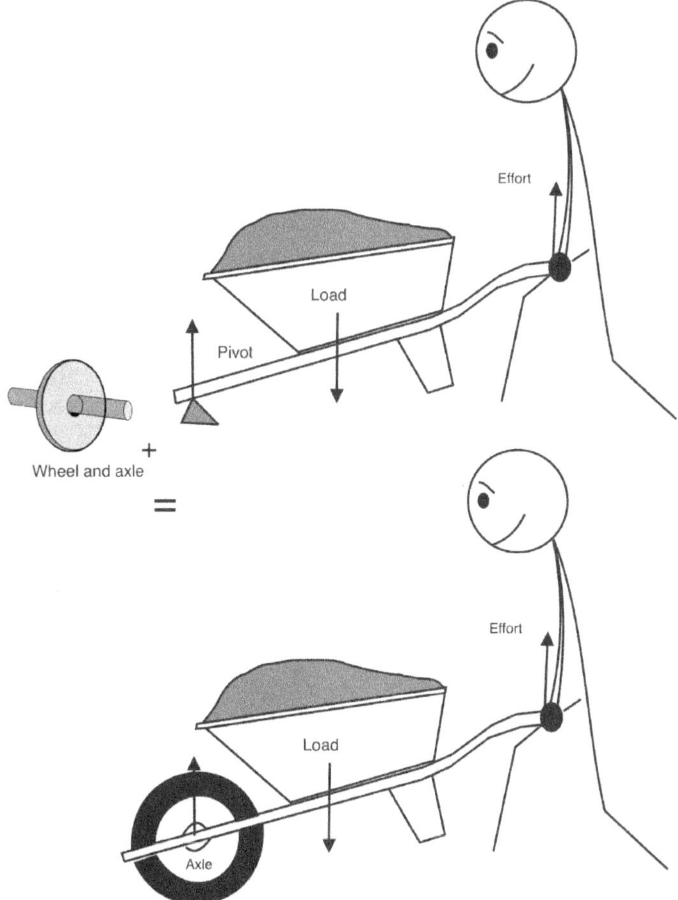

Figure 8.9 Garden wheelbarrow.

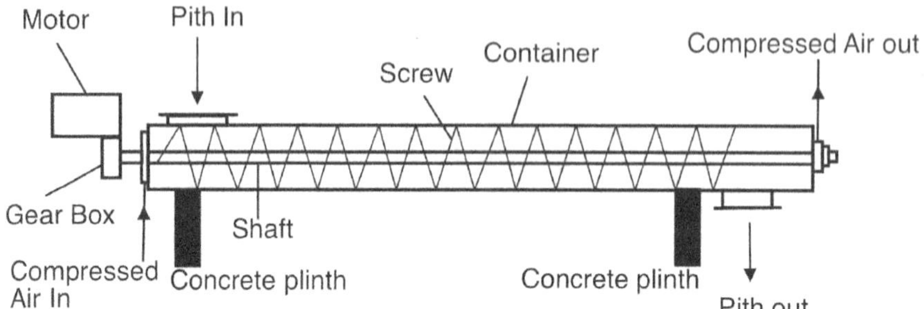

Figure 8.10 Screw conveyor transporting coconut pith.

Example 8.5: Design of a screw conveyor for transporting powder material

Earlier, Figure 8.5 shows raw coir dust and Figure 8.7 shows a design for a machine to produce these blocks using a class 2 lever. This example uses a screw conveyor to transport the loose coconut pith to the required position.

Answer

The scheme of the conceptual design shown in Figure 8.10 can be explained in the following way. A cylindrical tube is anchored to the floor. A screw conveyor is mounted inside the tube supported at both ends on ball bearings. The screw thus can rotate freely inside the stationary tube. The tube has a hole at the left end in Figure 8.10 through which pith to be transported, is fed to the conveyor. To keep the pith fluidized, compressed-air is passed through the left-hand side. The shaft of the screw is connected to the output shaft of the gearbox so that the screw can be driven by the gearbox. The input shaft of the gearbox is driven by a motor. The behaviour of the components can be expressed as follows:

1. When the motor is switched on it runs at high-speed producing low torque. This turns the gearbox.
2. The gearbox receives the high-speed input and produces its output shaft to run at low-speed and high torque. This turns the screw.
3. The bearings carry the screw and permit it to rotate easily.
4. When the screw is rotated, it transports the pith along the tube and
5. The transported pith is received at the hole at the right-hand end.

8.5 DESIGN METHOD 2: MORPHOLOGICAL ANALYSIS

Morphological analysis considers a product as a union of parameters in the functional or physical domain, enumerates the parameters and their possible instances and chooses one instance for each parameter and composes the compatible combinations, as conceptual designs.

Morphological analysis is a method for identifying and investigating the total set of possible concepts to solve the design problem considering all parameters. An artefact can be represented by a vector

$$\begin{bmatrix} \text{Parameter 1} \\ \text{Parameter 2} \\ \text{- -} \\ \text{Parametr } n \end{bmatrix}$$

In conceptual design, possible parameters can be components or the purpose functions. As an example, consider a wheelbarrow made up of the components: (i) lever frame, (ii) wheel and axle, (iii) bucket and (iv) the stand. Then all wheelbarrows can be represented by the vector

$$\begin{bmatrix} \text{Lever frame} \\ \text{wheel and axle} \\ \text{bucket} \\ \text{stand} \end{bmatrix}$$

If in a design, the parameters are components, a morphological chart of the product will show, components by name in the left column, and possible instances of the component for each one, in the cells in the right, in the corresponding row. On the other hand, if the parameters are the purpose functions, the morphological chart will show all purpose functions in the left column and possible function providers for each one, in the cells in the right, in the corresponding row. The approach begins by identifying and defining the purpose functions of the design problem. For each purpose function, a range of relevant function providers are identified, and the 'Morphological Chart' is prepared. A typical chart is shown in Table 8.1. This chart means that Function A can be deployed or provided by option A_1, option A_2, option A_3 or option A_4. A similar argument applies to Functions B, C, - - -X.

Table 8.1 Typical morphological chart

Purpose functions	*Option 1*	*Option 2*	*Option 3*	*Option 4*
Function A	A_1	A_2	A_3	A_4
Function B	B_1	B_2	B_3	
Function C	C_1	C_2	C_3	
Function X	X_1	X_2		

The concepts are vectors of the form

$$\begin{bmatrix} A \\ B \\ C \\ - \\ X \end{bmatrix}$$

where the elements of the vector are compatible options from the morphological chart. All possible vectors form the solution space. Consider the following Example 8.6.

The outcome from the morphological analysis is the development of a representative 'Solution Space'.

Example 8.6: Design of a discussion table

Consider a 'Discussion Table' as an example and establish a 'Solution Space' using the 'Morphological Method'.

A table is made up of two parts: the table top or the surface, and the legs. According to the morphological method, all possible legs and all possible table tops should be considered. Table 8.2 lists nine possible legs and Table 8.3 lists six possible table tops. In theory, each table top can have nine different types of legs and thus, a total of 54 two vectors can be derived from the combinations. Some of them may not be compatible. The method suggests that the design team identify good combinations (designs) and create a representative design space. The designs in this representative space are deemed to contain the optimal and near-optimal solutions. Table 8.4 shows eight designs that form the sample solution space derived from the morphological analysis. The design team then chooses one of these for further development.

Table 8.2 Morphology of table legs

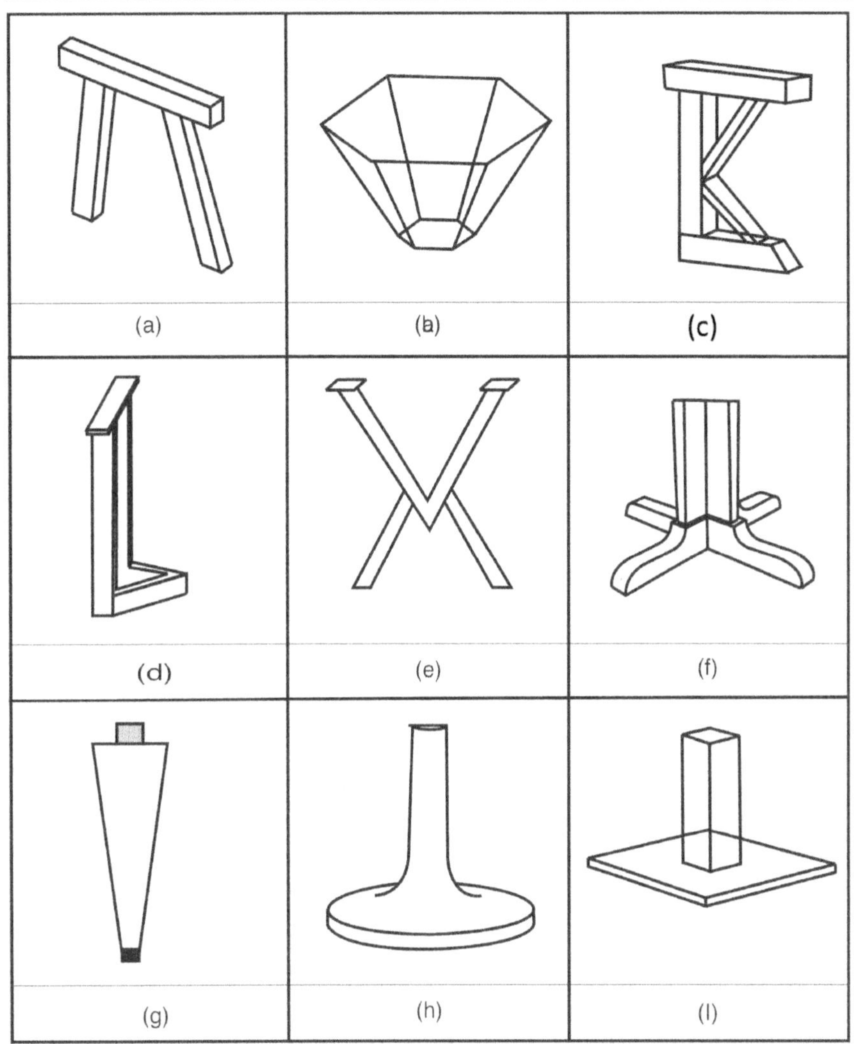

Table 8.3 Morphology of table tops

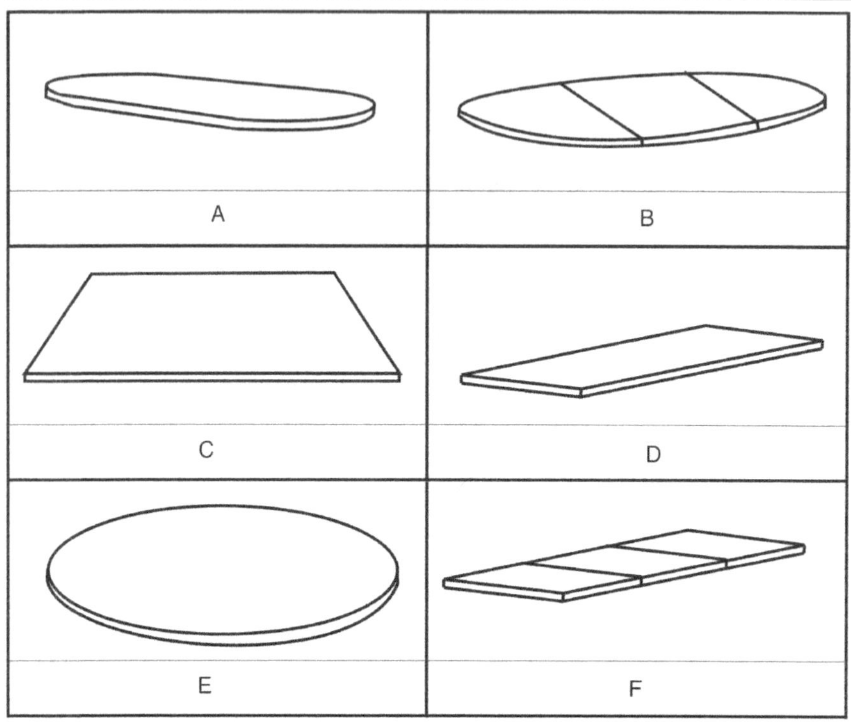

Table 8.4 Solution space developed from the morphological analysis

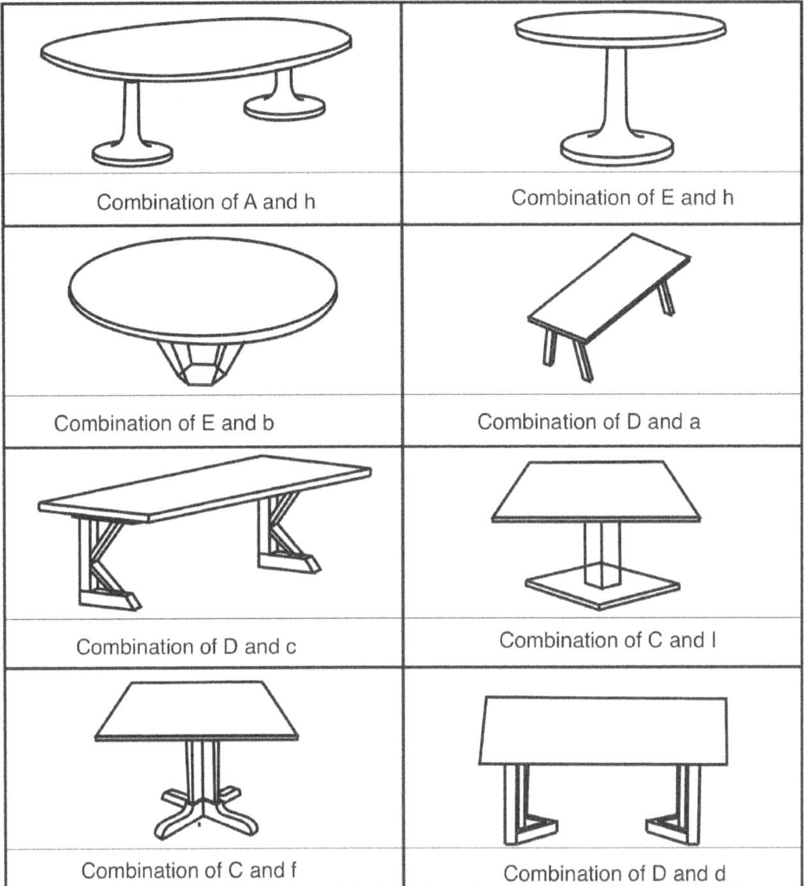

Example 8.7: Design of a table-top dessert server

In many parts of the world, dessert is an important element that concludes a meal whether it is a lunch or dinner. The course usually consists of a large variety of sweet foods, savoury items, cheeses and nuts, fruits and ice creams. Naturally, elegant cuisine requires high-quality tables and servers, to compliment the standard of the hotel or household. A 'Table-top Dessert Server' is required to present a variety of desserts on the table so that the guests can choose and access their preferred one from the dessert items. A 'Function Tree' is provided in Figure 8.11 to assist the design process. Establish conceptual designs using the morphological method.

The purpose functions from the function tree form the basis for the morphological chart. Consider Table 8.5 showing the morphological chart. The first row represents 'House and Hold the Structure'.

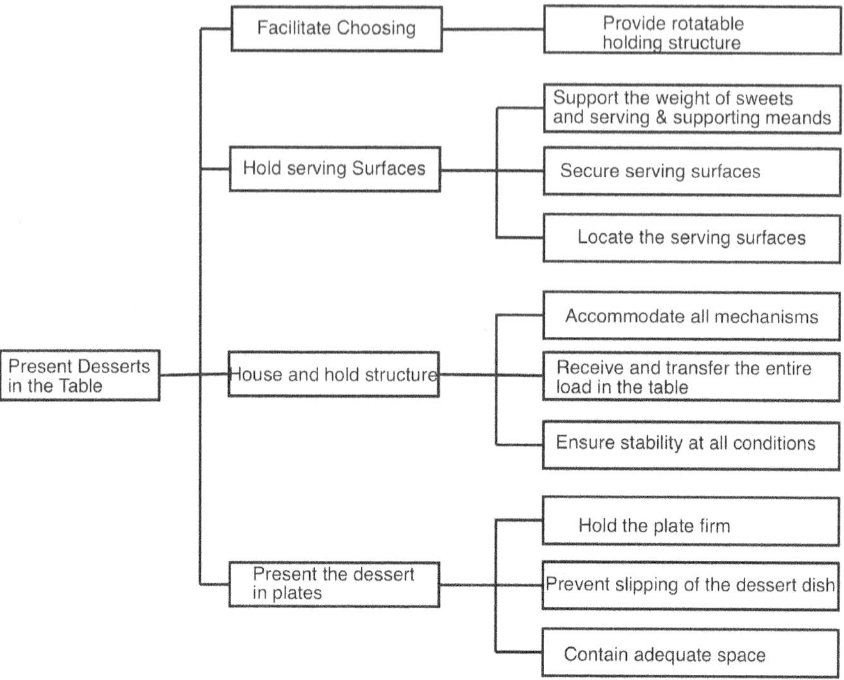

Figure 8.11 Function tree of a table-top dessert server.

This function can be achieved by five methods as shown in the second to sixth elements in the first row. Similarly, the second row has the first element showing the purpose function 'Present the dessert plate' for which there are three methods of achieving. In the same way, the third and fourth purpose functions and their methods of achieving are shown in the third and fourth rows.

To construct a conceptual design, all the purpose functions outlined in the function tree should be integrated. This is achieved by choosing one feature instance from each row in the morphological chart and integrating them. As a condition, the chosen instances should be compatible.

One such compatible combination is made by (i) the round plate to support the dessert plate, (ii) a grass structure supporting these plates, (iii) a thrust bearing to combine all these and (iv) a thick and small circular base to house and support the entire structure. This is shown as a concept (d) in Table 8.6. Similarly, three other conceptual designs were composed and shown in the same Table 8.6.

This example is adapted from a student project and the completed prototype is shown in Figure 8.12.

Table 8.5 Purpose-function-based morphological chart: Table-top dessert server

Hold the structure		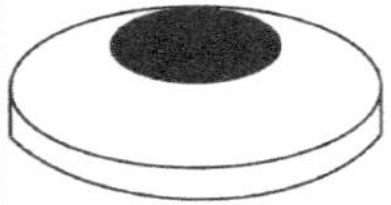		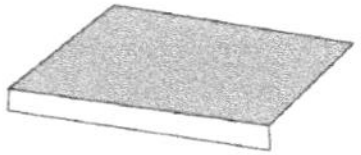	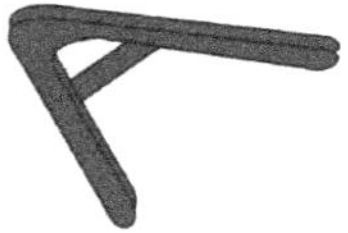
Present the dessert Plate			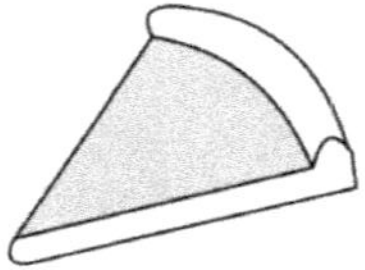		
Hold and Support the Dessert Plates		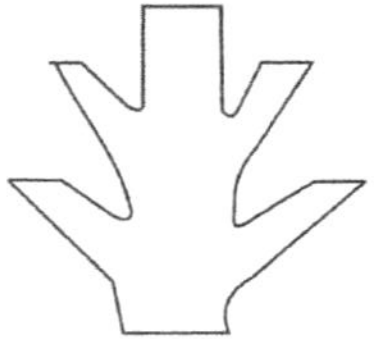	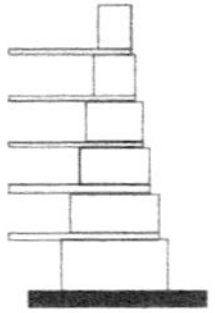		
Rotate Support structure to Facilitate Choosing	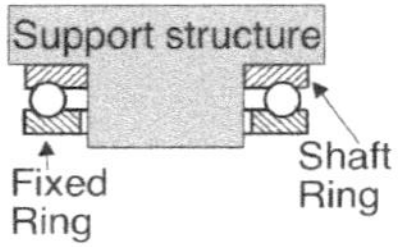 Support structure Fixed Ring Shaft Ring	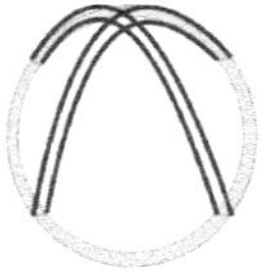			

Table 8.6 Generated conceptual designs of table-top dessert server

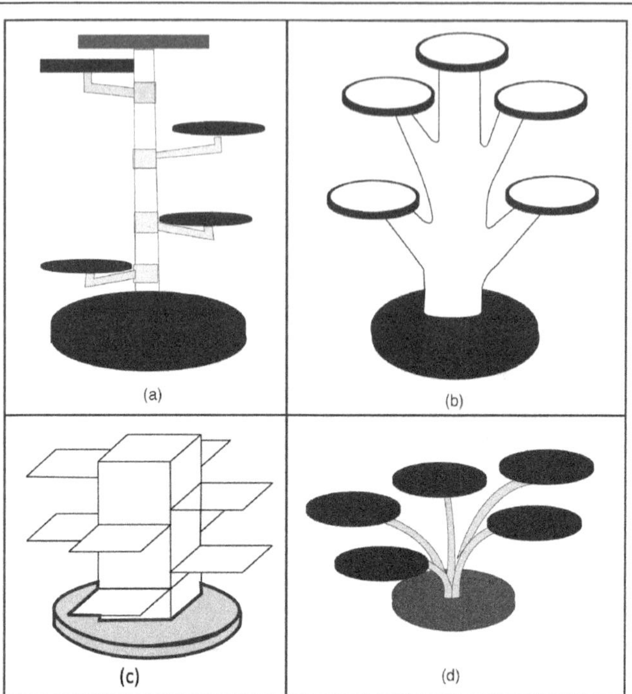

Figure 8.12 Completed table-top dessert server.

Example 8.8: Design of a display board

Big conferences normally would have several parallel sessions and participants go from one session to another in quick succession. To facilitate the participants, session programs are displayed at the entrance of the rooms where these sessions are held. Current poster stands suffer from several limitations. They are not attractive; the designs are in old styles and do not reflect the interests of visitors; they do not reflect the cultural inheritance of the local area. The aim of this project is to develop an attractive stand that can be used in conference halls when parallel sessions are conducted. Generate conceptual designs for a display board using morphological analysis as the design method.

Requirements and product concept

The requirements and the product concept consisting of the function tree and specifications have been established by establishing customer verbatim, needs, metrics and specifications in that order. A function tree of the product is also developed by the design team and is shown in Figure 8.13. It shows that there are three purpose functions, namely displaying the poster, supporting the display board and housing the components, and acting as storage to the items.

A morphological chart is prepared using the purpose functions, display the poster, support the display board and house and support the units in-use and the units not-in-use. Table 8.7 shows the morphological chart. Six designs were generated using the morphological chart.

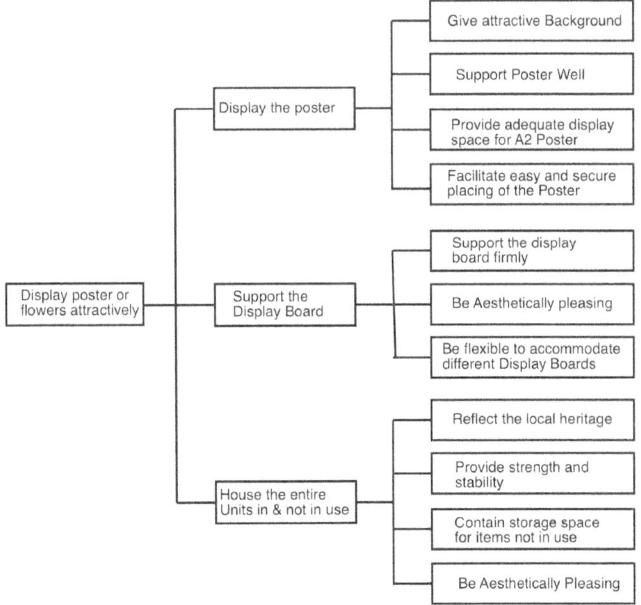

Figure 8.13 Function tree for the display board.

Table 8.7 Morphological chart for the display board

Display the Poster	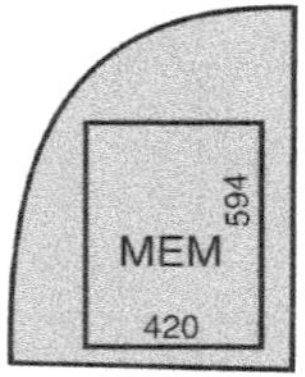 MEM 594 420 Non-symmetry	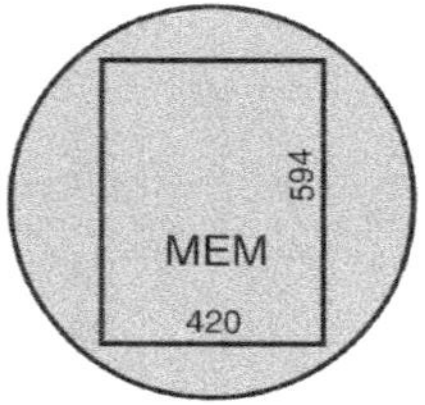 MEM 594 420 Circular	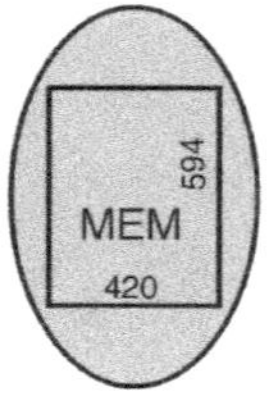 MEM 594 420 Elliptical	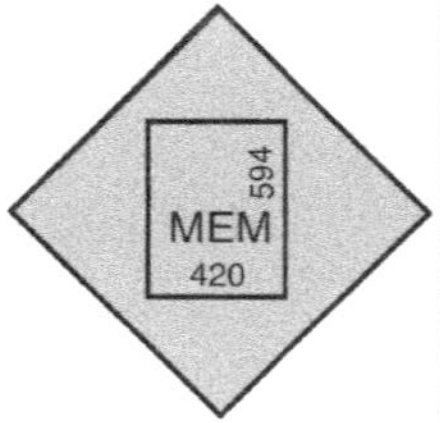 MEM 594 420 Diamond	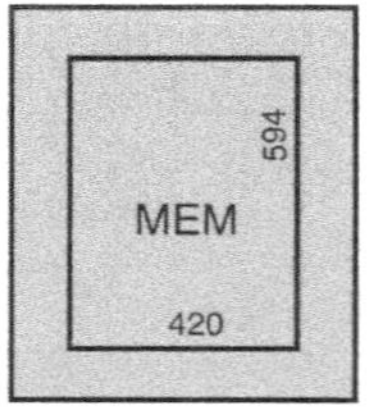 MEM 594 420 Rectangular
Support Display Board	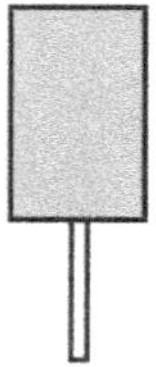 Single Pole	 Double Pole	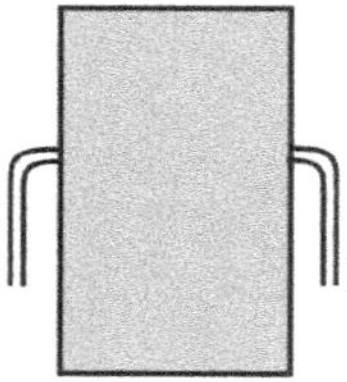 Side support	Wall Mount	
House and support Units in-use and units not-in-use.	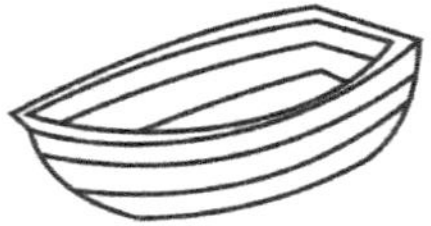 Traditional Fishing	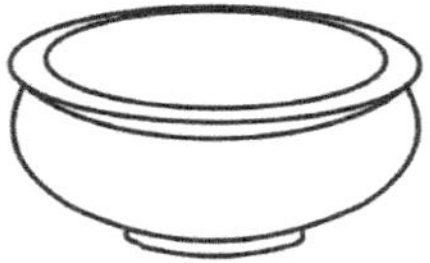 Traditional Clay Work		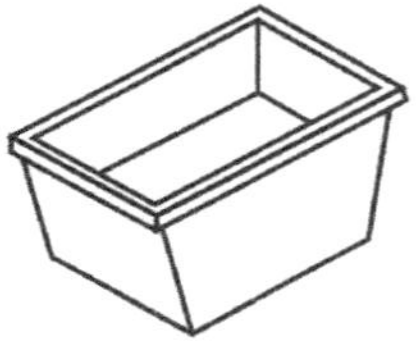	

Table 8.8 Solution space from the morphological chart: display board

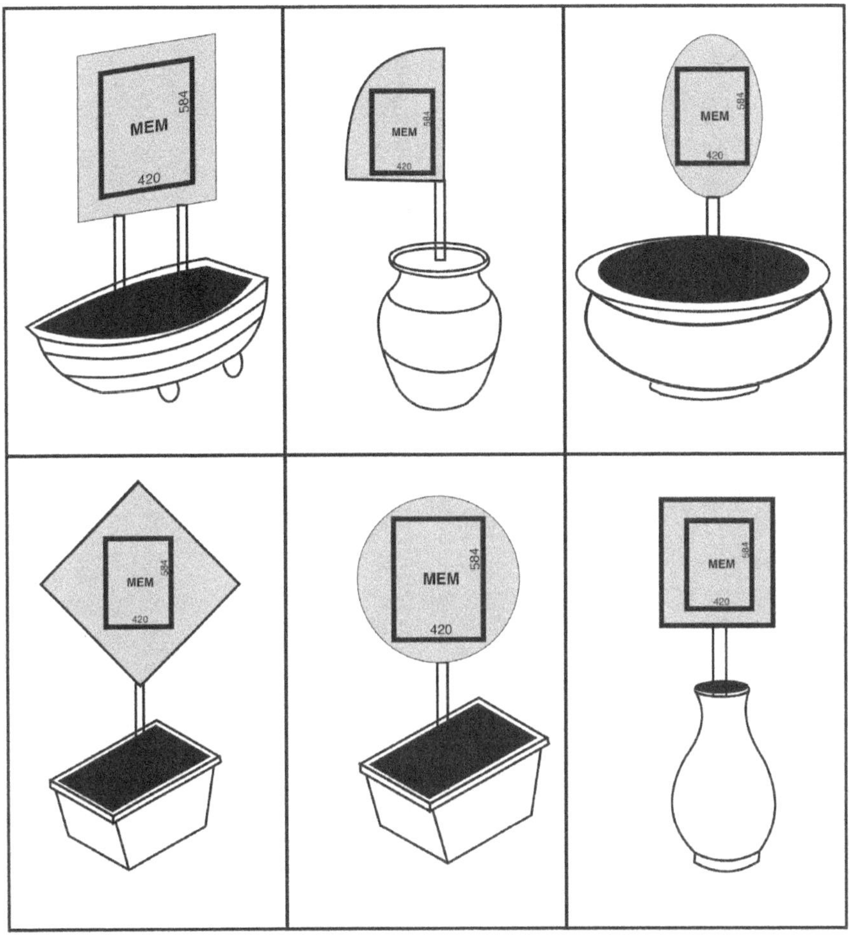

Each of them is a compatible combination of function providers chosen from the chart. The conceptual designs are shown in Table 8.8. Figure 8.14 shows the picture of the artefact made by a group of students.

Figure 8.14 Prototype of the display board.

8.6 DESIGN METHOD 3: C SKETCH

In the C sketch method, the designers are required to do only a graphical representation of how they intend to solve a problem, for a specified cycle time equal to, say 15 minutes for small projects. Each one of those designers is required to work by himself on his proposal, and after the cycle time is finished, each designer passes his proposal to the next designer. After that, each designer will work on how to improve the design that was passed to him, in his own way. Each one of them can add, modify or delete some elements of the proposed design. At the end, there will be several proposals, the number equals to the number of team members. This method is all about initiating original ideas from different minds and continue adding elements from different minds and to improve the original design ideas, by passing each design to others in a systematic way.

Consider the illustration of the C sketch method with four designers shown in Figure 8.15. Designer 1 starts his design and hands it to Designer 2 after a specified time. Designer 2 works on improving the concept for the specified time, and hands it to Designer 3. Designer 3 in turn works on improving it for the specified time and hands it to Designer 4. Designer 4 completes the design and hands it over to Designer 1 for the final look. This is represented by the outer collection of the arc arrows. The next inner set of arc arrows shows the progress of the design started by Designer 2. The representation continues like this until the innermost set of arc arrows represents the last designer, which in this case is Designer 4.

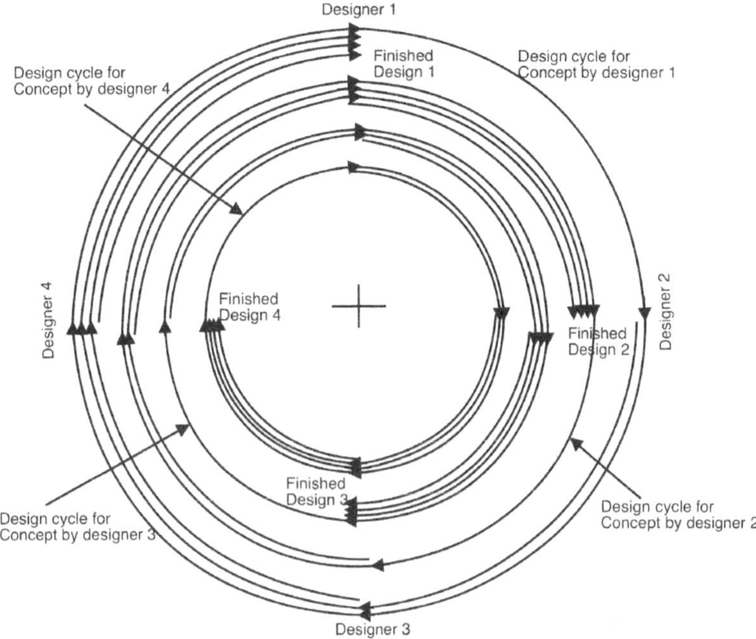

Figure 8.15 Illustration of C sketch design process with four designers.

Example 8.9: Design of a display stand for a high value-added component

There are several companies with their main theme centres on the phenomenon of precision, which is the designing and manufacturing of machines and other structures to achieve low tolerances and repeatability. These companies take pride on several aspects or values such as the types of machines, the offering and type of its services for many generations and the types and grades of materials handled. To present their values to customers, these companies tend to display their manufactured complex products for those passing by, to see. Special stands are designed and made to hold these items. In this example, 'C' sketch is used to develop the conceptual design of the stand for a company that was started by a father and son and continued by their succeeding generations. Figure 8.16 shows the picture of a stand designed and manufactured by a group of students using C sketch to develop the design.

Design process of the stands developed by them can be described in the following way. Four designers started their concepts and passed them through only two of the designers to result in four conceptual designs. This is illustrated in Table 8.9. The first designer visualized 'excellence' as the governing theme. He started with the earlier approach of craftmanship to excel excellence. The other two designers supported this theme. The second designer added advanced machines, to add better or improved processes to the excelling craftmanship as

Table 8.9 Progress of C sketching in the design of the stand for high value-added component

happened with the march of time. The third one introduced CAD/CAM instead of craftmanship and introduced strategic management to excel in quality. The thought process embraced the developments in excellence through generations, in precision manufacturing.

The second designer started with the theme of quality through dedication. The next designer emphasized the dedication and this was followed through, by four generations. The third one ended by comparing it with the development of the four dimensions of quality development with strategic quality management as the latest one. The third designer started with the theme of people as the driving force behind the company's development. He depicted the main persons in the company as pillars in the growth of the company. The other two designers supported the approach and added more pillars. The fourth designer highlighted every specialist characteristic the company has maintained with a closed fist and thumb up. The second and third designers added additional qualities in each generation. In all designs, a transparent rectangular plate was added on top to keep the component. The process is captured in Table 8.9 where the first row shows the starting themes. The second row shows the additions made by the second designer for the proposed themes and concepts. The third row shows the final designs from the entire exercise. The students then combined the good features from all designs to their final design they built. They combined the first two elements in the third row as their design, and this is shown in Figure 8.16.

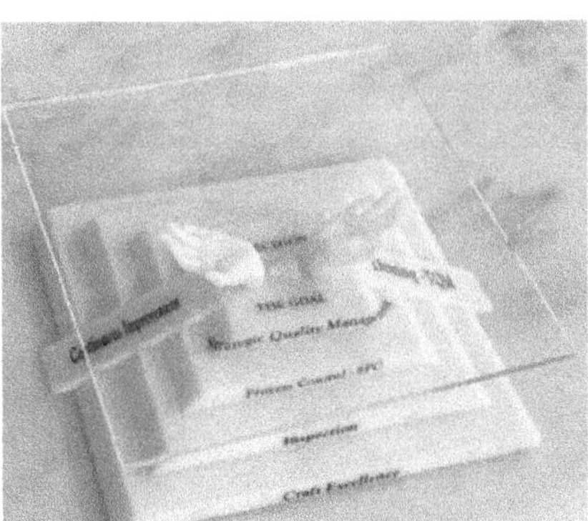

Figure 8.16 Stand for a high value-added component.

Example 8.10: Design of a wooden clock preaching noble ideas

An entrepreneur thinks that noble ideas have to be reminded often, and he thinks that a wall clock showing noble ideas will remind of them, every time someone looks at the clock for time. C sketch is seen as an ideal method to generate concepts for this clock. This is because the designer who starts the concept sets up the perspective or theme while others contribute to build the concept. The four themes chosen by the starting designers in this assignment are as follows:

1. Benefits of being seated amidst learned persons
2. Benefits of hard work
3. Wishlist by a person looking for progress
4. Merits of learned people.

The additions and deletions or alterations are textual in nature, and the contributions by individual designers in the C sketching process are not shown here. But the main point is that the need is for 12 ideas for each clock. Each of the subsequent designers in the C sketch method therefore had sufficient place for their contribution. The themes are the main area for creativity. The designs developed are shown in Table 8.10.

Table 8.10 Solution space for wooden clock preaching noble ideas

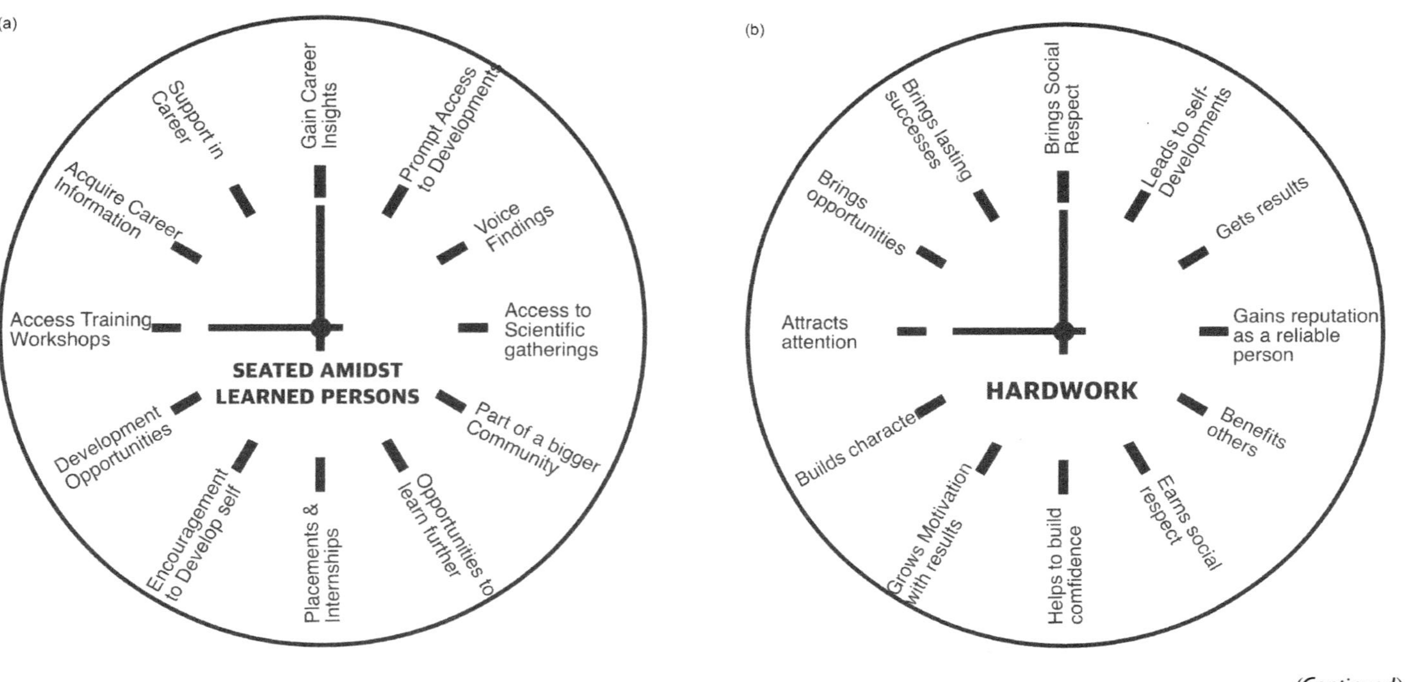

(*Continued*)

Table 8.10 (Continued) Solution space for wooden clock preaching noble ideas

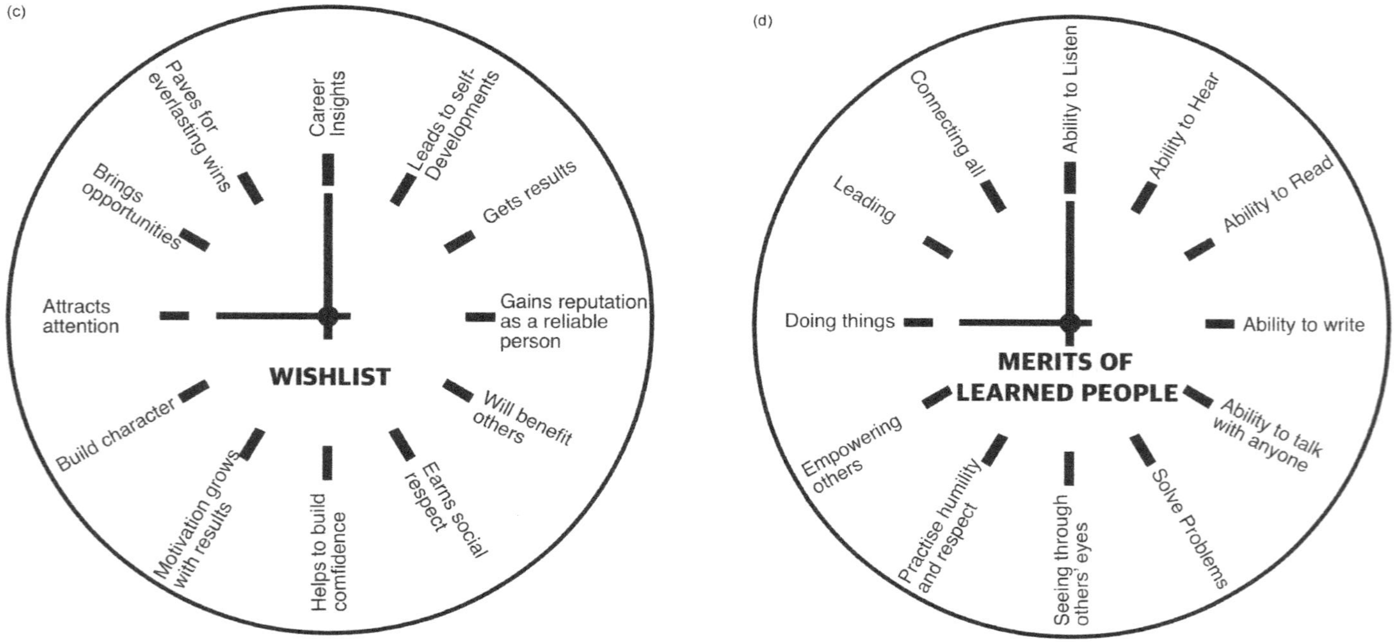

(c)

WISHLIST

Career Insights
Leads to self-Developments
Gets results
Gains reputation as a reliable person
Will benefit others
Earns social respect
Helps to build comfidence
Motivation grows with results
Build character
Attracts attention
Brings opportunities
Paves for everlasting wins

(d)

MERITS OF LEARNED PEOPLE

Ability to Listen
Ability to Hear
Ability to Read
Ability to write
Ability to talk with anyone
Solve Problems
Seeing through others' eyes
Practise humility and respect
Empowering others
Doing things
Leading
Connecting all

Example 8.11: Design of a dining chair

This example shows four designers developing four designs for a chair on a dining table.

Table 8.11 Progress of C sketching: design of dining chairs

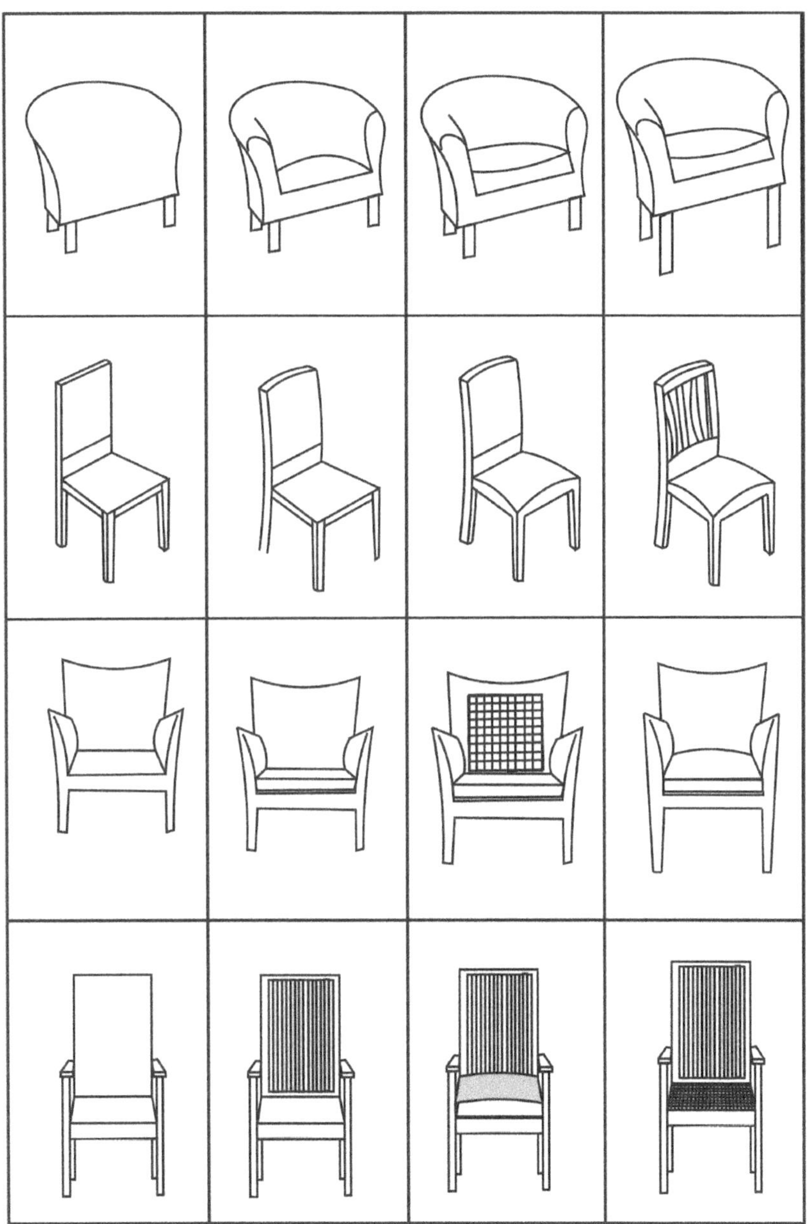

These four designs are shown along the four rows in Table 8.11. The first design was visualized by the starting designer as a luxurious chair like a chair in the living room. The second designer added details to the seat. The third designer added a cushion and the fourth designer added length to the legs and did the final touch-ups to the design. The second starting designer visualized a simple chair. He started with four legs and back with the seat. The second designer added curvature to the back-support and the hind legs. The third designer added cushion to the seat and the fourth designer added grill work to the back. The third starting designer visualized the chair with a comfortable back-support. The second designer erased some of the details and provided a cushion to the seat. The third designer added cushions to the side-support to the arms and a replaceable back cushion. The fourth designer extended the legs and completed the cushion work started by the third designer. But he removed the back cushion. The fourth Starting designer started with a simple chair with arm rests. The second added grill to the back. The third added a cushion to it. The fourth removed the cushion and replaced it with a woven seat. To compensate for the height variation, the legs were extended.

8.7 DESIGN METHOD 4: ATTRIBUTE LISTING

Attribute listing takes an existing product or system, breaks it into parts and then recombines them identifying new forms of the product or system. It establishes the attributes that the product should have and proposes alternative providers to provide the attributes, choose the providers of each attribute and combine them to form the final product. The attributes may include physical, mental, emotional and social, considerations depending on the complexity of the challenge. Attribute listing focuses on identifying attributes of a process or product and ways to improve one or many of them.

The method starts with the identification of the features or the components of it, that needs improvement. The important step is the process of breaking. In the process of 'breaking up' an existing product, the aim is to identify its elements that can answer questions like the following:

i. Is this element of an object or a structure necessary for its function?
ii. Do the elements have certain attributes (form, material, etc.) fundamental to the functioning of the product?
iii. Are there any other possibilities and ways to create this element? Or
iv. What happens when elements with changed attributes are put together?

In other words, conduct an in-depth analysis to identify the attributes. For this purpose, the elements and possible attributes are often listed in a table.

The steps in conceptual design using attribute listing are as follows:

i. Make a list of attributes of a product or a service and choose some of these attributes that seem particularly interesting or important to improve the design.
ii. Draw a morphological table using the important attributes as the first column.
iii. Identify alternative ways of achieving each attribute.
iv. Combine one or more of these alternative ways of achieving the required attributes and try to come up with a new product.

It is a good technique to use in conjunction with some other techniques especially brainstorming and morphological analysis. The method can be useful to come up with the next generation of products where advanced technologies can replace obsolete ones.

Example 8.12: Design of a memorabilia showing conceptual design methods

A table-top memorabilia item in memory of a conceptual design methods workshop is desired. To facilitate the design process, use a table-top globe as the starting product to collect and formulate the

Figure 8.17 Rotating globe.

required attributes. Establish four conceptual designs for table-top memorabilia, which enumerates the ten conceptual design methods discussed in the workshop using 'Attribute Listing' method.

Answer

Consider the table-top globe as labelled in Figure 8.17. The three parts and the identified attributes are given in Table 8.12.

Table 8.12 Attributes of different parts of the globe model

Part	*Attributes*
1 Globe	Displays lands and seas partitioned
	Displays boundaries as curving lines
	Can see needed detail
2 Axis locator	Carries the globe in axis
	Provides angle to the axis
	Facilitates revolving
3 Base	Ensures stable foot area
	Gives catch phrase area

The attributes established are used as the parameters to develop a morphological chart. The morphological chart developed is shown in Table 8.13.

Table 8.13 Morphological chart showing ways of achieving attributes

Attribute	*Option 1*	*Option 2*	*Option 3*	*Option 4*
Displays lands and seas as partitioned portions	Design methods as wedges or faces of a prism	Design methods as different sectors of circle	Design methods as slices of an orange.	Design methods as segments in overlayed prisms
Displays boundaries	As straight lines	As curves		
Can see needed detail	Can see names of the methods			
Carries the globe in axis	Support from one end	Support from two ends		
Provides angle to the axis	Supported by vertical axis	Supported by slanted axis		
Facilitates revolving	Resting in one position	Resting in any random position		
Ensures stable foot area	Covering projection	Balancing moment		
Presents catch phrase area	Presents a space for title	No space for title		

Table 8.14 Conceptual designs generated from the morphological chart

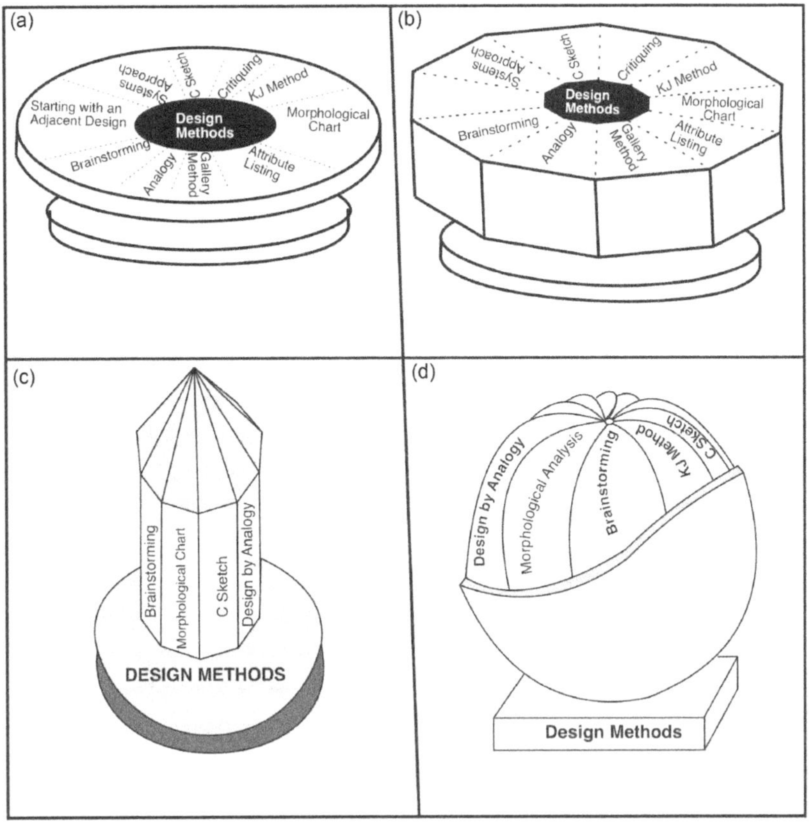

(a)	(b)
Design Methods; Systems Approach; C Sketch; Critiquing; KJ Method; Morphological Chart; Attribute Listing; Gallery Method; Analogy; Brainstorming; Starting with an Adjacent Design	Design Methods; Systems Approach; C Sketch; Critiquing; KJ Method; Morphological Chart; Attribute Listing; Gallery Method; Analogy; Brainstorming
(c)	(d)
Brainstorming; Morphological Chart; C Sketch; Design by Analogy; DESIGN METHODS	Design by Analogy; Morphological Analysis; Brainstorming; KJ Method; C Sketch; Design Methods

The designs are made by harmonious integration of one chosen item from each row. Four such designs composed by the integration are shown in Table 8.14. For example, design (a) is composed of the integration of the following:

1. Design methods as different sectors of circle
2. Separated by dotted lines
3. Can see the names of the design methods
4. Support from one end
5. Supported by vertical axis
6. Resting in one position
7. Covering projection
8. Presents a space for title

Example 8.13: Design of a bicycle for leg amputees

An entrepreneur identifies the opportunity to develop a bicycle for people who lost a leg. He feels that they can ride the bicycle by pedalling with a hand while the other hand can steer the bicycle. Using a road bicycle as the starting product, develop a bicycle for the handicapped described above.

Important components (attributes)of a road bicycle, as shown in Figure 8.18, are as follows:

1. Main frame
2. Saddle
3. Drive system (Pedal, Cog wheel, and chain drive)
4. Steering system (Handle bar and fork)
5. Wheel assemblies (Front and rear)
6. Front wheel hub
7. Rear wheel hub

Out of these the drive system, the steering system and the front wheel can remain the same and the remaining are the components that need change. The main attributes of the frame are as follows:

1. Interface to the steering system
2. Interface to the drive system
3. Interface the wheels
4. Interface with the rear wheel and Interface to the saddle.

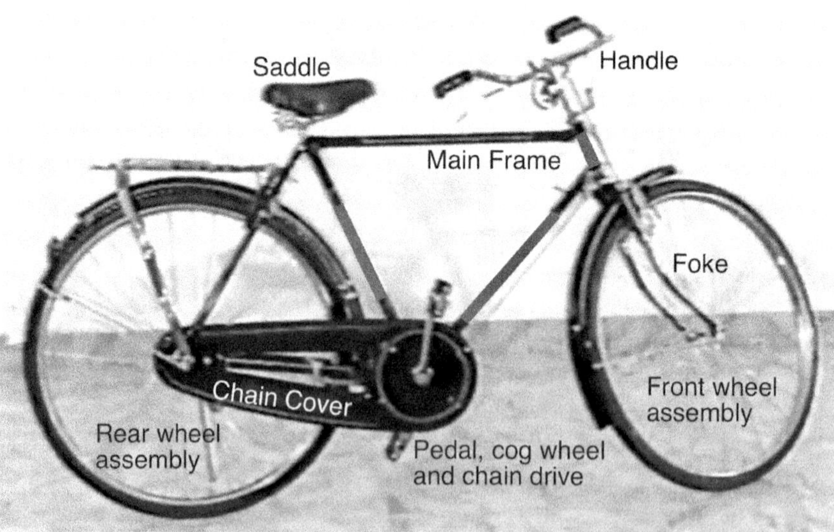

Figure 8.18 Parts of a road bicycle.

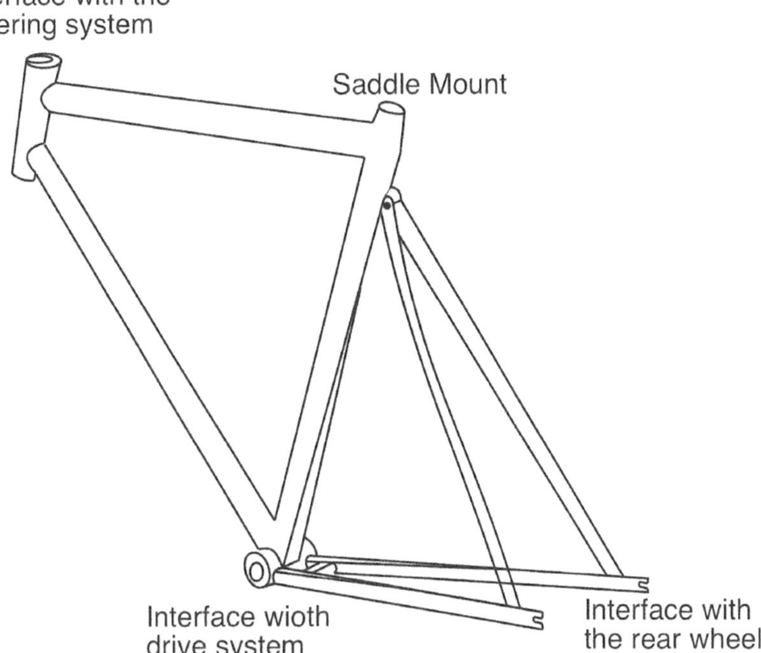

Figure 8.19 Frame of a road bicycle.

Figure 8.19 shows how the frame of a road bicycle has these attributes. The additional attributes that have to be included in a new frame are

1. Accommodation of a bigger seat
2. Drive system powered by one hand instead of the legs and
3. Two rear wheels instead of a single one. This is needed to provide stability at all times.

The rear wheels have to be powered by the user using the traditional pedal, cog wheel, chain and free wheel. Figure 8.20 shows the components that go into the traditional rear wheel assembly in a road bicycle. As indicated in the figure, the axle receives the load from the frame and passes to the wheels through the bearings to the hub, spokes rim and tyre.

Seating the user and deciding the entrance side are the first decisions. Once these are decided, the next decision is to fix the drive system accessible to the driving hand. The next decision is to fix the position of the handle bar. The final decision was to locate the rear wheels for transferring the load to the ground through the wheels.

Two concepts developed are shown in Figure 8.21. In the concept shown in Figure 8.21a, two separate forks hold two rear wheels with one of them fitted with the free wheel. The front wheel and the steering system are connected through a single bar. In the concept shown

Freewheel receives power and Supplies it in turn to wheel

Hub: Receives load and transfers it to the ground through spokes rim and tyre

Figure 8.20 Components in a traditional rear wheel system.

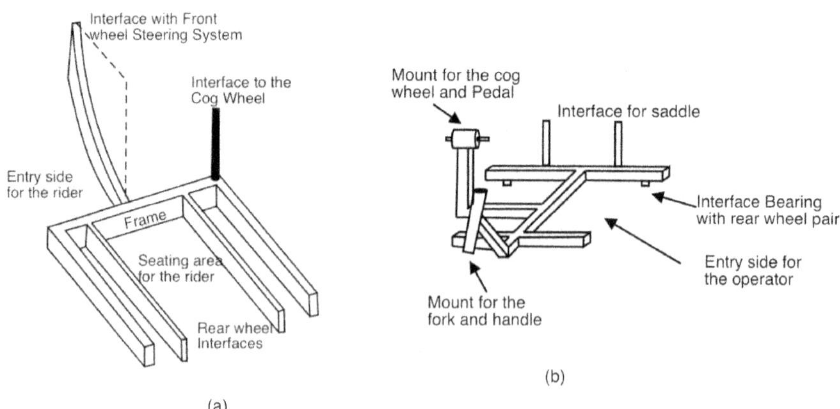

Figure 8.21 Concepts for a bicycle frame for leg amputees.

in Figure 8.21b, the hubs of the two wheels are firmly connected by a metal tube possibly through welding. The load coming from the frame is transferred to this tube by external bearings, and this tube becomes the rotating shaft. In the new rear wheel system, the attributes that needed change are two wheels instead of one, tube connecting the hubs as one rotating unit, and two bearings and housings to transfer load. The axle was not needed and hence eliminated. The free wheel and the hub connection are contained only in one wheel.

Figure 8.22 shows the exploded view of the parts to plan the embodiment design.

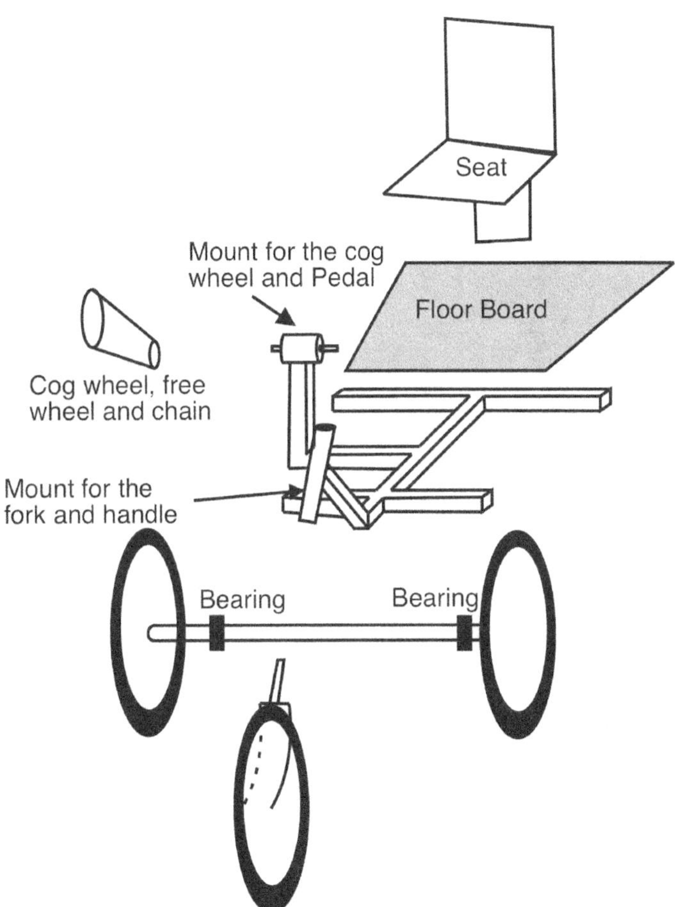

Figure 8.22 Exploded view for assembly planning cycle for leg amputees.

Example 8.14: Design of a memorabilia for MEM using Attribute Listing

Master of Engineering Management, MEM, program at United Arab Emirates University has been operating for more than 15 years. To celebrate this achievement, a conference is being organized. International experts are invited and are expected to give keynote addresses. To give them a small present that would remind MEM program every time they see it, a product has to be designed and fabricated using attribute listing. This is the task.

The first step is to identify a similar typical memorabilia item. A rotatable glass cube with the engraved features of the City of London was released as memorabilia to celebrate the London Olympics in 2012, and this was taken as one product for identifying the attributes. This product is made up of two parts: a base and a cube which stands on one of its vertices on the base. The product is shown in Figure 8.23.

Attributes of the two pieces, the base and the rotating cube were identified and listed in Table 8.15.

The internally carved features were identified as not suitable and dropped.

Figure 8.23 London Olympic memorabilia.

Table 8.15 Attributes of parts of the city of London memorabilia

Rotating cube: Part 1	*Base: Part 2*
Displays several features	Carries the cube in axis
Geometric shape boundaries	Facilitates revolving
Displaying needed features	Ensures stable foot area
Precision engineering	Presents catch phrase area
Internally carved features	
Right choice of material	
Innovative manufacturing	

The students developed the design shown in Figure 8.24. They established the design by combining the following options given in Table 8.16:

Figure 8.24 MEM memorabilia made by students.

1. Dodecahedron
2. Geometry boundaries as straight lines
3. Can see the names of the courses
4. 3D positioning and sharp straight faces
5. Sheet metal
6. Sheet metal net and precision welding
7. Resting in one point
8. Revolving
9. Covering projection
10. Presents a space for title

Table 8.16 Alternative ways of achieving attributes

Attribute	*Option 1*	*Option 2*	*Option 3*
Displays several features	Displays 11 courses, one in each face		
	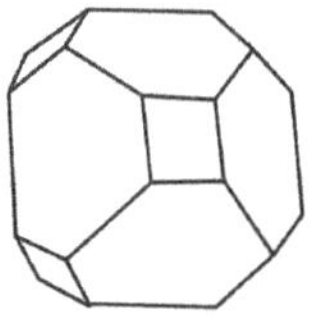	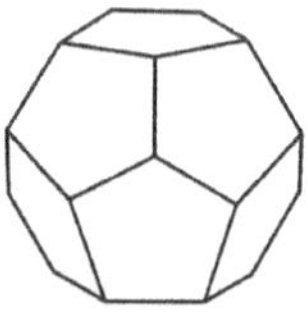	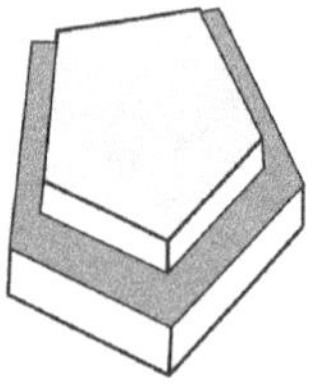
Geometric shape – boundaries	As straight lines	As curves	
Displaying features	Can see names of the courses		
Precision engineering	3D positioning & sharp straight faces	3D positioning & sharp curved faces	
Right choice of material	Sheet metal	Solid block	
Innovative manufacturing	Sheet metal net and precision welding	NC Machining	3D Printing
Carries the cube in axis	Resting in one point	Resting in two point	
Facilitates revolving	Revolving	Not revolving	
Ensures stable foot Area	Covering Projection	Balancing Moment	
Presents catch phrase area	Presents a space for title	No space for title	

8.8 DESIGN METHOD 5: START WITH AN ADJACENT OR UNACCEPTABLE SOLUTION

Professor Bruce Field from Monash University, Australia, originally proposed this method. The method is based on the following arguments. A design is deemed unsatisfactory because (i) it either does not do the expected function satisfactorily or (ii) because there are no provisions in the design to perform all the specified functions and additional provisions are needed to perform these functions. This kind of situation arises when someone working on a design leaves the job or when the current generation of the product is using an old technology, while better technologies with more capabilities have become available now.

A product can be visualized as a point in an n-dimensional hyperspace where each dimension is a feature. Then it is possible to identify the n number of desired features and a corresponding n-dimensional hyperspace of the desired or acceptable design. This can be established by referring to the stakeholder requirements. If this hyperspace, and the feature vector of the unacceptable solution are compared, two things (i) the unacceptable features present in the solution and (ii) desired features that are absent, can be identified. Now it is possible to adopt and produce concepts that have (i) none of the unacceptable features and (ii) all or most of the desired features present. The design process is shown by the flow-chart shown in Figure 8.25.

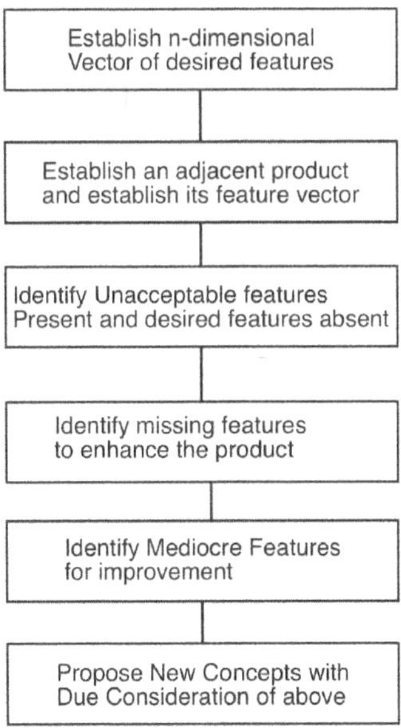

Figure 8.25 Design process from an adjacent design.

Example 8.15: Design of a display board/flower vase stand

The aim of this project is to develop an attractive display board for notices that can be used in conference halls when parallel sessions are conducted and can be used as a flower vase stand when single meetings are convened (Figure 8.26).

PRODUCT CONCEPT

Author's Note: This assignment, once was done by a group of students in the UAE. One of the requirements of the assignment was to reflect the nature of the local environment. In one of the early meetings, in their desperation to identify a starting point, one student exclaimed 'this is a desert, full of rocks and sand and some palms and camels, where can we start?'. They were advised to look at the properties or features of the aforementioned items. In the next meeting, they came up with the following: *Unwavering character is one of the main phenomena symbolized by a rock. In general, stones are strong, versatile and easily accessible and represent the ability to be grounded and connected with the earth. It represents hardness, strength and permanence of relationships. The people also have these properties and friendships or loyalty. Thus, a rock symbolizes humble and hardworking people living with conviction and rich cultural heritage. Therefore, we have decided to base our product on sand stone.*

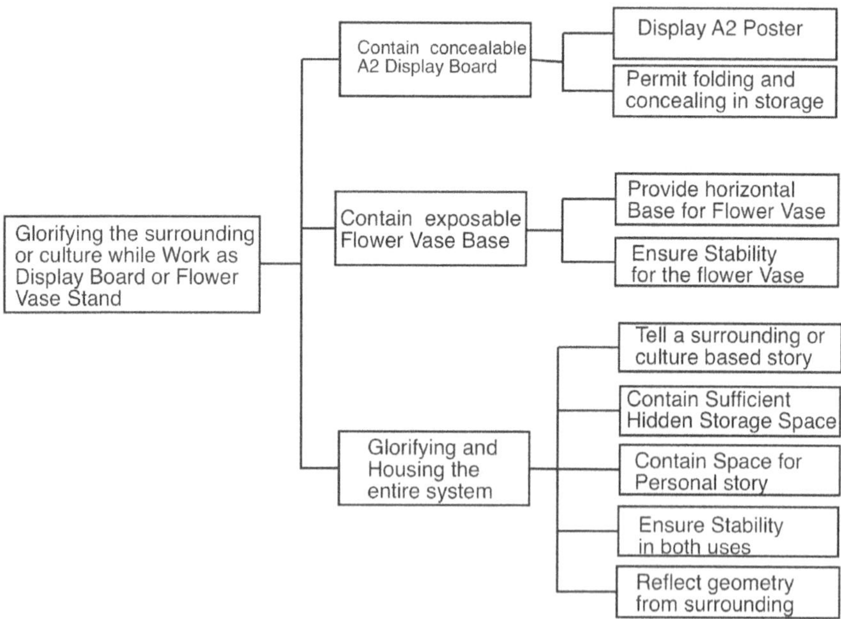

Figure 8.26 Function tree of display/flower vase stand.

Table 8.17 Options for material and form

Option 1: This transparent jar contains broken pieces of sandstone. While they are stones it get the shape from the jar and loses the magnitude, the very characteristic of the sandstone.	*Option 2*: This transparent jar contains broken marble stones. The broken nature does not provide the sense of the rock.	*Option 3*: This is a stone carving of a traditional coffee jug used by the local people. Though it is a carving made out of stone it will be seen as a valuable piece of art and not as a rock.
Option 4: This is a camel made out of stone by carving and will not be valued as a stone. Instead, it will be valued as a piece of craft.	*Option 5*: This is a boulder that is the representation of a hill or mountain. It is sandstone with layers and will be recognized as sandstone.	*Option 6*: This is another porous sandstone boulder found inside a desert and also will be recognized as a sandstone.

The decision of using rock as the material has already been made and the first decision was to decide the material form. The possible forms considered are shown in Table 8.17. The choice from these forms was the sandstone rock, option 6.

Solution concept

The first step in using this design method is to form the *n*-vector of desirable features. After consulting the stakeholder requirements the following nine-vector was identified as the desired feature vector for the display board/vase stand.

$$\begin{bmatrix} \text{Exhibitting the Magnanimity} \\ \text{Staying put in adverse weather} \\ \text{Structured life of the locals} \\ \text{Miniature of the local area} \\ \text{Light weight} \\ \text{Easy to handle} \\ \text{Easy to clean and maintain} \\ \text{Mouldable to different shapes} \\ \text{Catch Phrase for the Portrayal} \end{bmatrix}$$

The following nine-vector can be used to represent a local sandstone rock or boulder, which can be used as the adjacent or unacceptable design to start with.

$$\begin{bmatrix} \text{Magnifiencetly big Figure} \\ \text{Staying put in adverse weather} \\ \text{Porous Structured nature of rock} \\ \text{Can include features of local area} \\ \textit{Unacceptably heavy weight} \\ \textit{Difficult to handle} \\ \textit{Moderately difficult to maintain} \\ \textit{Needs carving for different shapes} \\ \textit{The catch phrase needs to be found} \end{bmatrix}$$

Comparing the two vectors, the appearance and porous structure of the boulder are desired properties of the rock that can be retained. The moderate feature on the inclusion of local details can be enhanced. The main features that have to be removed and replaced are (i) unacceptably heavy weight, (ii) handling difficulty, (iii) difficulty to maintain due to the filling of the whole volume, and (iv) difficulty in moulding shape to include local features. This demands a change in material and the choice was Flynn material and gypsum. Flynn can be seen as a woven cloth. This choice allowed the removal of the undesirable features and permitted the enhancement of the moderate feature. Concepts can be generated with a wooden frame covered by the Flynn mesh with a layer of gypsum. This can permit a hollow structure where unused items can be stored. The chosen and built design is shown in Figure 8.27.

Figure 8.27 Display board/flower vase stand by students.

Example 8.16: Design of a dessert trolley for a hotel

A star hotel needs a trolley to take the desserts to their guests during lunch and dinner. Design a trolley starting with a workshop trolley shown in Figure 8.28a as an adjacent unacceptable design.

The first step in using this design method is to form the n-vector of desirable features. To compile this vector, based on the nature of the product it was decided to consult the customer requirements.

Customer Requirements for a dessert trolley:

1. Cosy appearance and handle
2. Strong to carry large quantities
3. Easy to manoeuvre
4. Adequate space to display several items
5. Stylish and easy to clean shelves
6. Has multiple shelves
7. Has lower shelf for plates and cutlery
8. Has separate space for accessories (tissues, napkins, etc.)
9. Separate compartments for cakes, fruits, and ice creams
10. Can see all items while seated at the dining table

From the customer requirements, the vector of the desired features is identified as

$$\begin{bmatrix} \text{Cosy appearance and Handle} \\ \text{Structurally strong} \\ \text{Easy to manoeuvre} \\ \text{Adequate space} \\ \text{Easy to handle} \\ \text{Can see all items while seated} \\ \text{Multiple shelf structure} \end{bmatrix}$$

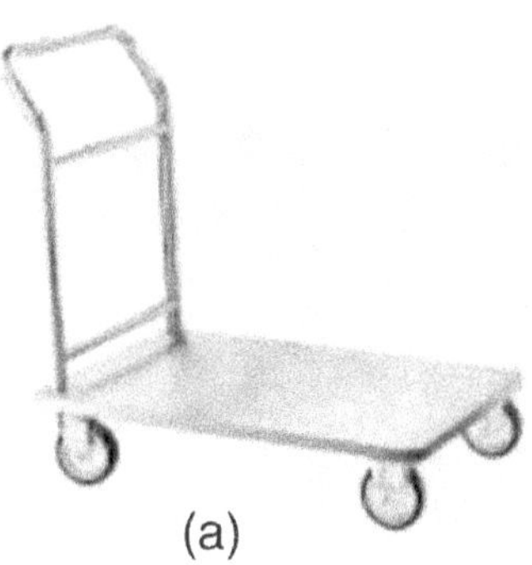

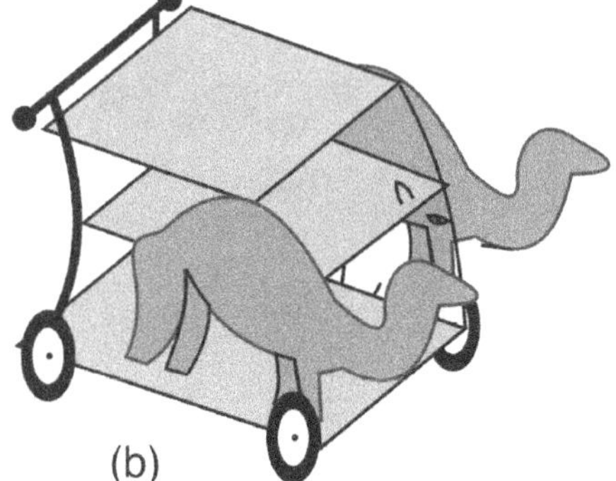

Figure 8.28 Adjacent design and the developed design.

The following seven vector can be used to represent a workshop trolley, which can be used as the adjacent or unacceptable design to start with.

$$\begin{bmatrix} \textit{Bland appearance and handle} \\ \text{Structurally strong} \\ \text{Easy to manoeuvre} \\ \textit{Limited space} \\ \textit{Difficult to handle (loading at low level)} \\ \text{Can see all items while seated} \\ \textit{Single shelf structure} \end{bmatrix}$$

The items that are in italics need changing while the items in normal letters can be retained. Space limitations can be rectified by introducing multiple layers. Difficulty in handling is rectified by the arrangement of items on the different shelves. The bland appearance is removed by introducing the arc shape from the sizes of the shelves and by the introduction of camel profiles on the sides. The proposed design is shown in Figure 8.28b.

Example 8.17: Design of a string hopper making machine

Rice noodles or string hoppers shown in Figure 8.29a is a Sri-Lankan food item made using a wooden kitchen press shown in Figure 8.29c. Rice flour with hot water and little oil is mixed thoroughly to make a dough. The dough is then filled in the cylinder shown in the figure, and the piston is engaged on the filled dough and is pressed with both hands.

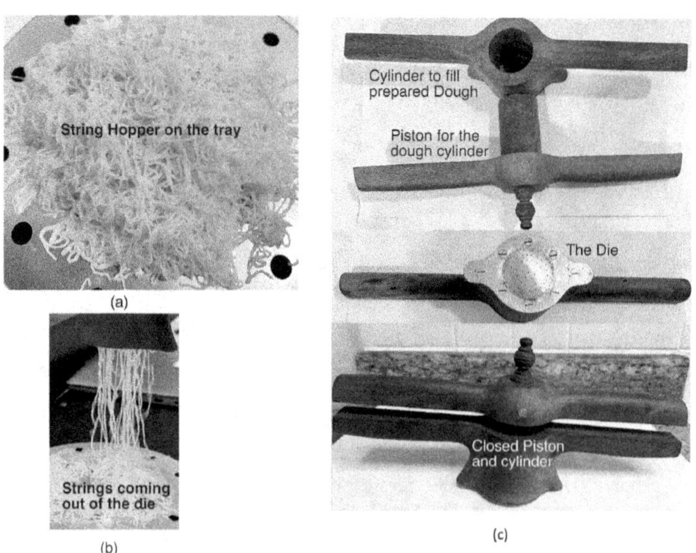

Figure 8.29 Manual kitchen press producing string hoppers.

The pressed dough is forced through the die as shown in Figure 8.29b to produce the string strands in the form shown. The strings produced by the kitchen press are received on a special tray by laying the strings in a circular form.

Producing and laying the string hopper on the tray has two operations done by the hand. In the first, it forces the piston to press the dough down the cylinder through the holes on the die. In the second, the hand circles and lays the string stream circling on the tray.

The task in hand is to design and develop a mechatronic machine to produce string hoppers from the prepared dough. It has to be a mechatronic system because the flow of the string stream should be started and stopped at the right time and the rotation of the tray should start and stop at the right time.

Answer

From the customer requirements the vector of the desired features is identified as the 9-vector

$$\begin{bmatrix} \text{Changeable Dies} \\ \text{Continuous flow of strings} \\ \text{Piston and cylinder arrangement} \\ \text{Adequate space for Dough} \\ \text{Mechanised forcing of dough} \\ \text{Variable flow rate of dough} \\ \text{Automated start stop of dough flow} \\ \text{Circular filling of Dough on the hopper tray} \\ \text{Automated starting of rotational placing} \end{bmatrix}$$

The following nine-vector can be used to represent a wooden kitchen press, which can be used as the adjacent or unacceptable design to start with.

$$\begin{bmatrix} \text{Changeable Dies} \\ \text{Continuous flow of strings} \\ \text{Piston and cylinder arrangement} \\ \textit{space for Dough enough for a single hopper} \\ \textit{Manual forcing of dough} \\ \textit{Manually Variable flow rate of dough} \\ \textit{Manual start stop of dough flow} \\ \text{Circular filling of Dough on the hopper tray} \\ \textit{Manual starting of rotational placing} \end{bmatrix}$$

The items that are in italics need changing while the items in normal letters can be retained. In effect, two actions that have to be mechanized are (i) extruding the dough through the die to create strings and (ii) laying the strings in the circular 'string hopper tray' in a uniform fashion. The control of it can be achieved through a mechatronic system.

The mechatronic system receives a signal to start the process consisting of the following activities:

1. Place the string hopper tray under the nozzle to receive the stream of string by turning the 'Placing Disc'.
2. Start the feeding mechanism by moving the piston down.
3. Start rotating the tray to fill the string in a circular fashion on the tray by turning the 'Filling Disc'.
4. Wait till the tray is filled completely by holding the placing disc stationary.
5. Stop feeding string and stop rotating the tray by holding the filling disc stationary.
6. Turn the placing disc to bring the next empty tray in position under the nozzle to fill the next tray.
7. Remove the filled tray for cooking and replace it with an empty tray.
8. Repeat the process until the desired numbers are made.

The scheme of the concept can be described as follows:

1. Prepared dough can be stored in a cylinder and, a power-driven and speed-controlled piston will force it through the die to make a stream of strings.
2. A rotating circular 'Placing Disc' is placed under the string stream so that the strings will be laid uniformly on the tray.
3. The piston driving the strings is stopped until the next tray is placed in position under the die producing the string stream. See Figure 8.30 for a schematic arrangement of the concept.
4. The required filling pattern of the string on the tray follows a circular path. This is governed by turning the 'Filling Disc'.
5. The motion of the piston to produce string stream should be stopped and started to match the movement of the tray. That is the empty tray should be under the nozzle and the turning of the tray (about its axis) should synchronize with the flowing of string.
6. The timed motions of the individual motors, feeding empty trays and removing filled trays would keep the manufacturing continuous.

The conceptual design of the machine is shown in Figure 8.30. The dough is prepared and filled in the top tank and the mechatronic controller controls the linear actuator that presses the dough through the dies. The controlled motion of the placing disc ensures that empty

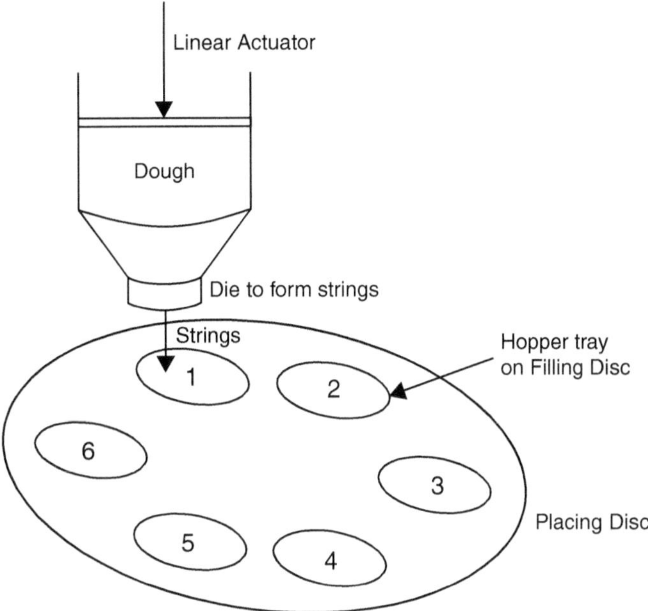

Figure 8.30 Conceptual design of the string hopper making machine.

trays are placed on positions 1, 2, 3, 4, 5 and 6 and string coming out of the die will fall on the tray. Rotating the tray simultaneously with the filling disc ensures that the filling is circular and uniform.

The function structure of the machine is shown in Figure 8.31.

The function structure can be explained in the following way. Two kinds of material the tray and dough are the inputs. Energy is available for use and the use is prompted by the signal. The process

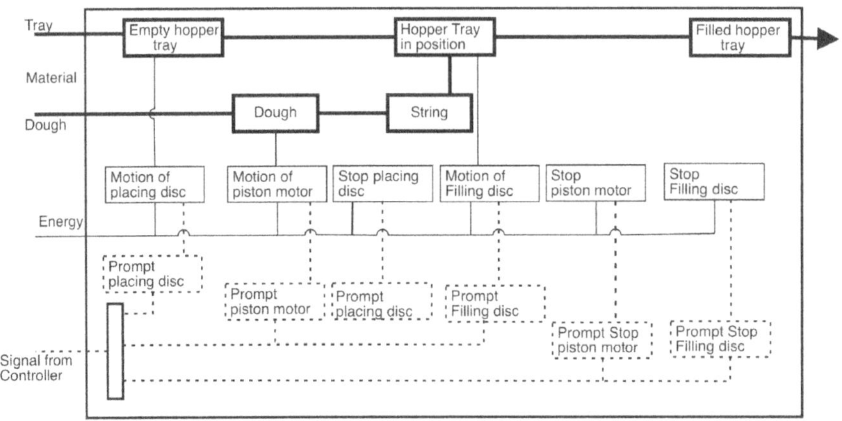

Figure 8.31 Function structure of the string hopper making machine.

starts with a prompt to the placing disc by a signal from the controller. The placing disc draws energy and moves until it is stopped by another signal. The dough is in the cylinder and the piston is moved to create string which falls on the tray under the nozzle placed in the filling disc. The filling disc starts rotating by a prompt until it is stopped. Now the tray is laid with the string on it becomes the output.

8.9 DESIGN METHOD 6: GALLERY METHOD

In gallery method, the designers work individually and collectively and, in this sense, it is a mixed method. In the first phase, individuals of a group begin sketching their ideas silently on sheets of paper. This is referred to as the phase of individual ideation. After a set amount of time, individuals display their sketches as a gallery and discuss their ideas. During the group discussion, the members of the group (i) present their ideas to the group (ii) critically evaluate each other's ideas and (iii) modify, eliminate or generate ideas as a group. This is followed by another round of silent individual idea generation and group discussion to choose a design. Pugh's method of Concept Evaluation is very suitable when the gallery is discussed.

Example 8.18: Design of a graduation gift

A gift on the graduation day for an engineering graduate was desired. The presenters want it to be unique and reflect the graduate's development during his life. This project was done by a group of graduate students.

The first step in conceptual design is collecting the requirements for the product which describe the desired behaviour of the product. They were classified into three main domains which are personal, academic and university domains. For each domain, there were specific requirements as given in Table 8.18.

The first phase of the method is to generate concepts individually and place them in the gallery for discussion. Four designs proposed by a team of four are shown in Table 8.19.

The first concept contains all the development activities in a clock, emphasizing the importance of time management. In this design, Clock's hands can be represented by mechanical tools like a screw driver and spanner which remind the student of his study days. Also, in this clock, each hour has a different sample or picture reflecting an element of the personal, academic and university domains. In the second concept, both hands join together to preserve the developments in the three domains, with the helmet which is a protecting equipment. The third design is a star pentagon shape with a mirror background containing the university logo. It has five sides; in the base, there are small models of the university memories, helmet, and graduation cap. The concept of this design is that the inner sides contain a picture of three domains, academic, university and personal, which reflect on

Table 8.19 Conceptual design gallery

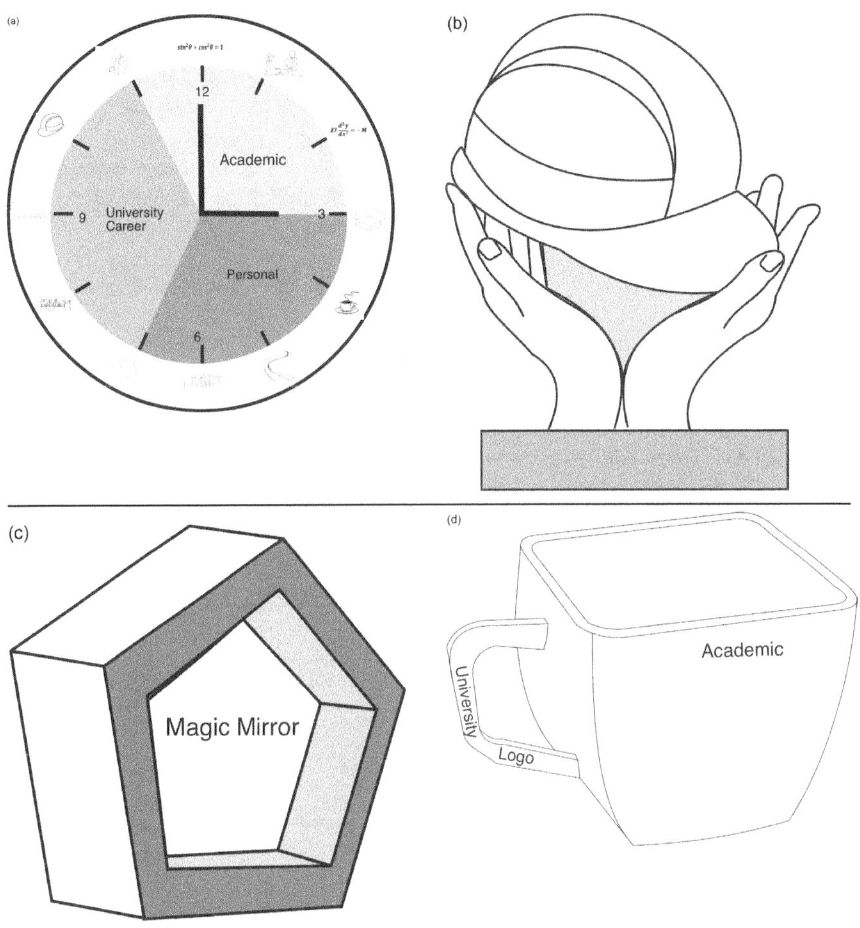

mirror only if the piece is held at a specific angle. The pentagon is a symbol that represents the spiritual domains advocated by many religions. It can be used as a desk mirror or trophy in the cabinet. The fourth concept came from the university café where the students spend their free time with friends. A four-sided cup with the University as the handle with three remaining sides to reflect the three domains.

During the discussion of the gallery, they identified the following:

1. Having pictures is a poorly presented good idea.
2. The topics shown in the pictures are good.
3. Couple of words underneath the pictures would make it even better.
4. They liked the idea of pentagon to represent the spiritual side.

Table 8.18 Characteristics to be displayed

Personal	*Academic*	*University*
Sport competitions and activities, & Social activity (hostel activities, events)	Graduation cap	Name and logo of the university
Sequence of basses	Graduated project and Industrial training	University achievements + Ranking
Relationships, new friends	Labs, experiments, tools, Equations, equipment, material	Special building
Canteen, café, food court	Hard working in different environments	Name of the town
Skills (communication and technical)		Travel to the university from home.

Table 8.20 Designs produced in the second round

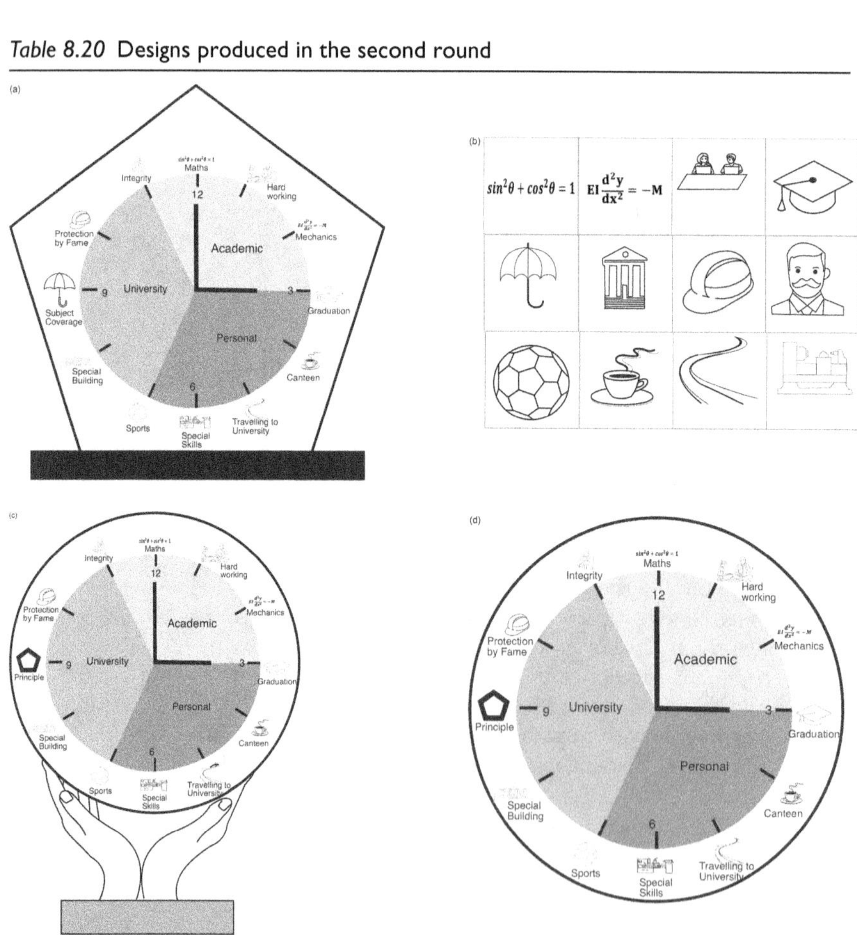

After the discussion in the gallery, the group went for a second individual session. Two students developed the clock further while one student put the clock inside a pentagon meaning that everything is under moral governance. The fourth student went on to develop the pictures inside the clock. Table 8.20 shows their individual designs.

After evaluating the designs and choosing one design, they made some more modifications and produced a design. The final design is shown in Figure 8.32.

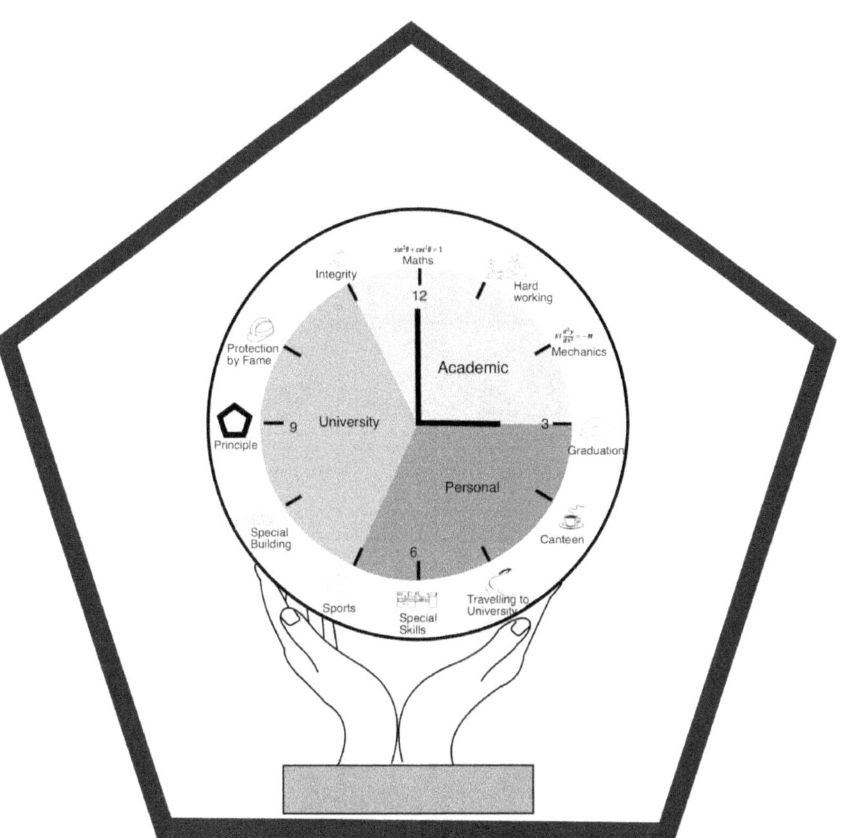

Figure 8.32 Final design of the graduation gift.

Example 8.19: Design of a Bread bicycle

In several places, bicycles are the main mode of transport and they adopt normal bicycle in several different ways to accommodate the load. These include trolleys, additions and changes to the handle bar, additional cargo space, carrier attachment, changes to the rear wheel system, changes to the front wheel system, etc. Each method has its own benefits and shortcomings. An entrepreneur from a village in an under-developed country is planning to open up a bakery. He believes that, taking the bread and other baked items to the door-steps, as the last step in the supply-chain, will greatly enhance the sale of his products. He plans to employ delivery boys to do this job on bicycles. He would like to build bicycles to do this job, but, the new bicycles should use as many of the standard parts as possible from a standard roadmaster bicycle. In this example, the requirement is to design a bicycle to transport about 75 kg of bread loafs and any other items using the gallery method. It may be assumed that the density of bread is 0.192 kg/litre and an average loaf is 11 cm × 11 cm × 25 cm in size.

Answer

The steps to follow are as follows:

1. Establish the Requirements for a bicycle to carry 75 kg of bread
2. Analyze what are the different approaches to accommodate the load and establish concepts.
3. Put the concepts in a gallery and discuss
4. Use the insights gained from the group discussion and have a second round of concepts
5. Choose one and consolidate the chosen conceptual design

Step 1: The requirements are as follows:

1. Should be able to carry 75 kg of bread
2. Should have 5 cm clearance between layers
3. Should have covered storage
4. Bicycle should use standard road master bicycle parts as far as possible
5. Minimal variation from standard bicycle
6. Should be operated by a single person
7. Flexibility in use (easily convertible to a bicycle)
8. Should not make the bicycle bulky

Step 2: In this era of the internet looking at the net to gather a broader idea of concepts can be useful in certain kinds of design problems. Cargo bikes have been in existence for a long period and the designs have developed in different facets. Looking at present-day cargo bikes the following six types can be easily identified. The requirement in this project is to develop a manual bike to be manufactured in small numbers, the first gallery of designs is chosen from these categories and was evaluated [3].

Type 1 – Long John Style Cargo bikes: These are heavy-duty front loaders that traditionally have bucket style storage space in the front. They are used for carrying bulky objects and children since the rider can have an eye on them all the time. It provides plenty of loading space. There are plenty of websites that show the long John style cargo bikes [4,5]. In this bike, it is easy to protect the cargo easily in adverse weather.

Type 2 – The Trikes: Cargo tricycles are just three-wheeled variants of the Long John. Three wheels make them extra stable when riding [6,7]. The bucket has to be modified to suit the cargo.

Type 3 – Cargo bike Trailer - This is an additional trailer attached to the bike. These can be detached whenever they are not needed. Typical ones can be seen on the internet [8,9].

Type 4 – Three wheelers with basket at the rear: This type of bike has two wheels at the rear with a compact basket. Sometimes this is called the elderly person's bike because of its compactness and stability. Typical ones can be seen on the internet [10,11].

Type 5 – Long Tail bikes: Long tail cargo bikes look exactly like a normal bike, just with a slightly longer tail. They transport cargo or passengers on an extended tail section behind the rider. They are the better options for those having mixed uses. When they are not carrying any cargo, they would look like a normal road bicycle. Typical ones can be seen on the internet [12,13].

Type 6 – Road bicycles with Extra reinforcements carrying large baskets: There is a large variety of reinforced attachments that have been developed to go with the normal roadmaster bicycles enabling them to carry large boxes at the front and rear. Typical ones can be seen on the internet [14,15].

Step 3. Gallery discussion and observations made are summarized in Table 8.21.

The analysis summarized needs a detailed calculation of the cargo container. Consider the calculations in the following way:

Volume of the loaf $= 11 \times 11 \times 25 = 3{,}025\ \text{cm}^3$

Each bread loaf is 11 cm tall and if trays are arranged inside a clearance is needed and let that be 5 cm.

Therefore, the effective height of a loaf is 16 cm and the width of the loaf will be 13 cm if 2 cm clearance is assumed.

If the height of a container is assumed to be 1 metre there can be six layers.

If 75 kg of bread has to be stacked in six layers each layer has to have

$$= \frac{75}{6} = 12.5\ \text{kg}$$

Each layer should have $\dfrac{12.5}{3.025 \times 0.192} = 21.5$ loafs

Assuming 21 loaves per layer, the layers will be arranged in the following way as shown in Figure 8.33.

Table 8.21 Gallery discussion and observations

Type	*Discussion*	*Observations*
Type 1	This bike is having special construction and can carry accommodate bulky load very well.	Not suitable for the project.
Type 2	The trikes again have special features to accommodate bulky cargo. Though it is well suited for the task, it does not meet the requirements 4, 5, 6, 7 and 8.	Not suitable for the project.
Type 3	Cargo trailers can accommodate bulky loads in the trailer when needed and can be detached whenever they are not needed. Hence a standard Road master can be used.	This is a very good option for consideration. The main point for consideration is the size of the trailer to be accommodated.
Type 4	Three wheelers with the container for cargo in the rear is a very stable bike. However, it does not meet the requirements 4, 5, 6, 7 and 8.	Not suitable for the project
Type 5	Long tail bikes can accommodate bulky cargo when needed and can be used as a normal bike when not transporting the cargo. Its disadvantage is the length of the bike.	This can be an alternative if the volume of the cargo is large.
Type 6	For medium size loads, having extra reinforcements and attachments to a normal road master bike is a good option. Minor attachments with the ability to easy attachment and detachment is a good idea.	This is a very good option for consideration. The main point for consideration is the size of the cargo container to be fitted and the easiness of attaching and detaching.

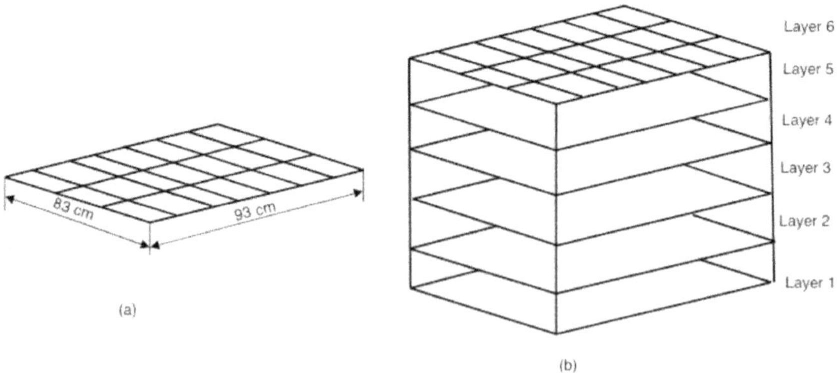

Figure 8.33 Arrangement of bread in layers.

Table 8.22 Gallery of the second round

Description of concept	*Sketch of the concept*
A front basket is attached to the bicycle carried by special frames attached to the front axle and the frame.	
A rear-basket is attached to the bicycle carried by special frames attached to the rear axle and the frame.	
A detachable trailer carrying the bread basket is attached to the bicycle at the rear axle.	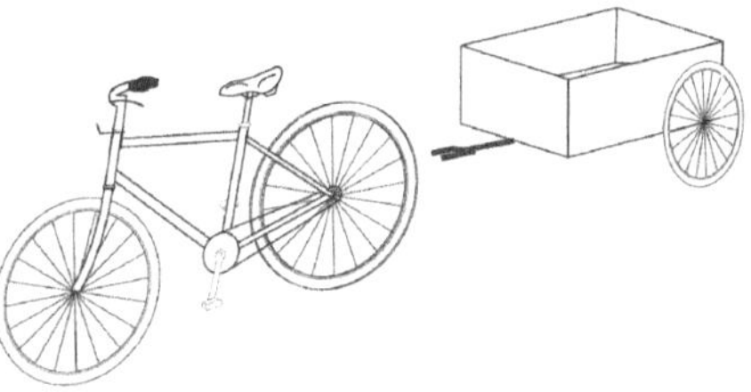

Getting back to the conceptual design of the bicycle, it should accommodate a box of 85cm × 95 cm × 100 cm.

A second individual phase of concept generation was carried out with a focus on designs using many components from the normal roadmaster bicycle and the volume of the cargo as calculated above. The second set of designs are shown in Table 8.22.

The group then jointly evaluated the concepts and chose a concept with two smaller baskets to accommodate four layers of bread in each. This gives some space for other products from the bakery. The concept is shown in Figure 8.34.

Figure 8.34 Final design by the gallery method.

Example 8.20: Design of souvenir placemats

A placemat is a small mat underneath a person's dining plate, used to protect the table from the heat of the plate and food. Unlike the larger tablecloth that covers the entire surface of the table, a placemat protects an individual place setting. Modern placemats serve several secondary purposes: to protect, decorate, entertain or even advertise. Placemats add a touch of contemporary glamour to the dining table, making an occasion to feel special. Placemats are made of several materials such as plastics, glass, natural fibres, wood and even marble. With the increase in environmental consciousness, an entrepreneur thinks that placemats can be used as a means of spreading environmental awareness. The task in this example is to design placemats with emphasis on environmental promotion using the gallery method.

Answer

The steps to follow are as follows:

1. Establish the requirements for a placemat set
2. Analyse what are the different approaches to placemat designs available in the market
3. Put the concepts in a gallery and discuss
5. Use the insights gained from the group discussion and have a second round of concepts
6. Choose one and consolidate the chosen conceptual design

The requirements are as follows:

1. Should have dimensions to cover a 25 cm diameter circle or a square of 25 cm sides.
2. Should have a minimum of 0.7 cm thickness of the mat

3. The temperature on the table top should not exceed the room temperature by more than 10°C.
4. Should have aesthetically pleasing appearance
5. Should be easily cleanable
6. Attractive art work on the top surface
7. Most importantly should contain succinct message on environmental preservation.

Table 8.23 Gallery of placemats available in the market

As the second step a search in the market for placemat designs was conducted and sample placemats were collected as candidates for the gallery. Table 8.23 shows the gallery.

Looking at the placemats collected from the market they fell into four categories (i) fibre table mats (ii) marble table mats (iii) woven plant-based table mats and (iv) wooden table mats as shown in Table 8.23. Of course, there can be several others, but these give a feel on the variety available in the market.

Looking at these mats and treating them as gallery and evaluating them, it was observed that there are only a few printed placemats and there are few placemats that deliver a message. The opportunity therefore is to produce printed placemats. The major creativity that can be introduced here, is with prints or images and words. Inspiring images and words can be introduced in appropriate colours and shapes. Table 8.24 shows some designs of images and words.

Table 8.24 Concepts in round 2 placemats with a message

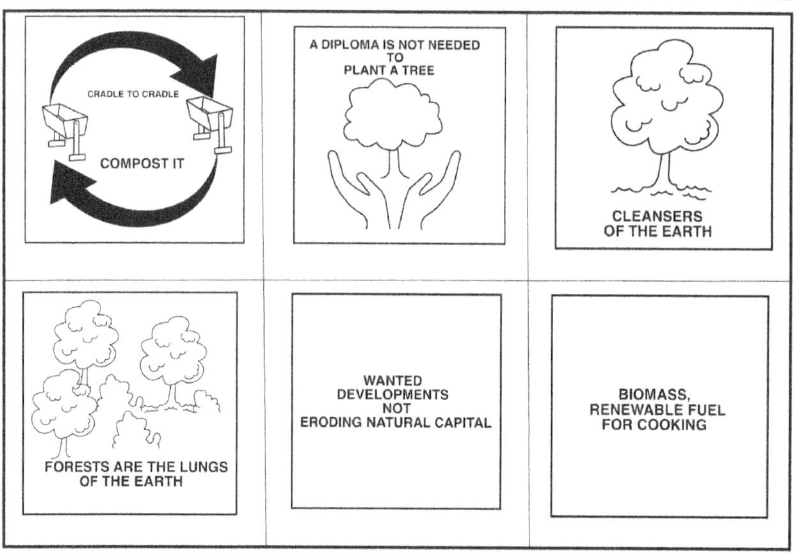

8.10 DESIGN METHOD 7: DESIGN BY ANALOGY

Design by analogy identifies an analogous base product, which is reasonably well known and has well-understood and established characteristics, and maps a relational structure from base to target. Steps in design by analogy are (i) understanding the design problem and identifying the key characteristics needed (ii) identifying a base product that exhibits those characteristics (iii) carrying out a mapping between the base and the target and (iv) evaluating and consolidating the design.

Haik et al. [2] recommend the following steps for design by analogy:

1. Understand the design problem, and identify the 'Key Product Descriptors' or KPDs needed.
2. Identify a 'Base System' or 'Source' that exhibits those characteristics.
3. Carry out a mapping between the base and the target.
4. Evaluate and consolidate the design.

Example 8.21: Design of a stand for precision component in the display

Precision Engineering is the designing and making of machines, fixtures, and other structures that have exceptionally low tolerances, are repeatable and are stable over time. There are several thousands of companies that are engaged in precision engineering and they pride themselves as leaders in varied aspects or values. They normally display complex artefacts or components they have designed and/or manufactured so that the visiting customers can be impressed with the capabilities of the company. A display exhibiting them can accommodate 12 such components, which are placed on individual stands. The plate on which the stand is placed is 160 mm × 160 mm and the stand should be within a foot area of 150 mm × 150 mm and should be of 50 mm high. The task is to design this stand using the design method, 'Design by Analogy' to design the solution concept generating stage.

Step 1. Established KPDs

1. Reflect the achievement of the company for three generations.
2. Multi-functional where the company is equipped with different specialities.
3. Robust performance revealing the power of several individually capable men and machines in the company.
4. The company can handle all types of tasks no matter how small or big the job is. It won't fail under pressure by handling big tasks.
5. Well-planned internal process from inception means that all tasks are done in sequence and in stages.
6. caring undertaking is an essential key problem descriptor that refers to the smooth progress from start to finish.

Table 8.25 Mapping of the characteristics of the gear and the company

Gear wheel	*Company*
Designed to be capable of handling regular and irregular loads including impacts.	Equipped with different required specialties
Several individually capable teeth work in harmony.	Several individually capable men and machines
Ability to handle varied size of loads	No job is small
Unassailable by the load	No job is big
Smooth loading	Caring undertaking from start to finish
Smooth unloading	Satisficing completion
Smooth transfer from tooth to tooth	Well-planned internal process from inception
Durable	High fliers for generations

Step 2. After identifying the six key problem descriptors the second step is to identify a base system that exhibits these characteristics. A gear wheel is chosen as the appropriate analogous product to reflect the key characteristics.

Step 3. The mapping

'Strong by Design' is the characteristic that the company wants to display and reflect, and the gear wheel reflects this characteristic very well in the way shown in Table 8.25.

Step 4. The design

The design team had novel designs that reflected their inspirations and feelings. The first concept was a circular base with two steps carrying a shaft and a strong gear. The second concept had a square base and steps with sharp and defined corners reflecting the precision manufacturing processes. At this point displaying the history of the company and the strong characteristics shown through generations was considered. The three layers at the base represented the three generations that built technical expertise of the company. An engraving of the 'falcon' was included to reflect the wisdom, victory and leadership the company is possessing. The process of developing concepts continued in this manner and the developed concepts are shown in Table 8.26. Figure 8.35 shows the manufactured piece of the chosen design.

Table 8.26 Conceptual designs

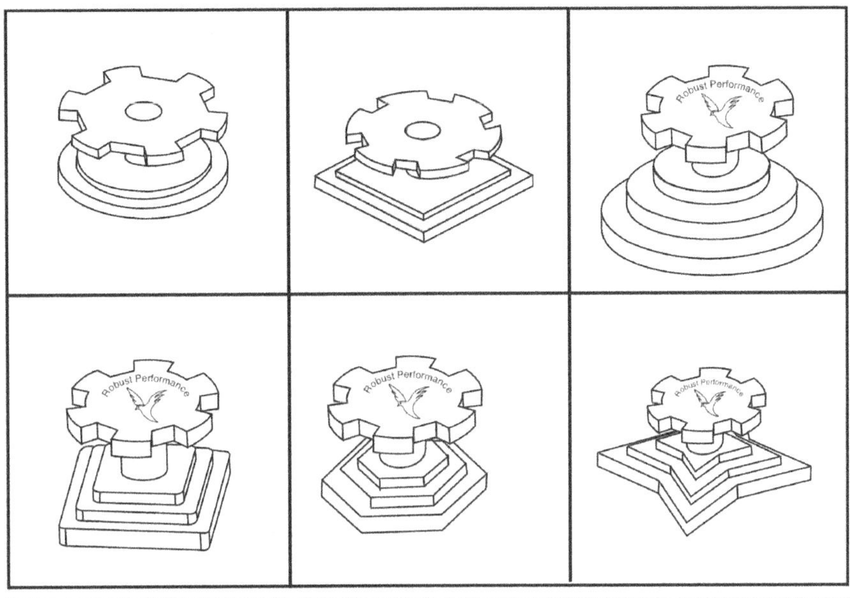

Figure 8.35 Manufactured stand.

Example 8.22: Table-top dessert display

In this example, a table-top dessert server is designed using the analogous method. As explained in Example 8.7, a 'Table-top Dessert Server' is required to present a variety of desserts on the table so that the guests can choose and access the dessert items, they prefer to have. The table-fan blade shown in Figure 8.36 was chosen as the analogous product for developing the display.

Step 1. Established KPDs

1. Should be able to support up to four dishes.
2. Oriented in different angular positions
3. Each presenter (supporting the dish) functions independently
4. Aesthetically pleasing appearance when supporting dishes or otherwise.
5. The contents of the dishes should not fowl (i.e. knocking down) each other.
6. Any dish should be accessible by rotating the presenter.

Step 2. After identifying the key problem descriptors, KPDs, the second step is to identify a base system that exhibits these characteristics. A fan impeller is chosen as the appropriate analogous product to lend itself reflecting the key characteristics.

Step 3. The mapping

'Harmonious Integration' is the fundamental characteristic that is fundamental to this design. How the fan impeller reflects this characteristic very well in the way shown in Table 8.27.

Figure 8.36 Table fan.

Step 4. The design

The design team integrated the identified characteristics and produced a design. The product manufactured is shown in Figure 8.37.

Table 8.27 Mapping of the characteristics of the impeller and the dessert display

Fan impeller	*Dessert display*
Fan can have varying number of blades such as 3,4, 5 and 6	Can have capability to display 3, 4, 5 or 6 presenters for the dessert
The fan blades do not have any preferred angular position	The presenters do not have any preferred position
Each fan blade takes a share of the air and handles the load and give it to the main shaft	Each presenter takes the load of the dessert and pass it to the main shaft
The blades have pleasant shape and have fixed radial positions	The presenters have aesthetically pleasing shape like the fan blades, with ability to assume any radial positions
The blades are at fixed radial positions so that there is no overlap and fowling	The blades are at different fixed heights so that there is no overlap and fowling
The blades rotate about the axis and thus any blade can be accessed without physically moving	The presenters rotate about the main vertical shaft and thus any presenter can be accessed without physically moving

Figure 8.37 Dessert server developed analogous to fan blades.

Example 8.23: Design of a table-top reading assistant

A reading assistant assists the reader by providing more than one book in open condition for the reader to cross-refer. The user can sit in his place and access different books with the help of this reading assistant. Better and more enhanced reading assistants that can handle many books in open conditions and present them to the seated reader is a need for the research community. The task in this example is the design of attractive, decorative and professional reading assistants, that satisfy the needs of the market, for placing at the top of the table.

Step 1. Established key product descriptors, KPDs

1. Several planar surfaces to hold the books in open condition.
2. The surfaces should be supported independently and integrated as a single product.
3. Ability for the reader to switch to the next book without getting from his chair.
4. The product should be portable.

Table 8.28 Analogous products

5. Space for placing sundry items like pens, pencils, markers, etc.
6. Can be considered as a decorative piece of furniture.

Step 2. Identification of the base system

Four base systems have been identified for this intended reading assistant as shown in Table 8.28.

A table lamp made up of planar faces, a rubber plant having almost planar leaves, pyramids having planar faces and drawer set that has drawers with planar bottoms are identified here.

Step 3. In Step 3, their properties are mapped as shown in Table 8.29.
Step 4. In Step 4, the final designs are made and are shown in Table 8.30.

Table 8.29 Mapping between the base product and the desired design

Properties of the base system	*Intended reading assistant's properties*
Lamp shade	*Reading assistant*
Several planar faces	Each face can hold a book
Hexagonal head meaning six faces	Hexagonal head can hold six books
Bottom edge has a thick retention lining	The bottom can be a retention piece
Broader bottom edge or base	Broder bottom part can hold note paper
Portable	Portable
Circular base and stem	Circular base can be enlarged and the stem can be a rotational shaft
Rubber plant	*Reading assistant*
Leaves have large near-planar surfaces	Each surface can hold a single book
Leaves are near horizontal	The book holding surface is horizontal
Leaves are connected to stem	Holding surfaces are connected to stem
Each leaf is supported by the stem	Stem-shaft supports holding surfaces
Stem is supported by the base	Shaft is supported by the base
Pyramid	*Reading assistant*
Made by planar faces	Each face can support a single book
Made by four planar faces.	Can support four books
The faces are triangular in shape	Can make a trapezium through frustrum
If it is made a frustrum it will have narrow top and broad bottom	Top can accommodate the book width
Base is supported by the columnar shaft.	The bottom can accommodate book and note paper
	Add a separate matching rectangular base.
Drawer cabinet	*Reading assistant*
Drawers have horizontal surfaces	Each surface can accommodate a book
Cabinet has several drawers	Can accommodate several books
Has flexibility in opening directions	Book supports open in several directions
Can open and close drawers at will	Can open and close book supports

Table 8.30 Base products and conceptual designs

Base Product	Conceptual Design
	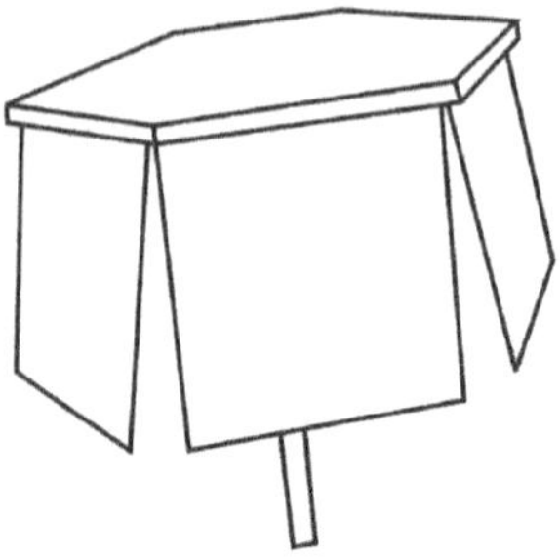

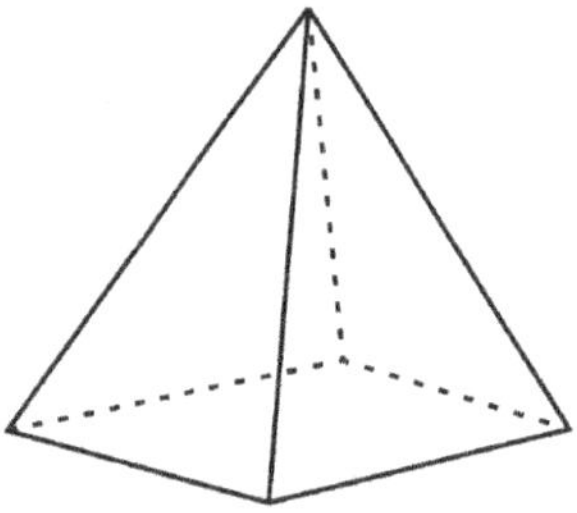	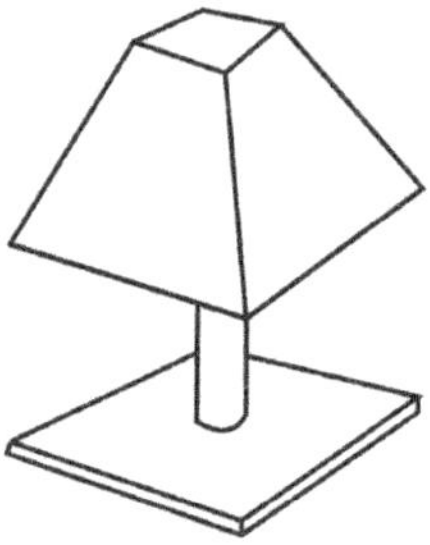
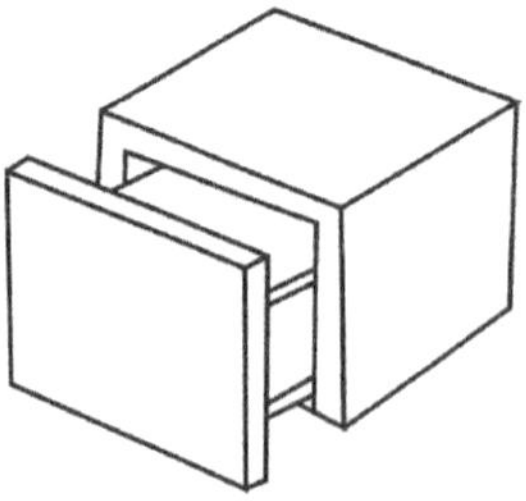	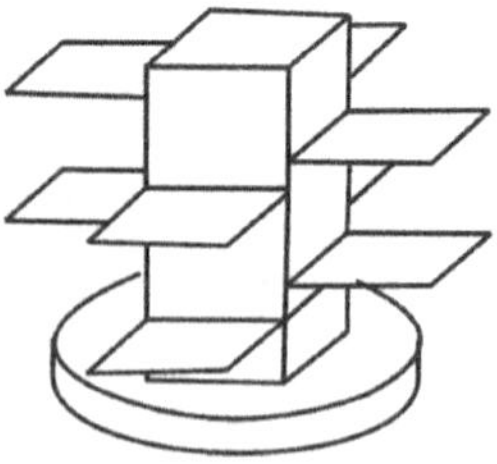

8.11 DESIGN METHOD 8: KJ METHOD

The KJ method to generate ideas requires, all relevant facts and information written on individual cards which are collated, shuffled, spread out, and read carefully. The cards are then reviewed, classified, and sorted based on idea similarity, affinity, and characteristics. Using the titles of the individual groups as the basis, a morphological chart is prepared and conceptual designs are formed or proposed. KJ method can be divided into six steps. They are

1. Put opinions (or Data) onto sticky notes one item per sheet
2. Put sticky notes on the wall
3. Group similar items – Clustering
4. Name each group
5. Vote for the important groups and rank them
6. Mixing and matching: Develop conceptual designs mixing and matching themes.

The KJ method is very useful when there are a wide variety of opinions or data. Figure 8.38 explains the process. Steps 1 and 2 result in the recording of all relevant data as shown in Figure 8.38a. These notes are moved around to form clusters of related ideas as shown in Figure 8.38b. In the next step, the clusters are given a heading as shown in Figure 8.38c. In the next steps, the groups are ranked and important groups are identified. These prioritized relevant points are used to generate designs.

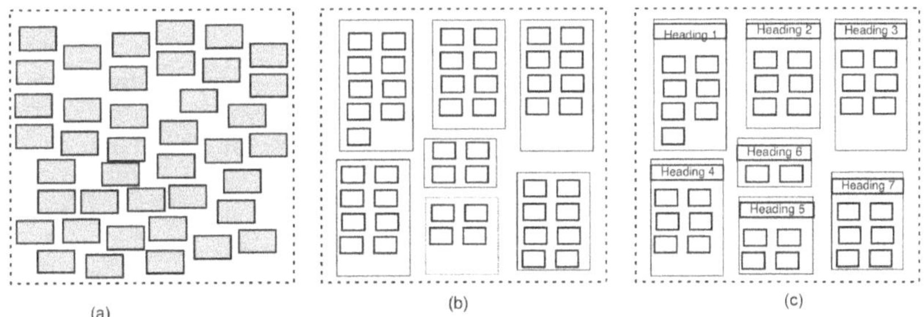

Figure 8.38 Arrangement of post-it stickers.

Example 8.24: Design of a memorabilia item for a country

A memorabilia item stirs recollection and is collected for its association with a particular interest. Memorabilia items are produced from time to time to celebrate events. Assume that a national event is coming and a memorabilia item is needed. The memorabilia item has to reflect the country's wealth which has several facets. This example describes the development of a memorabilia item for the United Arab Emirates.

The points are collected as per the KJ method and the clusters are identified and named, as shown in Table 8.31.

Table 8.31 Headings of clusters and the entries in sticky notes

People Open minded High ambition Leadership	*Clothes* Kandora Thobe Burqa
Features of UAE Sheikh Zayed – Founding Father UAE Map UAE Flag and colour Religious traditions adherence Restore hope to poor countries museums Traditional Landmark buildings Modern landmark buildings	*Technological frontiers* Burj Khalifa Burj Al Arab Expo2020 Qasr Al Hosn Dubai Eye Sheikh Zeyed Mosque Skyscrapers AI University
Technological advancements Oil Derricks Palm Islands Dubai Metro Atlantis Al Barajeel Oasis Complex	*Celebrations* Dubai fireworks Weddings
Interesting locations Dubai Opera Dubai Frame Louvere Museum Abudhabi Dubai eye	*Nature in UAE* Palm Trees Falcon Desert Ghaf Trees Beaches
Space mission Emirates Mars Mission Hazza Al Mansouri – First Emirati in Space Emirates Astronomical Observatory Satellite	*Aviation industry* Emirates Etihad Fly Dubai Air Arabia Wizz Air
Petroleum industry Refineries ADNOC	*Nuclear energy* Barakah Nuclear Power Plant Masdar
Traditional food Balaleet Ragag bread Harees (Cruhed wheat) Luqaimat	*Practices in animals* Raising camel and sheep Falcon hunting Horse racing Camel racing
Inherited practices Traditional dance Farming Herding Weaving Spinning	*Places* Tents Mud houses Villas Palaces
	Inherited traditions Dallah coffee pot Pottery making

Combining the details, a clock was designed and built. The clock is shown in Figure 8.39.

Figure 8.39 A memorabilia clock for United Arab Emirates.

Example 8.25: Design of a display for precision items on stands

In Example 8.21, a stand for a single precision item for display has been made. In this example, a display unit that can accommodate 12 such stands and components is developed. The idea is to produce a display, that carries the items placed on individual stands as a group. It is expected that a component can be contained in a 10 cm enveloping cube. Use the KJ method to develop conceptual designs for the stand.

Answer

Different approaches and design ideas are collected as per the KJ method. They are analysed and arranged into clusters as named in Table 8.32.

Some conceptual designs based on the above information groupings are shown in Table 8.33. The fabricated display based on concept 2 is shown in Figure 8.40.

Table 8.32 Headings of clusters and the entries in sticky notes

Pollards of different heights Pollards with circular heads Pollards with triangular heads Pollards with rectangular heads Heights from 1.0 to 1.5 m	*Tiered displays with horizontal boards* Similar to stairs Single compartments Multiple compartments Partitioned and non-partitioned
Tiered displays (a shaped stands) Single compartments Multiple compartments Partitioned and non-partitioned	*Full round displays similar for plants* Different pot sizes Single tier Multiple tier
Tiered semi-circular horizontal boards Single compartments Multiple compartments Partitioned and non-partitioned	*Tiered podium like stand* Single side Both sides
Star-like horizontal discs – moveable 2 tiered 3 tiered 4 tiered	*Tiered stands with full-circular boards* Single compartments Multiple compartments Partitioned and non-partitioned
Horizontal circular stands able to turn like Lazy Susan Single tiered Multiple tiered Star-like board	*Vertical boards with horizontal placing protrusions* Triangular boards Rectangular boards Circular boards Elliptical boards
Shape of protrusions Rectangle Semicircle Semi-elliptical	*Vertical circular board with space for stands hanging on protrusions* Reflecting various characteristics of the company
Mobile stands Fitted with wheels	*Poles with radial arms* Hang the plates for placing stands and exhibits Single tiered Multiple tiered
Cylinders Different heights Diameters Colours	

Table 8.33 Conceptual designs of the display

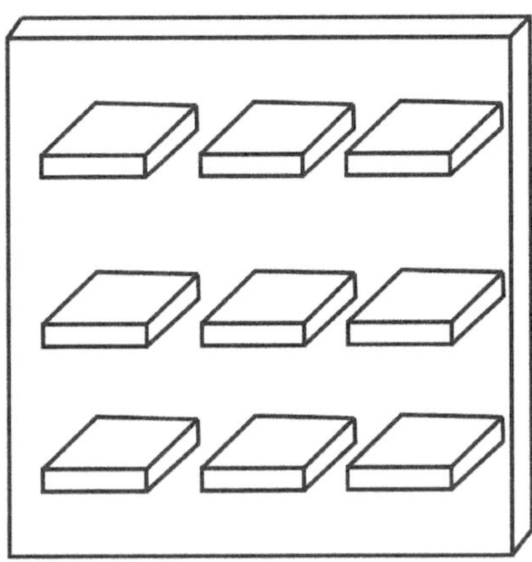 This is a simple display having spaces for nine stationary stands on which individual precision items can be placed.	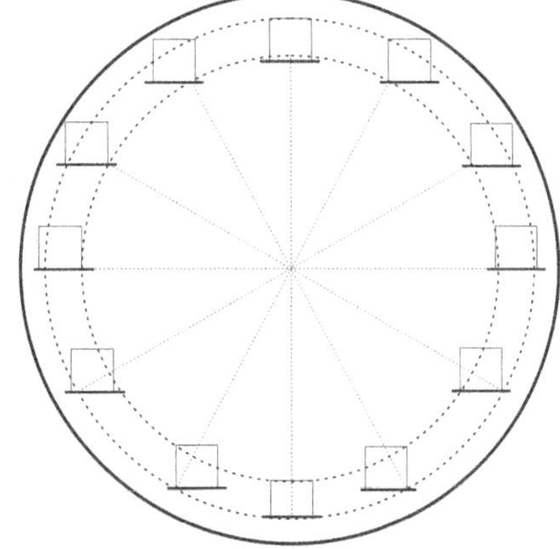 This is a simple circular display with stationary spaces on which the stands and precision items can be placed.
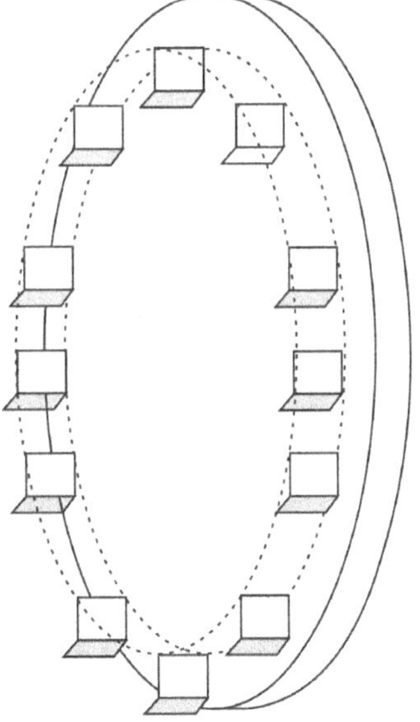 This circular display carries bolts to hang up to 12 spaces for placing the stands and components. The spaces are hung on bolts normal to the circular disc.	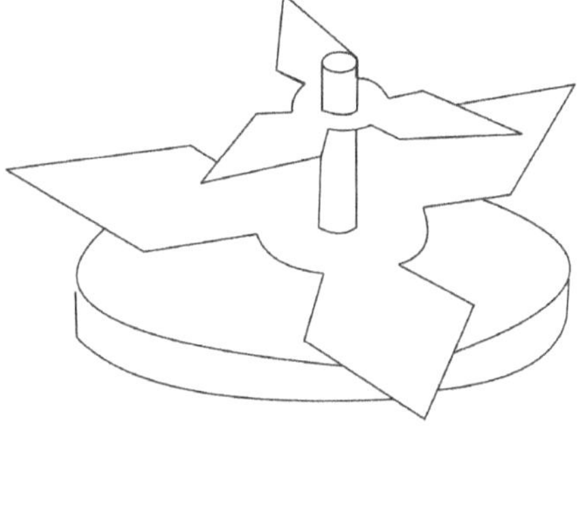 In this display rotatable stars are placed to provide space for the stand and the components. Number of stars can be increased as per need.

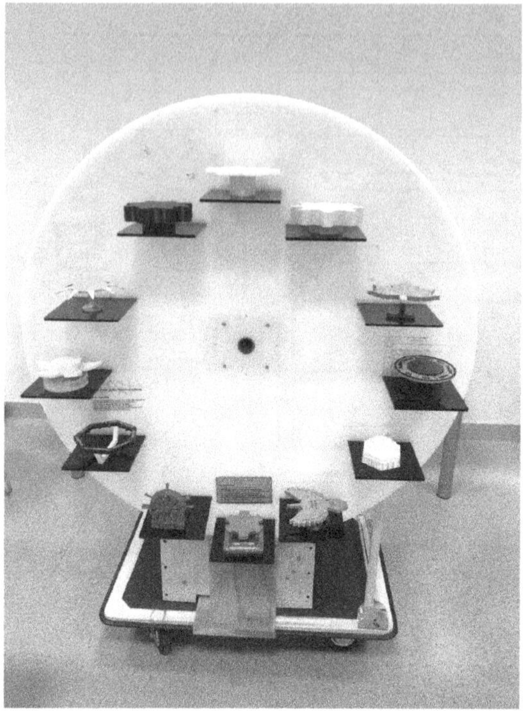

Figure 8.40 Display for precision components.

8.12 DESIGN METHOD 9: CRITIQUING

Critiquing is the process of giving feedback for improving, by bringing expert insight into existing interfaces, prototypes, brands, services, design approaches and processes or technical difficulties in implementing a design. It is usually run by a group of 3–7 people and can involve designers, developers, marketing analysts or businesses. The aim is to learn from the design and to explore where it works well and, where it could be improved, in different perspectives. Critiquing refers to receiving input on current design ideas and can be especially relevant when next generation products are developed. Critiquing can provide a complete change in approach as new technologies can provide openings for new concepts and better manufacturing processes. It can make previously infeasible concepts feasible.

The steps in critiquing involve the following:

a. Establishing the goal of the product and the criteria for evaluation, with respect to the specifications.
b. Establishing the list of components (can be parts or sub-assemblies)

c. Evaluating each component for its contribution towards the overall goal, and
d. Coming up with suggestions (results of critiquing) for improvements or actions addressing the inefficiencies.

The suggestions should be focussed in a helpful manner and objective. They may include the following:

a. Positives and Negatives of the design (What are the positives and negatives with respect to a set of criteria and how they are aligned with the goal)
b. Additional considerations (What are the possible considerations from different perspectives, particularly from customers' perspective)
c. Suggestions on alternative solutions (Why not try it another way)

The success of critiquing relies on the fact that someone does lead the discussion, define what questions should be discussed and facilitate the conversation. The leader should keep in mind that design critiquing is focussed on evaluating and converging on a design that has already been prepared, by directing specific paths and changes. The question 'why is this design achieving or not achieving our project objective?' can refocus the team towards the overall goal of the design.

Example 8.26: Design of a step ladder

The design of a step ladder is shown in Figure 8.41. The goal of the step ladder is to provide access to small heights. A stepladder is a portable, self-supporting, A-frame ladder with two front side rails and two rear side rails. Generally, steps are mounted between the front side rails and braces are mounted between the rear side rails. Health and Safety Executive (HSE) of the United Kingdom observed high number of accidents and failures of the stepladder and commissioned Loughborough University to investigate it. Their work is described as critiquing in this example.

Ergonomics evaluation into the safety of stepladders, by Loughborough University, for the HSE, United Kingdom [3].

Several guidelines have been given on the correct use of step ladders. Three important guidelines from them, on the safe use of step ladders are

a. Do not use a stepladder with spreaders (stay) unlocked.
b. Do not use the top step or cap as a step.
c. Do not use a step ladder to access high places

Records show a substantial number of step ladder accidents where people are injured. Failed ladders show the following failures:

a. Rail (or stile) failure with aluminium ladders (42%)
b. Bracing (support between rung and stile) failure with aluminium ladders (21%)

Figure 8.41 Design of a step ladder.

c. Unstable condition with aluminium ladders (16%)
d. Unstable condition with wood ladders (5%)
e. Cuts from sharp edges with aluminium (5%)

While the above failures are recorded, investigations on accidents show that tipping and overbalancing are the most common causes of accidents on stepladders. A fresh evaluation on the design or critiquing is needed.

Answer

Part 1

The question 'why is this design achieving or not achieving our project objective?' is a good starting point.

Accident statistics suggest that user demands, especially for stability, are not being met. The user often does not accurately assess the relative hazard levels and make a valid judgement and this wrong judgement leads to accident. Therefore, there must be an understanding of what the user reasonably anticipates that they can do with the stepladder, and the stepladder must provide adequate strength and stability to allow this use. In other words, the design should be strong enough to a certain amount of 'misuse'.

This data illustrates the limited potential that the safety information has for changing behaviour. Even when the users read and comprehended safety instructions, the majority of users were prepared to ignore it to get the job done. This suggests that effective stepladder safety measures must be inherent in the design, rather than at the user's discretion.

Part 2: Additional considerations

Investigating accidents and failures of stepladders reveals that misuses were the main causes of the majority of them. The misuses are explained as follows:

1. *Overreaching*: This mode of misuse is unsafe since the current stepladder design permits individuals to place themselves sufficiently outside the footprint of the stepladder that their centre of gravity can cause the stepladder to tip over.
2. *Unorthodox postures*: When users are unfamiliar with stepladders, they may adopt unorthodox postures to either improve their perceived security or to gain advantage in their reach.
3. *Stepladder cantilevering*: Quite often users will reach for support from objects near them to either stabilize themselves, improve their kinaesthetic or to provide mechanical advantage. More serious is the use of the stepladder itself to cantilever the body mass.
4. *Standing on inappropriate structures*: One of the most common misuse modes is standing on inappropriate structures within the stepladder such as top cap or top step. This is even more dangerous in parallel operation.
5. *Inappropriate use of the guard rail*: When participants used the top platform to stand upon, they found that extra security could apparently be provided by leaning against the guard rail which was supplied with some stepladders.
6. *Vertical over-extension*: One of the most undesirable actions is for the user to vertically overreach from a shorter stepladder, instead of obtaining a taller stepladder. This activity places the user's mass very high up relative to the stepladder feet, and this can allow the centre of gravity to move outside the footprint, since shorter stepladders tend to have smaller footprints.

Part 3: Suggestions on alternative solutions

The key geometric variables which contributed to the overall form of stepladder are established as the seven specific parameters, identified as A through G in [3], as defined below.

A – Width of the ladder feet in the x-axis.
B – Width of the ladder feet in the y-axis.
C – Width of the second step from the top of the ladder.
D – Distance from the second step from the top to the ladder foot in the x-axis.

E – Distance from the second step from the top to the ladder foot in the *y*-axis.
F – The height from the ground to the upper surface of the second step from the top of the stepladder.
G – The distance from the point vertically below the ladder's centre of gravity to the ladder feet on the y-axis.

The suggestion is to *permit some reasonable misuses and redesign these parameters in such a way that the members of the ladder can withstand the extra loadings and still retain stability.*

Example 8.27: Design of a wet umbrella holder

Umbrellas protect people from rain, but when they reach the destination, wet umbrellas pose a problem in storing them while the user goes inside. The water that comes with the umbrella spills everywhere and makes a wet zone at the entrance to big offices and hotels. A wet umbrella holder is a device where wet umbrellas can be stored and where the spilling water is collected. This limits the wet zone. An umbrella holder designed for a big building is shown in Figure 8.42. Critiquing the design is required.

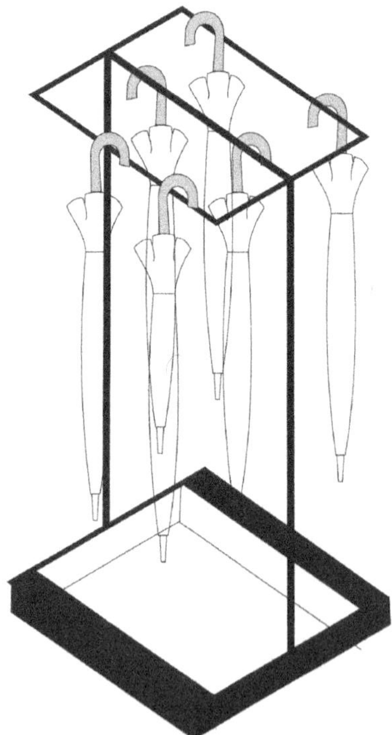

Figure 8.42 Umbrella holder.

Critiquing this design requires the criteria which in this case can be the stakeholder requirements. The requirements for the umbrella holder are established as follows:

1. Should be able to hold several umbrellas
2. Should collect the remaining water dripping from the outer surface of the umbrella canopy.
3. Should limit the wet zone
4. The collected water should be easily disposable
5. The Umbrellas should be kept upright so that there are no chances for water to go to the inside of the umbrellas
6. The holder should be easily moveable
7. The umbrellas in the holder should not get entangled

Part 1

As explained earlier the question 'why is this design achieving or not achieving our project objective?' is the starting point.

This design is a good one meeting several of the requirements. It can store several umbrellas, and the water dripping is collected in a basin at the bottom or base of the holder. The wet zone is limited and the umbrellas are kept upright, and there is limited chance for water to get to the inside of the umbrella. Since the canopy of each umbrella is wrapped and tied there is little chance for the umbrellas to get entangled. However, disposing the collected water has to be manually emptied by lifting the collecting basin. This is an avoidable physical task that can spill water on the adjoining floor. The holder is not easily moveable and therefore has to be included as permanent furniture even during summer.

Part 2: Additional considerations

(None)

Part 3: Suggestions on alternative solutions

Four suggestions on additions are made at this stage. They are as follows:

1. *Fixing wheels*: The base of the holder can be modified to have wheels attached to it. This will enable it to be easily moved and the holder can be brought out from the store room only when it is required.
2. *Fixing the tray and a draining tap*: The current design has a removeable tray to collect the dripping water. A better-quality tray with a draining tap can be fixed permanently.
3. *Better material*: better material such as stainless steel can be used to make the holder attractive and rich.
4. *Separate compartments*: Separate compartments for each umbrella can be considered when separation and issuing tokens become necessary.

Example 8.28: Design of a mechanical turntable

The turntable shown in Figure 8.43 makes a 60° rotation, waits there for some time, then makes another 60° turn, waits there for the same time as earlier and continues in this manner. As shown in Figure 8.43 the motor continuously drives the input shaft of the worm gear. The output shaft of the worm gear turns the drive wheel of the Geneva mechanism. The driven Geneva wheel is driven by the pin in the driving wheel. The Geneva wheel turns during one sixth turn of the driving wheel and remains stationary during five sixths turn of the driving wheel.

Part 1

It is convenient to start with the question 'why is this design achieving or not achieving our project objective?'

This design is a good one in meeting the requirement on timing and repeatability. Since it is connected to the angular rotation, any fluctuation in speed will not affect the geometries. For operations with small

Figure 8.43 Intermittently rotating table.

masses, there will be small momentums and hence there won't be much jerks. However, if there is sufficient mass involvement, there will be jerks during the starting and stopping.

Part 2: Additional considerations

If the speed of the input motor can be varied, different time slots can be introduced for moving and staying.

If there is a possibility to control the acceleration and deceleration of the turntable then the jerks can be eliminated.

Inclusion of a belt between the gearbox and drive to the mechanism can be considered to dampen the jerks.

Part 3: Suggestions on alternative solutions

Three suggestions on additions are made at this stage. They are as follows:

1. *Replace the drive system*: Replace the continuous motor, worm gear box and the Geneva Mechanism with a stepper motor.
2. *Introduce a timing belt drive*: Introduce a timing belt drive between the worm gearbox and the Geneva Mechanism. This elastic material can reduce the jerks.
3. *Control time for acceleration and deceleration*: Control the acceleration and deceleration time to keep the magnitudes of inertia forces within acceptable limits.

8.13 DESIGN METHOD 10: SYSTEMS APPROACH

In this method, the product is treated as a system which is an assemblage of sub-systems. Andreassen [16] divides the system or product into six classes of sub-systems namely:

1. *Working system*: the sub-system which influences the transformation of the object and changes it to a new form.
2. *Prime mover or energy system*: which delivers the energy for the transformation
3. *Transmission- or energy distribution system*: that distributes energy to the system component
4. *Control system*: Composed of sub-systems for controlling, supervising and inspection of other systems' state and performance, creating man/machine interface, and establishing the human safety system
5. *Frame system*: that keeps the whole system together in space, and
6. *Helping system(s)*: that solves different necessary helping tasks.

The design process using this method starts with the identification of the sub-systems needed for the product and is followed by proposing various alternatives for these sub-systems as in the morphological chart method. This method is very useful in mechatronic system design because it will mostly have all six sub-systems.

Example 8.29: Design of a table-top dessert server

As described earlier a dessert concludes a meal and the course usually consists of a large variety of sweet foods, savoury items, cheeses, nuts, fruits and ice creams. A 'Table-top Dessert Server' is required to present a variety of desserts on the table so that the guests can choose and access the dessert items, they prefer to have. Establish conceptual designs using the Systems approach. This non-mechatronic design helps to understand the concept of 'needed sub-systems.'

Answer

A system can have six constituent sub-systems: working system, prime mover system, transmission or energy distribution system, control system, frame system, and helping system. The table-top dessert server is made up of three sub-systems from the above six. They are (i) working system, (ii) frame system and (iii) helping system as shown in Figure 8.44. Proper systems approach selects components from mechanical (or material), electrical (energy) and information (for control) databases to build the systems. There were no such databases available for this project and hence the components needed for the product were identified and hand-picked as shown in Table 8.34. These were used instead of the databases.

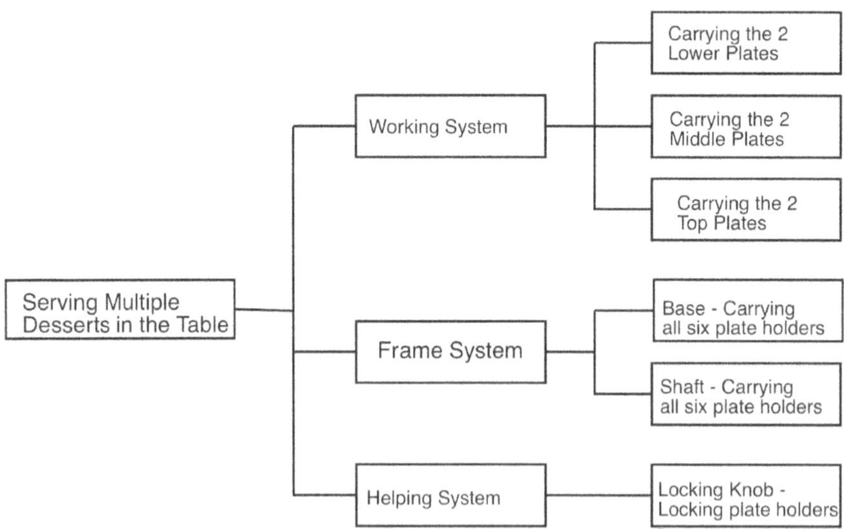

Figure 8.44 System structure and functions.

Table 8.34 Possible components

Single plate presenters (mats or coasters)	*Feet*
Central pole	Spacers in the pole
Single layer board supporting presenters	Bearings for layer boards
Two layer board supporting presenters	Locking knob for the pole
Three-layer board supporting presenters	Locking pin
Base board	

The solution space for this system can have a single piece of wood that can accommodate all six dishes, two pieces each with three locations or three pieces each with two locations as shown in Table 8.35.

Table 8.35 Different layer boards and chosen concept

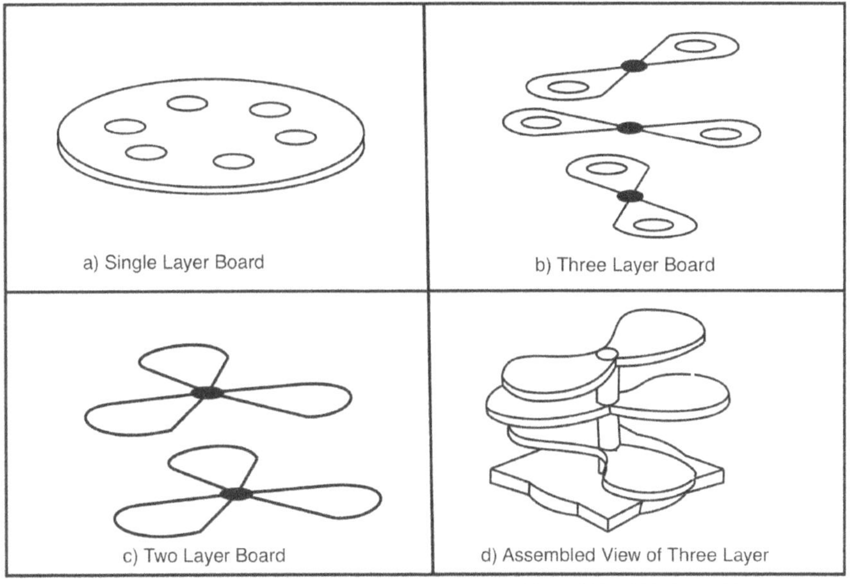

During manufacturing the wooden pieces were cut in the shape of leaves. The three-layer concept was chosen and developed. The student project developed is shown in Figure 8.45.

Figure 8.45 Finished table-top dessert server.

Example 8.30: Design of a mechatronic turntable

This example describes a project, where a turntable using electronic components to control the rotary motion of the table is designed.

The sketch of the turntable is shown in Figure 8.46 for easy comprehension and conceptualization of the problem. The desired function or goal of the system is to place the hole under a fixed nozzle in turn. First the hole 1 is placed under the nozzle. It stays there for some time. Then it moves again and brings hole 2 under the nozzle. It stays in this position for some time and then moves to bring hole 3 under the nozzle. Thus, the work of the turn table is to bring the holes under the nozzle and keep them under the nozzle for a specified time and continue the process. The turn table is powered by a motor which is driven by power from the national grid. A shaft and bearing assembly, connect the motor and the turn table and distribute the power. A cross plate is employed to support the table shaft assembly and another plate is employed to support the motor and the electrical and electronic elements.

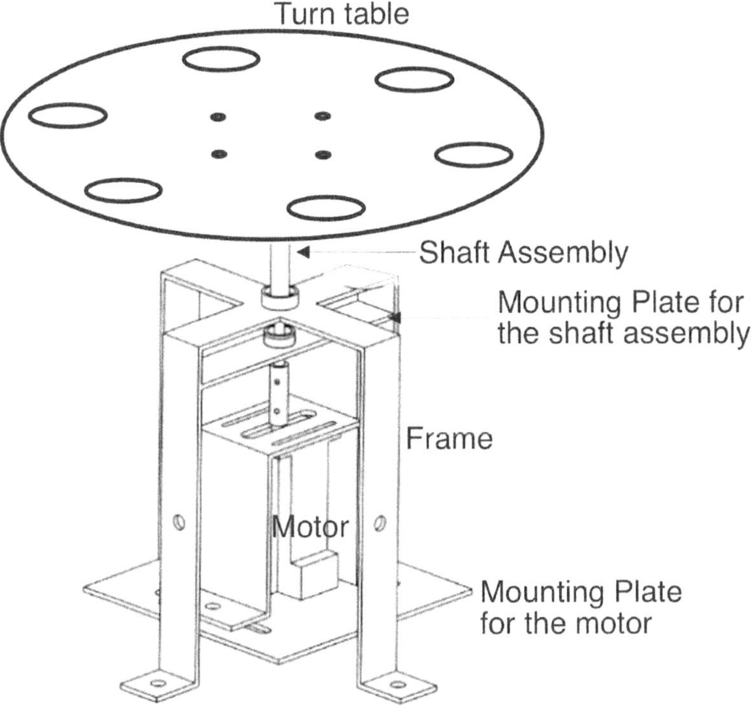

Figure 8.46 Schematic of a turn table.

According to Andreassen's classification they can be arranged in the following way:

1. *Working system*: the expected change here is the aligning of the hole in the turntable under the fixed nozzle in turn.
2. *Prime mover or energy system*: The motor which is the system that delivers the energy for the work.
3. *Transmission – or energy distribution system*: The shaft assembly that distributes energy to the system components.
4. *Control system*: This is the electronic system that monitors and issues various signals associated with the correct functioning of the system.
5. *Frame system*: Keep the whole system together in space.
6. *Helping* system(s): The two support plates that perform different necessary helping tasks.

This deciding is the equivalent of formulating the 'scheme' concept described earlier in the book. Choosing these sub-systems is the main task of conceptual design. The remaining task is to decide the components that form the various sub-systems.

In the next step, the components that build these sub-systems have to be identified. This is somewhat similar to the embodiment design as described in this book. To assist this Andreassen [16] proposes a method. He classifies components into three categories: (i) mechanical (ii) energy and (iii) information. If a database on the details of these is available then embodiment of the sub-systems can be built by asking the following questions:

i. What are the components that are needed to build each of the sub-systems?
ii. How they are interfaced?

For the turntable, the classification of components shown in Table 8.36 was used.

Table 8.36 Classified components

No	*Mechanical*	*Energy*	*Information*
1	Bearings	Motors	Processors
2	Springs	Actuators	Memory
3	Gears	Controllers	Interface
4	Chains	Power Supply	Code
5	Belts	Knobs	PC
6	Shafts	Relay	PLC
7	Bars	Display	Arduino
8	Bolts	Solenoid Valve	Rosebery
9	Screws	Compressor	
10	Coupler	Turbine	
11	Hub		
12	Nut		
13	Washers		
14	Keyway		
15	Dampers		
16	pulley		
17	Weights		
18	Railway		
19	Cart		
20	Valve		

CONCEPTUAL DESIGN

The conceptual design has the sub-systems as shown in Table 8.37. Figure 8.47 shows the conceptual design of the turntable. Figure 8.48 shows the system structure and its functions.

Table 8.37 Sub-systems in the mechatronic turntable

Sub-system	*Description*
Working system	Turntable and the power interface turning the table
Prime mover	Stepper motor
Transmission	Shaft and bearing assembly
Control system	Arduino microcontroller and the power drive
Frame system	Frame that carries and houses the components in position
Helping system	Nuts and bolts and the wheels

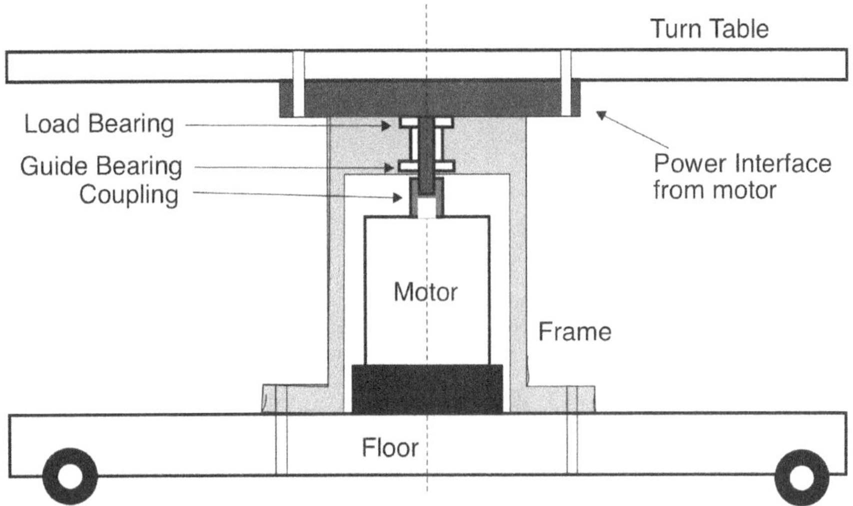

Figure 8.47 Conceptual design of the Mechatronic turn table.

The scheme in the 'system structure and functions' can be explained in the following way. The mechanical assembly starts the motor turntable shaft connected to the coupling and the coupling is connected to the shaft. The shaft is carried and guided by the load and guide bearings in the frame. The shaft in the bearings is connected to the interface which is connected to the rotating table. The assembly is now in the frame ready to run. A manual switch switches the power on and energizes the microcontroller. This triggers the microcontroller and the power supply. The microcontroller switches the power on and off as per the parameter settings and this in turn rotates the turntable and stops for specified periods.

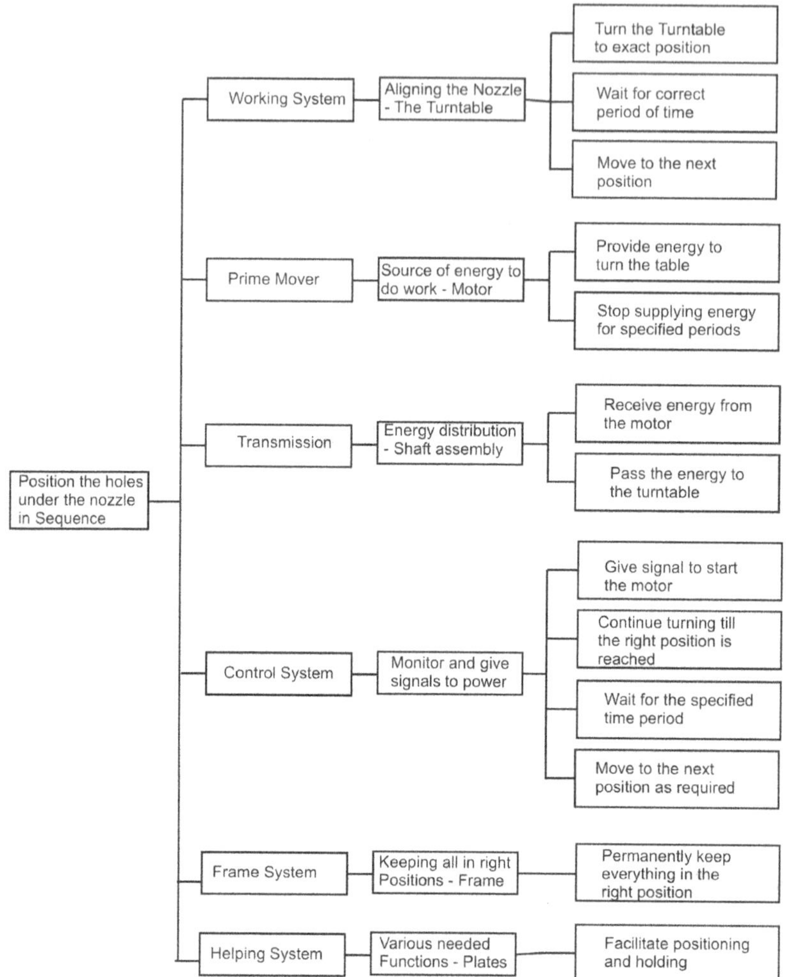

Figure 8.48 System structure and functions.

Example 8.31: Design of a wear monitoring test rig

An observation in a cement factory is described as follows: The drive of the grate cooler attached to the kiln was connected to the drive through a journal bearing. The bearing gets worn out in short intervals. When the wear increases beyond a certain limit the play in the bearing creates a 'knock', which shakes the cooler plate assembly. This and the resulting vibrations loosen the bolts that fix the cooler plates to the chassis and the cooler plates get dislodged causing a breakdown stopping the kiln.

Experience identified, journal bearing as the candidate for observation and improvement. The wear of the bearing is identified as the characteristic for continuous monitoring to choose a material that would last longer, to reach the critical quantity of wear when the bearing has to be changed. The task is to design a wear monitoring rig to monitor the wear of the bush.

THEORY

From a theoretical point of view, wear is influenced by various factors such as the ambient temperature, dusty condition of the air, condition of lubrication, etc. Shigley's 10th Edition [17] explains wear in the following way:

Consider a rectangular block of contact area A and subjected to a pressure P. The friction force acting against the motion is $PA\ f$.

The work done when it moves by a distance $S = PA\ f \times S = PA\ f\ Vt$

If the material wear is w then the volume removed is wA and this is proportional to the work done. Hence

$$wA = KPA\ f\ Vt$$

where K is the proportionality constant.

For Uniform pressure wear $w = KPfVt$

Introducing a modifying factor f_1 depending on motion type, load, and speed and an environment factor f_2 to account for temperature and cleanliness conditions the equation becomes

$$w = f_1 f_2 KPfVt$$

Make a note that the key parameter pressure is constant.

But when the pressure is applied on the journal bearing by the load-carrying shaft, the pressure between the shaft and bearing varies. Thus, investigation into the wear of the bearing is necessary. A half journal bearing that can be optically monitored is desired.

CONCEPTUAL DESIGN

Following the systems approach the conceptual design has the sub-systems as shown in Table 8.38. System structure is shown in Figure 8.49.

A test rig carrying two similar settings driven by two motors connected to worm and wheel gearbox and universal joints was constructed as shown in Figure 8.50.

Table 8.38 Sub-systems in the test rig for journal bearings

Sub-system	*Description*
Working system	Loaded shaft carried by two half journal bearings driven by motor
Prime mover	Motor that can change its speed (ideally)
Transmission	The belt transmitting the load
Control system	Arduino Microcontroller and the power drive
Frame system	Frame that carries and houses the bearing housing assembly, the motor and other components
Helping system	Nuts and bolts and the wheels

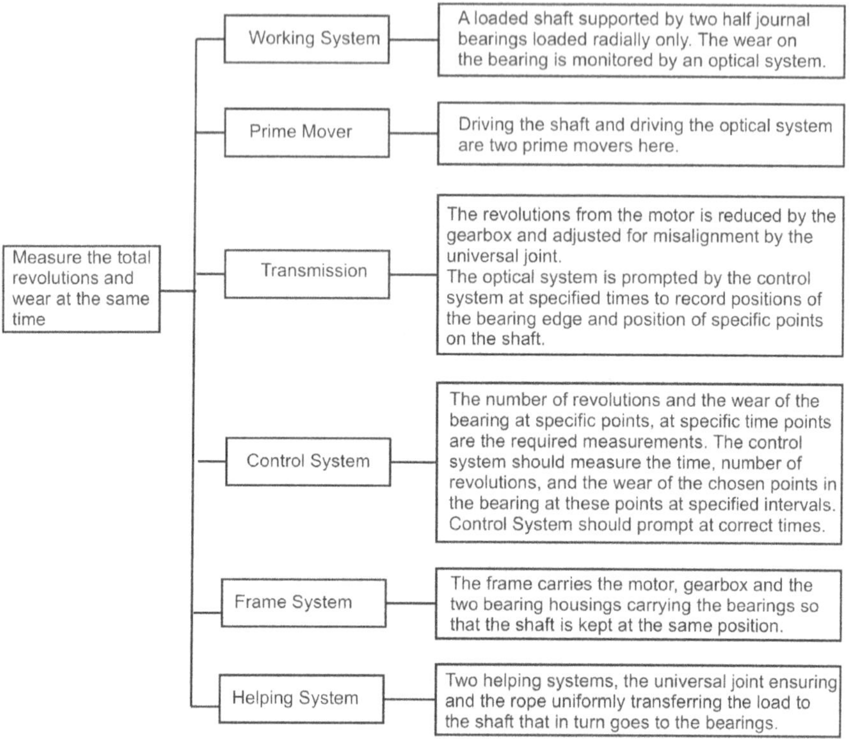

Figure 8.49 System structure of the wear measuring system.

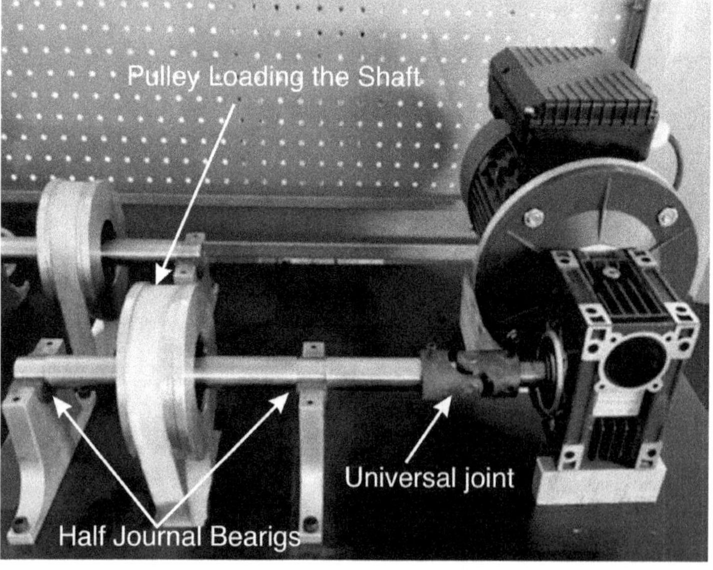

Figure 8.50 The test rig built for testing the wear of journal bearings.

8.14 DESIGN METHOD II: BRAINSTORMING

Brainstorming is a group discussion method to produce ideas for ideation that is familiar to the majority of novice designers. Several authors have specified steps to conduct the brainstorming activity. One such identifies seven steps in brainstorming as (i) write down a statement of the challenge so it is visible to all (ii) remind the group of the Divergent Thinking Guidelines (iii) set a quota of ideas (options) and keep going until you meet it (iv) gather and record concise and specific ideas; ideas should be stated in 'headline' form and be recorded in written form so that all participants can see and read them (v) record ideas as they are stated (vi) periodically (every five ideas or so) check with the client or the group to make sure the ideas are going in the right direction and (vii) proceed until the quota is met, or there are enough ideas to answer the challenge.

Example 8.32: Design of a flower vase stand

Conference halls in big cities have rooms of various sizes. It is intended to have a flower vase at the entrance of each of these rooms. In this project you are required to design a flower vase stand that should store the vase when not in use. The height of the flower Vase to be displayed is about 700 mm. Design the flower vase stand using brainstorming.

Answer

All vase stands have some core blocks like cylinders, hexagonal prism, etc. to give height and storage and this was at the centre of the initial discussion. Subsequently, the discussion was focussed on additional features.

The main point here is to receive all ideas with respect and record all of them in a systematic manner for consideration later. A cloud diagram is used to record them. Figure 8.51a shows the different core blocks, proposed by participants 1, 4, 5 and 6. The ideas are recorded in the clouds shown by the cloud diagram.

The features that can be included as proposed by participants 1, 2, 3, 4, 5 and 6 are recorded in the clouds in Figure 8.51b. Recording the source of every feature idea is good and would facilitate getting more information if needed. The conceptual designs generated from these ideas are shown in Table 8.39.

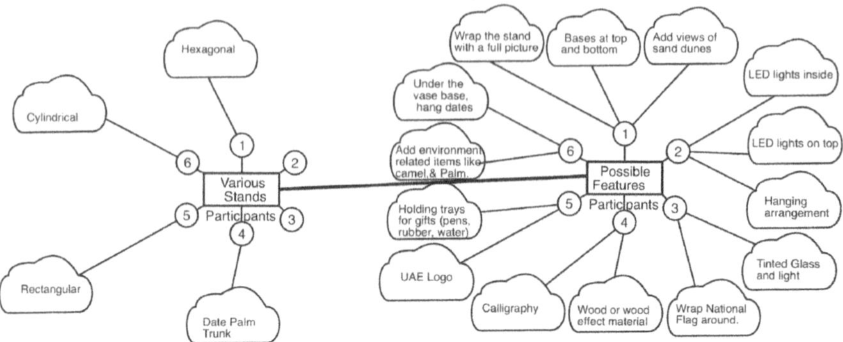

Figure 8.51 Ideas came out of brainstorming.

Table 8.39 Conceptual designs

Example 8.33: Design of a reading desk in a reference library

A reference library plans to have islands of four desks separated by two screens at right angles as shown in Figure 8.52, in their reference floor. They believe that this will provide adequate floor area and privacy for the users. The example requires conceptual designs for the desks, using brainstorming as the design method.

Answer

The process started with the description of the problem and the aim at the centre of the recording of the brainstorming session. The first design was for a conventional setup. The concept was proposed by one participant and others added features to improve it. A computer to take notes, space to keep books, work space, comfortable chair with wheels and access to a large number of printed books are features added by other participants. In a similar fashion, three other conceptual designs were also proposed and details added. For example, the third concept was proposed by another participant for a library that caters for a large number of researchers. The library has a large collection of latest books and journal papers. He proposed a 'U' type table. The details added by other participants include reading assistant to keep many books in open condition, space for a laptop computer, more working space and access to a large number of books and journals. The session has recorded five concepts and details of each concept are recorded and shown in Figure 8.53.

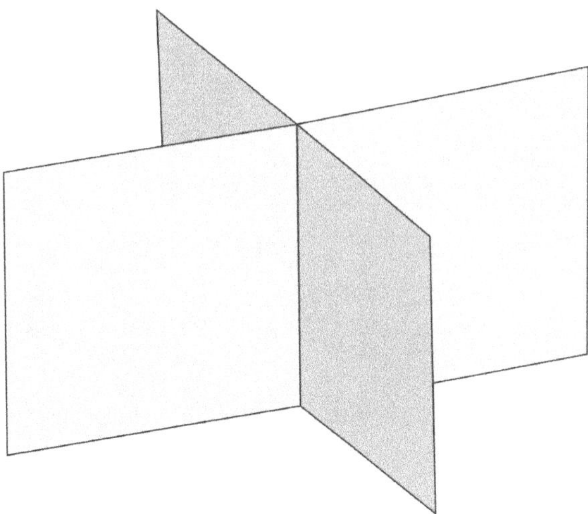

Figure 8.52 Screens isolating the reference tables.

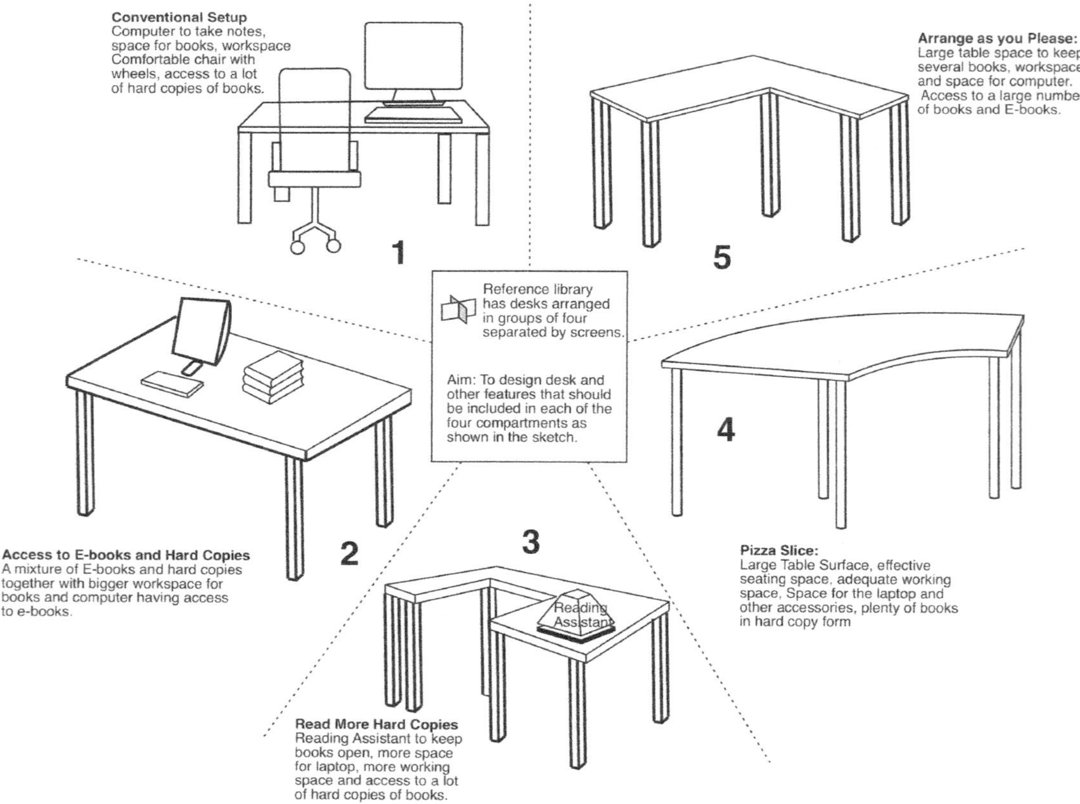

Figure 8.53 Recording of the brainstorming session – reading desk in a reference library.

8.15 DESIGN METHOD 12: CHECKLIST

Checklists are formulated by experienced persons working in the design in general or on the product portfolio of a company for a long time. Thus, the checklists can be generic and applicable to many products or specific to the properties of a company's own products. The checklist for concept generation is best applied when developing an idea into a concept. The technique needs a point of focus. This point of focus should be a product idea, already with material features, shape and dimensions. Applying a checklist in concept generation can yield a product concept which is developed further than just its initial idea state.

Osborn's [18] checklist is a comprehensive list of questions about ideas and problems that aims to encourage creativity and divergence in concept generation. This series of simple, generic questions are designed to support creative and divergent thinking when faced with a design problem. The questions should be taken one at a time, to explore new ways and approaches to the problem. In a brainstorming session, it can be useful to write each statement on a card, and randomly select a card when discussing alternative solutions. Alternatively, paste the questions onto a board and place them in the design team's environment. Osborn's checklist is given in Table 8.40.

A variation of this list is sometimes referred to as SCAMPER – Substitute, Combine, Adapt, Modify/Magnify/Minify, put to other uses, Eliminate, Reverse/Rearrange [18].

Possible procedure:

1. Define a product idea in detail, including features such as shape, dimensions, etc.
2. Search for and select a checklist for concept development. Using more than one checklist is recommended here.
3. Systematically work through the checklist by answering the questions on the checklist.

   Note: Improving the product idea is the aim here and if it is not achieved try another list.

4. Improve the idea by answering the questions on the checklist over and over again. This iterative attempt may be needed to improve the concept.
5. Present the developed idea in an explanatory sketch.

Table 8.40 Osborn's checklist

Other uses?	*New ways to use as is? Other uses if modified?*
Adapt?	What else is like this? What other idea does this suggest? Does past offer parallel? What could I copy? Whom could I emulate?
Modify?	New twist? Change meaning, colour, motion, odour, taste, form, shape? Other changes?
Magnify?	What to add? More time? Greater frequency? Stronger? Higher? Larger? Longer? Thicker? Heavier? Extra value? Plus ingredient? Duplicate? Multiply? Exaggerate?
Minify?	What to subtract? Smaller? Condensed? Miniature? Lower? Shorter? Narrower? Lighter? Omit? Streamline? Split up? Understate? Less frequent?
Substitute?	Who else instead? What else instead? Other ingredient? Other material? Other process? Other power? Other place? Other approach? Other tone of voice? Other time?
Rearrange?	Interchange components? Other pattern? Other layout? Other sequence? Transpose cause and effect? Change place? Change schedule? Earlier? Later?
Reverse?	Transpose positive and negative? How about opposites? Turn it backward, upside down, inside out? Reverse roles? Change shoes? Turn tables? Turn other cheek?
Combine?	How about a blend, an alloy, an assortment, an ensemble? Combine units?

Example 8.34: Design of a human powered lawn mower

A push lawn mower [19] is a manual lawn mower where the cutter is attached to the wheels so that whenever the wheels turn the cutter also turns. The wheel motion provides the linear movement that traps the grass for cutting. Starting with this push lawn mower concept, apply the checklist method to enhance the concept.

The checklist contains the questions shown in the left column of Table 8.41.

The concept of the designed human-powered lawn mower is shown in Figure 8.54.

Table 8.41 Questions and relevance to lawn mower

Questions	*Relevance to the lawn mower design*
Other uses?	The mower can be developed as an exercising tool.
Adapt?	Can adapt the cutter as part of the energy sink in exercise.
Modify?	Modify the push bar as the recipient of drive energy
Magnify?	–
Minify?	–
Substitute?	Substitute the handle with the exercise bicycle
Rearrange?	–
Reverse?	–
Combine?	In short combine lawn mower and exercise bicycle

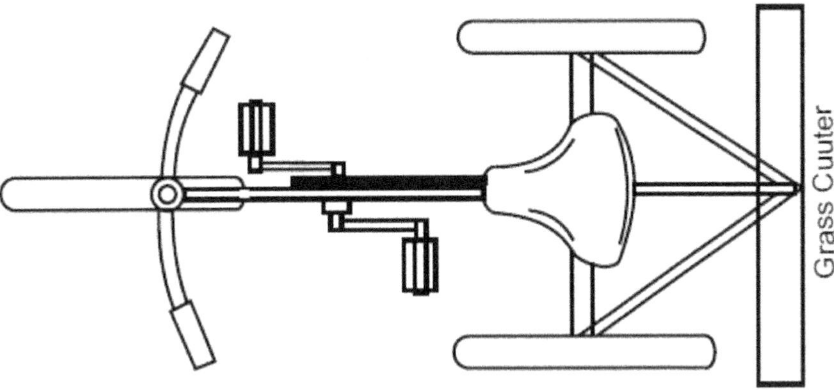

Figure 8.54 Human-powered lawn mower.

Example 8.35: Design of a canteen table

The 'Food Village' in a university has a common eating hall with a large number of tables. The students take a tray from the tray stand, get their food items from various vendors within the village and go to these tables. The trays help them to carry all their chosen food items to the table in an easy way. Once they go to the table, they take all the items from the tray and put the tray aside. The tray that was very helpful up to this point becomes a nuisance at this stage. Figure 8.55 shows a typical four-seater table in a canteen. Apply the checklist method and improve this design.

Figure 8.55 Starting focus: A simple table and chairs.

Table 8.42 Questions and relevance to canteen table

Questions	*Relevance to the canteen table design*
Other uses?	–
Adapt?	Can adapt the table to store the trays outside the table-top surface.
Modify?	Modify the table to have a space for storing empty trays
Magnify?	Can magnify; but it will have lot of waste space
Minify?	–
Substitute?	–
Rearrange?	Rearrange the table parts incorporating a space for empty trays
Reverse?	–
Combine?	In short include a space within the current design to store empty trays

Answer

The checklist questions are systematically considered and the answers are recorded in Table 8.42. The answers were considered to get an improved design. The improved design of the table for the canteen is shown in Figure 8.56.

Figure 8.56 Table with a deck to park trays.

Example 8.36: Design of a dessert display from Lazy Susan

After lunches and dinners, a dessert is served and in luxurious situations, a variety of them are offered for the guests to choose. When multiple dishes are served, a lazy Susan as shown in Figure 8.57 is used. The main benefit of the lazy Susan is the rotating table that gives access to all dishes while being seated. Using this as the starting focus and using the checklist method establish the design of a table-top dessert server.

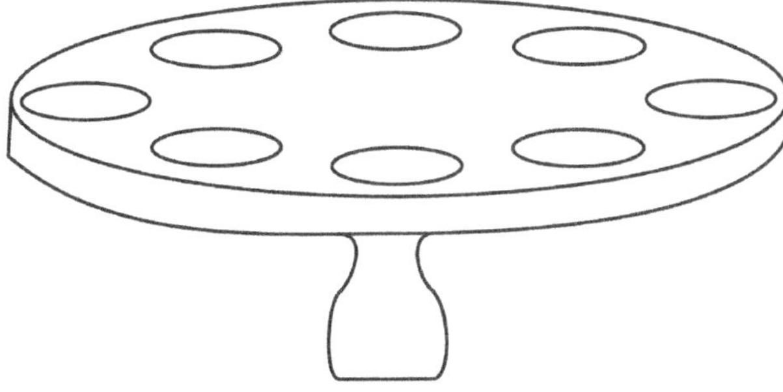

Figure 8.57 Starting focus: Lazy Susan food presenter.

Answer

The checklist questions and answers are shown in Table 8.43.
The designed table-top dessert server is shown in Figure 8.58.

Table 8.43 Questions and relevance to table-top dessert server

Questions	*Relevance to the table-top dessert server*
Other uses?	–
Adapt?	Can adapt lazy Susan presenter to have smaller diameter
Modify?	Modify lazy Susan to have more decks and modify the rotating table to make all the food items visible.
Magnify?	–
Minify?	–
Substitute?	Substitute the rotating table with rotating star-like arms
Rearrange?	–
Reverse?	–
Combine?	In short change the rotating table with a deck of rotating arms to carry various dessert items.

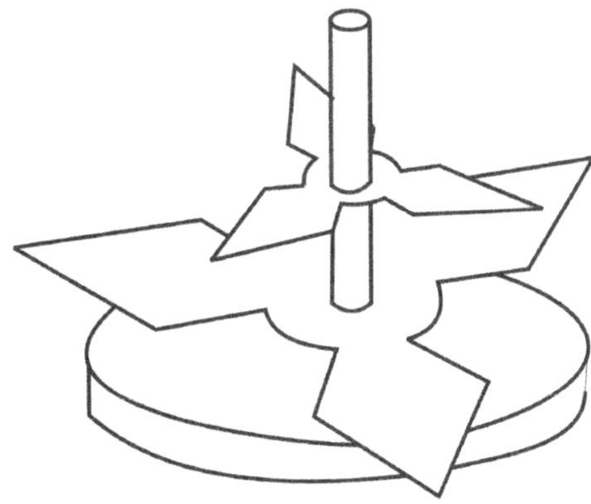

Figure 8.58 Table-top dessert server with rotatable layers.

8.16 DESIGN OFFICE FOR A MANUFACTURING COMPANY

Any organization or business needs an office as the centre of all business activities and needs a certain amount of information in order to function properly. The office acts as an intermediary between the outside world or public and the organization. Depending on the types and magnitudes of the organization there are several kinds of office such as records management office, human resources office, purchasing and stock control office, sales and marketing office and so on. The role and purpose of the office are to provide rooms, labour and other facilities which are used to organize and coordinate various activities of the organization. Functions of an office include (i) distribution and exchange of goods and services; (ii) generation, collection, processing and preservation of data; (iii) dissemination of information and (iv) organizational management.

One of the fundamental abilities or enhancers for designing is the potential to ask questions. A poem from Rudyard Kipling, the Nobel Prize winner's poetry reads

> I keep six honest serving men
> (They taught me all I knew);
> Their names are What and Why and When;
> And How and Where and Who

A checklist can easily be formed with these six questions: what, why, when, how, where and who.

Using Office as the focus and the checklist using the six questions as the basis, develop the design for a design office for a local industry.

Note from the author: This question is the summary of a project undertaken by a group of students from the MEM program. The findings were presented in the Annual Conference of the American Society of Engineering Education in 2017 [20].

Answer

The answer to this question is reached in two stages: (i) literature survey and (ii) apply the checklist method to create the conceptual design.

Literature Survey is carried out under four titles as shown below:

1. *Role of an Engineering Manager*: Leading from present to the future is one of the requirements of a modern engineering manager and in this connection, contribute to continuously upgrade the current operations in the short term, and in the long term, development and introduction of new generation of products in a timely manner and contributing to new company strategies related to technologies.

2. *Role of Research and Development*: An effective R&D department provides
   a. R&D strategy to position its innovation efforts internally and externally by defining where to place emphasis and the direction for R&D.
   b. R&D process ensuring that the right inputs and outputs are available to support functions such as product development, research, technical service, marketing and manufacture.
   c. Resources for developing the capabilities to encourage innovation; this includes tools, people, techniques and facilities.
   d. Organization by selecting the right structure for R&D allowing processes and resources to work as efficiently as possible. Structures can be based on competencies, products, services or disciplines.
   e. R&D culture representing values and behaviours that contribute to the unique social and psychological environment of an organization.
   f. Information systems ensuring that the right information is collected, sifted, analyzed and communicated. The team may include partners, universities and technology consultants.
   g. R&D metrics with Key Performance Indicators or KPIs measuring the performance of the R&D activities
3. *Design Process*: In systematic approach to design, the design model is the description of the stages or the sequence of activities the process goes through. There are several such design models. Design methods are tools and techniques that are being used at different stages in the design process. With these in the background, companies choose a design stage model and add the required design methods at different stages to suit their needs to create their design model. Design Council has a model with four distinct phases, Discover, Define, Develop and Deliver, mapping the divergent and convergent stages of the design process. It shows the different modes of thinking that designers use. In this book, a five-stage design model is adapted.
4. *Office Management*: 'How the functional, technical and financial factors should be organized to maximize the chances of a successful product development?', is the question that has to be addressed here. In general, the overall design is broken into parts for development and combined again to form the final solution. In this context providing written material, multimodal supports and multiplatform environments, support sense-making and common understanding in meetings. Minutes, images, networks, shared documents, and digitally shared information, all contribute to extend the process of information construction and exchange.

Table 8.44 Questions raised

Question	*Questions raised*
What	What details the design office should consider?
	What platforms are adopted by the sector of the company?
	What are the capabilities needed to embrace e-opportunities?
	What are the steps needed to ensure short-term profitability?
	What is the strategy for the R&D activities?
	What outputs and outcomes are expected from R&D?
	What are the tools needed to assist R&D activities?
	What outcomes are seen as achievements of R&D?
	What phase of the design process is the main concern?
	What are the tools, software etc needed in the identified phase?
	What are the tools to discover the problem (potential) area?
	What facilities have to be provided to enhance creativity?
	What is the best arrangement of the working office station?
	What are the main requirements of a design office?
	What are the roles the meeting rooms have to play?
	What are the main items that have to be preserved?
Why	Why the company should upgrade its current operation?
	Why the design process is important?
	Why the design process needs many software tools?
	Why the outputs of the design at various stages are preserved?
When	When should the design project be divided into tasks?
	When should the results of the design tasks be combined?
	When should software tools be given to support design process?
	When can past designs records help?
	When can design be stored as records?
	When should software tools be given to support design process?
	When can past designs records help?
	When can design be stored as records?
How	How to ensure smooth operation of the company?
	How design office can help to keep attention to details?
	How to develop new generation of products and services?
	How developing visions for the future be integrated?
	How e-transformation will improve current condition?
	How the new technologies will be required?
	How the company can encourage innovation activities?
	How R&D should be structured and organized?
	How the right information on past and future designs and processes can be made available for immediate accesses?
	How to ensure availability of the good practices established methods and the lessons learned?
	How the design project is divided into tasks for individuals?
Who	Who allocates the integration and division tasks in design?
	Who is responsible for the overall activities in Design Division?
	Who are the designated office bearers in a design division?
	Who decides the software tools needed?
	Who plans and executes activities to achieve goals?

Hypothetical model of a design office

The questioning technique and the diagnosis gave a clear insight and formed the basis for a hypothetical model. It starts with the identification of the goals of the design division as its top layer. Several questions in Table 8.44 can be grouped under this category. In the next layer, the activities that have to be performed to achieve the goals are established. This again is based on the questions in Table 8.44. In the next layer, the resources that are necessary to perform the above activities are specified. The next layer defines the personnel and their organization to best carry out the activities and achieve the goals. Finally, the key performance indicators necessary to monitor and control are defined. Figure 8.59 illustrates the hypothetical model.

Specific application sample

Design division for a high-tech manufacturing company, which manufactures composite components for leading international companies. They have gained clear knowledge and experience in manufacturing and the implications of manufacturing in the production of high-quality, end products. They have understood the implications of manufacturing defects that can have compromising effects in the design.

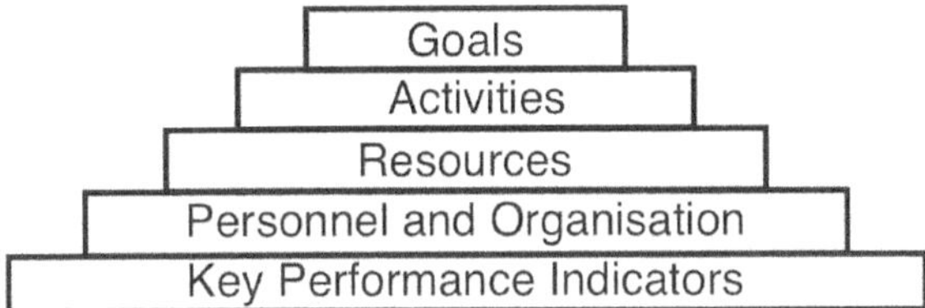

Figure 8.59 Hypothetical model of a design office.

A. Goal

They identified the goal as 'a design division that can (i) act as a reliable and efficient repository for designs with associated experiences (ii) use very fine analytical software and verification facilities so that the effects of manufacturing defects can be reliably estimated (iii) established good practices and examples that can be used in future activities (iv) a design model to suit their needs, and (v) a structured division with designated positions to carry out the functions and develop the capability to receive technology transfer from experts as and when appropriate'.

B. Activities

1. *Problem identification* – To clearly understand and define the problem, knowledge of composites, structural design and application-related theories are used. Activities involve (i) constructing defect concentration diagram; (ii) literature survey; (iii) referring relevant standards; (iv) review the execution of specific projects and (v) discussion with the customers. A definite problem statement and defined success criteria are the outcomes of this activity set. Several meetings and discussions may be part of this process.
2. *Analysis of the problem* - Activities are (i) establishing the current knowledge in terms of published data and established good practices of both internal and external categories and (ii) establish the purpose of the design and the design rules to be applied using the knowledge gained. Here again meetings and discussions form an important part of the activities.
3. *Design process* – The design process is implemented as a change management system consisting of (i) Engineering Change Request (ii) Engineering Change Order, and (iii) Change Order Closure in accordance with the change management system. Stage gate management is employed to manage the process from stage to stage with defined outputs from each stage. This is one of the areas where a lot of engineering efforts are made individually and integrated. Meetings of various kinds and sizes are needed here.
4. *Verification and validation* – The design is subjected to various analyses using advanced computer-based tools to verify and validate the designs. This is a specialist activity and requires a lot of resources in terms of e-facilities and skilled personnel. Analyses required for validation take a lot of time and new technologies are introduced in the market on a frequent pattern. Thus, acquiring new techniques and technologies is a constant additional requirement for this section. The verification part should ensure that the new design meets the specification in full.
5. *Implementation* – DA is an independent assessment to capture any omissions at the design stage before its committal to manufacture. It is a tool to check the adequacy and efficacy of the design. DA ensures that the end product meets or exceeds customer expectations. DA works with the team and being a team of experts, provides expert insights and additional support to achieve the end goal.

C. Resources

The resource requirements for this design division are mainly in the electronic or software format. They can be classified under the following categories:

Approved and common tools – This includes software packages like Microsoft Office, list managers, capture tools, brainstorming tools, adobe, and many others.

Robust Design Guides: Being in the cutting edge of the high-tech sector the design process and the proving of the designs are governed by design guides. These guides are often updated and a tracking of them and making them available and accessible is a fundamental task. This is one of the main requirements. Another important item is the proven design manual which outlines the set-menu procedures for standard problems within the sector.

Robust configuration and data management system – The importance of this needs no mentioning. The system should provide access to the design guides, design manual and standards at the press of a button. It should manage the data generated at all phases of the design in an orderly fashion so that all data generated can be made available for use at a later stage.

DA requirements – Information about assessment to verify and validate various requirements and stipulations specific to the sector. This may include standards, test procedures and the like.

Demonstrated technology readiness and manufacturing readiness – These are essentially case studies of various kinds that can give insights and understanding of complex procedures. This should also provide facility for viewing these case studies and self-brainstorming or dreaming.

Hardware support for effective meetings – The entire set of activities in the design division outlined above needs several meetings both with internal personnel and external experts. The meeting rooms should be equipped with multi-screen facilities and strong hardware and software support to facilitate meaningful discussions. There should be many meeting rooms with different capacities.

Space Allocation – Another important aspect is the space management in the design office. The first and foremost is the allocation of space for several well-equipped meeting rooms of different sizes. These facilitate meetings and discussions of high quality. The next important item is the working station for individual engineers. The importance of the facilities and space cannot be over-emphasized.

D. Personnel and organization

The fundamental and most valuable resource for a design division is the competence of the personnel. The division should have a leader who knows the way, walks the way and leads the way. The division should have capable intermediate managers who can lead their teams towards the goals. They should have authority on par with their responsibilities. A clearly defined organization structure is fundamental to the design division. Another fundamental requirement for a young company is the provision of regular training and knowledge-building programs. Being in the high-tech sector this should include training by international experts.

E. Key Performance Indicators

KPIs tell the management how effective and efficient the design division is. KPIs should be developed to meet the objectives of the company, which forms the Design Division. The work on this project suggested the following four KPIs.

6. *First time quality of the design* – This evaluates whether the design team has fully comprehended the design problem, developed the solution, carried out necessary and sufficient analyses and carried out the required verification and validation checks so that the DA passes the design with minimal additional requirements in the first time. This is a measure of the competence of the division.
7. *Client Satisfaction Ratings* – The improved design at the end has to be approved by the client. Hence the client satisfaction rating is a clear indication of the effectiveness or quality of the design.
8. *Development of repeatable procedures* – Design processes in the high-tech sector are very special and companies develop in-house good practices or methods. The number of reusable procedures and techniques developed within the design division is a good indicator of the quality of the division.
9. *Activity follow-up* – Each project will have activities such as meetings and evaluations. A record and analysis of these could give an indication on the active nature of the project.
10. *Comparison between budgeted or estimated time and actual time* – Finishing the task with reliable and satisfactory results in the budgeted time is a clear demonstration of competence.

REFERENCES

1. French M.J., *Conceptual Design for Engineers*, Springer, New York, 1990.
2. Haik Y., Sivaloganathan S. and Shahin T.M., *Engineering Design Process*, International Edition 3rd edition, Cengage, Noida, 2017.
3. Ergonomics evaluation into the safety of stepladders, Loughborough University for the Health and Safety Executive, United Kingdom. https://www.hse.gov.uk/research/crr_pdf/2002/crr02423.pdf
4. https://lh4.googleusercontent.com/di-GDwf_XHNrFSfpZKH0I2LJRENjxnsfzcqzUGouhILu76BB8KUw0wwGlVUjjx38Zn19ByJtiAUZ2ciOLpK6TbV0yWg-ol8UtFQaMx2fw6nb9xcS8bTGcxFcngJejqSpvwo3s3mV Visited on 7th November 2023
5. https://www.babboecargobike.com/blog/a-long-john-cargo-bike-that-is-perfect-for-urban-cycling Visited on 7th November 2023
6. https://www.cargobike.co.uk/product/tamar-cargo-trike/ Visited on 7th November 2023
7. https://i.pinimg.com/736x/e2/34/ff/e234ffe6e8c4de110431ca0ab5e4d63f.jpg Visited on 7th November 2023
8. https://en.vidaxl.ae/e/vidaxl-bike-cargo-trailer-black-and-yellow-65-kg/8718475701361.html Visited on 7th November 202
9. https://downtown-mag.com/wp-content/uploads/sites/7/2020/07/Cargo-Lastenrad-Intro-Test-Review-2020-366-1140x806.jpg Visited on 7th November 2023
10. https://ae.pricena.com/en/product/home-adult-three-wheel-tricycle-24-inch-6-speed-cargo-cruiser-trike-bike-with-rear-basket-for-shopping-exercise-bicycle-color-black-price-in-dubai-uae-134000935 Visited on 7th November 2023
11. https://www.industrialbicycles.com/images/Worksman-Mover1.jpg Visited on 7th November 2023.
12. http://www.portalbikes.org/long-tail Visited on 7th November 2023.
13. https://www.cargobike.ca/bike-types/long-tail Visited on 7th November 2023.
14. https://bikepacking.com/bikes/clandestine-shepherds-midtail/ Visited on 7th November 2023.
15. https://www.carousell.sg/p/woman's-26"-aleoca-signora-blue-1121263171/ Visited on 7th November 2023.
16. Andreasen M.M., *Functional Reasoning in Mechatronic Design, 41030: Mechatronic Engineering Design*, Design and Innovation Division, Technical University of Denmark
17. Budynas R. and Nisbet K., *Shigley's Mechanical Engineering Design*, 10th edition, McGraw-Hill Education, New York, 2015.
18. Osborn's Checklist. www.ifm.eng.cam.ac.uk/research/dmg/tools-and-techniques/osborns-checklist/, accessed on 28th May 2022
19. Push Lawn Mower. https://www.pinterest.co.uk/pin/learning-new-things-444308319479374372/, accessed on 8th June 2022
20. Sangarappillai Sivaloganathan, Salah Burhan Al Omari, and Aysha Al Ameri development of a design division for an industry: a capstone project in a master's of engineering management program. In: *ASEE Annual Conference and Exposition*, Salt Lake City, 2018.

Chapter 9

Concept evaluation and selection

9.1 INTRODUCTION

Developing a sample solution space containing proposed conceptual designs, evaluating them to establish that they are of high grade, and choosing a high-grade conceptual design from them, are the strategical steps for meeting the challenge of '*choosing a design from the unknown solution space*'. The chosen design is deemed as the optimal or a near optimal solution from the solution space. Evaluation of the sample therefore is crucial to ensure that the sample contains high grade designs. Chapter 8 described how a sample solution space can be developed. To evaluate and choose a design the following steps are taken:

i. Establish a set of criteria and their weightings
ii. Use the criteria and evaluate the designs in the sample developed earlier in the conceptual design stage
iii. Improve the concepts if needed, and decide whether the sample contains high grade designs. If the sample does not contain high grade designs go back to conceptual design.
iv. If the sample contains high grade designs, choose a conceptual design for further development using the decision matrix.

This chapter describes how the criteria are established and rated, and how evaluation and choosing a design are carried out using the Decision Matrix method [1]. Another good method to deal with the sample designs is Pugh's [2] `Controlled Convergence' method. This method is also explained here. Examples are provided to experience both methods.

9.2 SELECTION OF CRITERIA

Selection criteria can be established in two different ways: (i) by considering the design brief and customer needs or (ii) by considering the objective of the product through the objective tree. The following sub-sections describe them.

DOI: 10.1201/9781003484950-12

9.2.1 From design brief and customer requirements

The starting point for product development is the design brief that explains the design team what the product is, and the benefits to be delivered. The main criteria used to satisfy the objectives of the company or client are the benefits to be delivered. On the other hand, the customers rate the products according to what the product can do. For this reason, the degree of satisfying the customer requirements should also be part of the basis for concept selection. Thus, the criteria for concept selection should be the combination of benefits to be delivered from the design brief and important customer requirements from the list.

9.2.2 From the objective tree

Design objectives are goals for what a design is meant to achieve or the functional requirements that the design should strive for. For example, consider a bench vice. The objective of a bench vice is to hold an object with grip, thereby permitting work to be done on the object. This objective is achieved by generating counter forces and moments in the 1, 2 and 3 directions, as shown in Figure 9.1. The figure shows balancing forces generated by the applied force components and interactions between the jaws and the object in all three directions. In the same way, applied torques and reactions

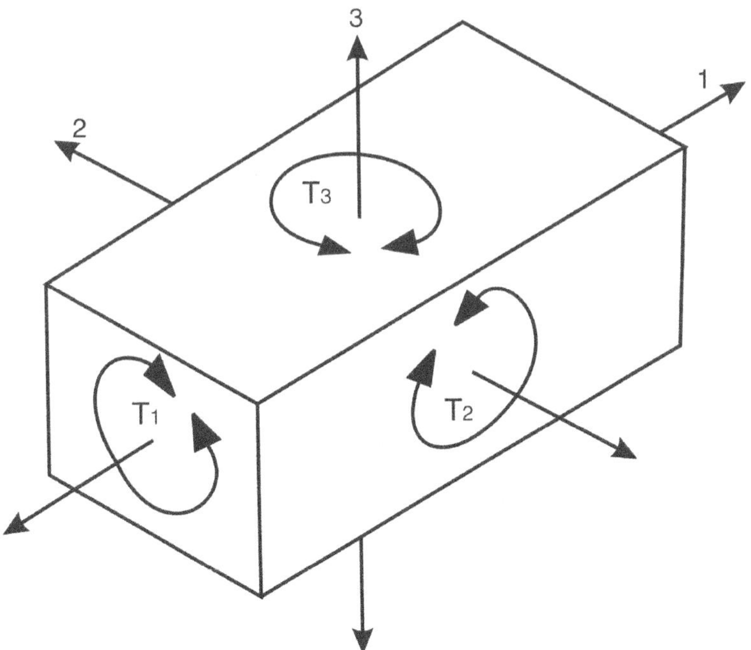

Figure 9.1 Free body diagram of a piece in grip and subjected to 3D load.

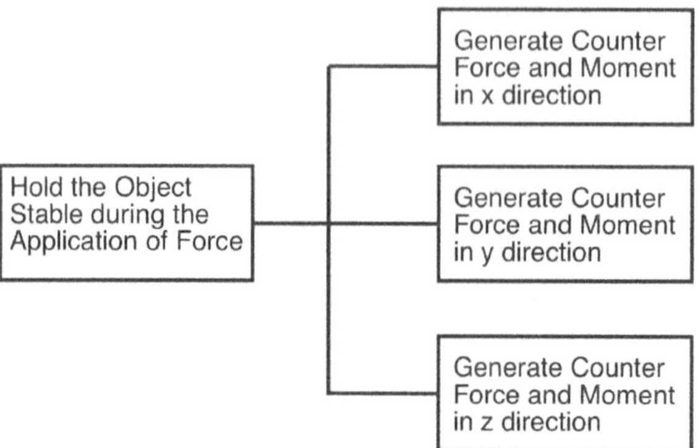

Figure 9.2 Objective tree for a bench vice.

are also marked in all three directions. Thus, the objective tree can be as shown in Figure 9.2.

From the objective tree, the criteria for concept selection can be the following:

a. Generate Counter Force and Moment in x Direction
b. Generate Counter Force and Moment in y Direction
c. Generate Counter Force and Moment in z Direction

Whichever the method of establishing the criteria, they have to be given weights to make concept selection. Allocation or calculation of the weight for each criterion can be made in several ways. The first and easy method is the random allocation by the design team. The next one is the ordering method which yields the same results as the tabulation method.

9.3 WEIGHT OF INDIVIDUAL CRITERION IN A LIST

9.3.1 Ordering method

Ordering method gives highest weight for the criterion that is more important than all others and gives a lesser value for the factors that are in the descending order of importance.

Steps in the ordering method:

1. Arrange the criteria in the descending order of importance.
2. Assign values to each criterion values, in the descending order of importance, starting from n where n is the number of criteria. The maximum value allocated would be n and the minimum value would be 1.

3. Find the sum of all the values allocated for each criterion.
4. The weights for the factors can then be the fraction obtained by dividing the Factor Value by the sum.

Example 9.1: Weights for factors arranged in the descending order of importance

Consider a situation where there are seven criteria A, B, C, D, E, F and G. Find the weights for the factors using the ordering method. Assume that the descending order of importance is A, C, E, B, D, G and F.

Answer

The steps in the ordering method are:

i. Arrange the criteria in the descending order of importance starting with the most important criterion at the top
ii. Allocate points to each of the criterion starting at the top and decreasing by 1 each time, while give the top one receives the number equal to the number of criteria.
iii. Now take the sum of the points.
iv. The weights for each of the criterion is its points divided by the Sum

Since the number of criteria is 7, A is allocated 7, C is allocated 6, E is allocated 5 and so on. The assigned values are shown in column 3 in Table 9.1.

The sum of the assigned values is 28.

The weights and their calculations are shown in Table 9.1.

Table 9.1 Example calculation on ordering method

No	*Criterion*	*Assigned value*	*Weight*
1	A	7	7/28
2	B	4	4/28
3	C	6	6/28
4	D	3	3/28
5	E	5	5/28
6	F	1	1/28
7	G	2	2/28
Sum		28	28/28

9.3.2 Weights of the individual criterion using tabular method

Steps in the tabular method:

1. The tabular method starts with the creation of a table with $(n+4)$ columns and $(n+2)$ rows.
2. On the first column, enter the names of the criteria one in each cell starting from the second row.
3. On the first row starting from the second column, enter the names of the criteria one in each cell.
4. After naming all the criteria, in the next column, add 1 to all cells.
5. The next column is entitled as score and the last column indicates the weight for each criterion in the respective row.
6. Fill the diagonal cells with a '–' (dash) to cancel them.
7. Now consider the criterion in the second row (the first criterion) with others starting from column three onwards. Enter a '1' if the row criterion is more important than the column criterion. Otherwise enter a '0'.
8. Fill each of the subsequent rows in the same manner.
9. Now take a row sum for each criterion and enter them in the column 'Score'. This is the number of criteria which are less important than the one in row with an additional 1.
10. Find the total of the score and divide each row sum with the total. Enter the answers in the next 'weight' column.
11. The weights for individual criterion in the row are given in the weight column.

Example 9.2: Criteria weights for a table-top dessert server

The design of a table-top dessert server is under way. The first four items shown below are from the design brief and the remaining are important customer requirements. Using them as criteria for concept selection, estimate the weight for each criterion using the tabular method.

a. Rotating server
b. Minimum of six different deserts
c. High-class product
d. Easy to assemble and dismantle
e. Easy to access any dessert
f. Height of the server is below 0.5 m
g. Monster cycles to failure >1,000
h. Pass stability industrial test
i. Cost

Table 9.2 Tabulation for weight calculation

	a	*b*	*c*	*d*	*e*	*f*	*g*	*h*	*i*	+1	*Score*	*Weight*
a	–	1	1	1	1	1	1	1	1	1	9	9/45
b	0	–	0	1	0	1	0	1	0	1	4	4/45
c	0	1	–	1	1	1	1	1	1	1	8	8/45
d	0	0	0	–	0	1	0	1	0	1	3	3/45
e	0	1	0	1	–	1	1	1	1	1	7	7/45
f	0	0	0	0	0	–	0	1	0	1	2	2/45
g	0	1	0	1	0	1	–	1	1	1	6	6/45
h	0	0	0	0	0	0	0	–	0	1	1	1/45
i	0	1	0	1	0	1	0	1	–	1	5	5/45
Total											45	1.00

i. There are nine criteria. Hence the table should have 13 columns and 11 rows.
ii. The top most row and left column have the columns and rows earmarked for individual criterion as shown in Table 9.2.
iii. Considering criterion 'a' in row 2, it is more important than all other criteria and all cells are marked 1.
iv. The (row sum) score is entered in the cell for score which tells the number of criterion that are less important than '*a*'+1.
v. The process is repeated with other criteria such as *b*, *c*, *d* - - -.
vi. Now take the sum of all scores and divide each score by it, to get the weights of individual criterion as shown in Table 9.2.

Note: It can be seen that the weights obtained by the tabulation methods is the same as the ordering method.

Example 9.3: Weights with some selected values

Sometimes certain criterion demand specified values as weight. For example, in an eight criteria a, b, c, d, e, f, g and h situation, selection need criteria 'g' to have a weight of 0.25 and 'h' to have a weight of 0.15. The order of importance for the remaining criteria, is given as a, c, d, f, e and b. Allocate the weights using the ordering method.

Answer

Since the weights for g and h add up to 0.4 the remaining weights should add up to 0.6 to make the total of weights equals 1. This 0.6 is divided among a, b, c, d, e and f. Since there are only six factors the assigned values should start from 6. The assignment and weight calculations are shown in Table 9.3.

Thus, the weights for the eight criteria are

$$\begin{bmatrix} 0.17 & 0.03 & 0.14 & 0.11 & 0.06 & 0.09 & 0.25 & 0.15 \end{bmatrix}$$

Table 9.3 Weight allocation using ordering method

No	*Criterion*	*Assigned value*	*Weight*
1	a	6	$\frac{6\times 0.6}{21}=0.17$
2	c	5	$\frac{5\times 0.6}{21}=0.14$
3	d	4	$\frac{4\times 0.6}{21}=0.11$
4	f	3	$\frac{3\times 0.6}{21}=0.09$
5	e	2	$\frac{2\times 0.6}{21}=0.06$
6	b	1	$\frac{1\times 0.6}{21}=0.03$
Sum		21	0.6
7	g		0.25
8	h		0.15

9.4 DECISION MATRIX

9.4.1 Construction and rating calculation

The decision matrix is sometimes called the weighted sum method because of the method employed by the decision matrix. Here, the designer rates the conceptual ideas against criteria that are deemed relevant to the task. In a design situation, they come from the design brief and important stakeholder requirements. The conceptual ideas or designs go on the rows, and the design criteria on the columns. The weighting factors (WF) are determined for each criterion as shown earlier. For each concept, the designer tries to see how well it achieves each design criterion on a rating from 1 to 10, 10 being the best. These numbers are referred to as the rating factors (RF). A grid is prepared in the following way, for constructing the decision matrix:

i. In the grid, the first column contains all the different concepts from which the choice has to be made.
ii. The top row enumerates the criteria one by one in each cell.
iii. The second row enumerates their weight factors.
iv. The rest of the cells are divided into upper triangles and lower triangles.
v. Each concept is rated against the criterion represented by that column on a 1 (low) to 10 scale and the rated factor is entered in the upper triangle.

vi. All concepts are rated in this way and the upper triangles are filled.
vii. The lower triangles are filled by the product of the column's weighting factor and the rating factor.
viii. The row sums of the products (lower triangles) are entered in the last column. This is the weighted sum of the rating scored by that design.

Example 9.4: Explain the given decision matrix

Office chairs have several designs and thus the solution space for the designs will have several members. Five conceptual designs P, Q, R, S and T for an office chair are proposed based on a brainstorming session. A decision matrix has been constructed to evaluate the designs as shown in Figure 9.3 with the following criteria and their weights. Explain the decision matrix.

a. Proper lumbar support: 0.12
b. Equipped with a headrest: 0.15
c. Robust armrests: 0.20
d. Upholstery Material: 0.20
e. Adjustable height and inclination of the backrest: 0.13
f. Frame with 360-degree swivel: 0.10
g. Wheels and number of wheels: 0.10

Answer

The decision matrix shown in Figure 9.3 can be explained in the following way:

	Criterion a	Criterion b	Criterion c	Criterion d	Criterion e	Criterion f	Criterion g	Total
Weight	0.12	0.15	0.20	0.20	0.13	0.10	0.10	1
P	6 / 0.72	5 / 0.75	6 / 1.20	5 / 1.00	6 / 0.78	5 / 0.50	7 / 0.70	5.65
Q	9 / 1.08	6 / 0.90	5 / 1.0	6 / 1.20	10 / 1.3	5 / 0.50	5 / 0.50	6.48
R	7 / 0.84	7 / 1.05	7 / 1.4	7 / 1.4	7 / 0.91	7 / 0.70	7 / 0.70	7.0
S	5 / 0.60	6 / 0.90	5 / 1.0	5 / 1.0	6 / 0.78	6 / 0.60	6 / 0.60	5.48
T	6 / 0.72	7 / 1.05	6 / 1.20	6 / 1.20	6 / 0.78	7 / 0.70	8 / 0.80	6.45

Figure 9.3 A typical decision matrix.

Consider the first row

This records that designs are rated with respect to criteria $\begin{bmatrix} a & b & c & d & e & f & g \end{bmatrix}$ as shown in the first row.

The second row shows their weights as $\begin{bmatrix} 0.12 & 0.15 & 0.20 & 0.20 & 0.13 & 0.10 & 0.10 \end{bmatrix}$

The rating of the design P is $\begin{bmatrix} 6 & 5 & 6 & 5 & 6 & 5 & 7 \end{bmatrix}$ and these are entered in the upper triangle of each cell. This is shown in the third row of the decision matrix.

If the ratings are subjected to the weights of the criteria, the overall rating for P can be found as

$$\begin{bmatrix} 6 & 5 & 6 & 5 & 6 & 5 & 7 \end{bmatrix} \begin{bmatrix} 0.12 \\ 0.15 \\ 0.20 \\ 0.20 \\ 0.13 \\ 0.1 \\ 0.1 \end{bmatrix}$$

$$= [6 \times 0.12 + 5 \times 0.15 + 6 \times 0.2 + 5 \times 0.2 + 6 \times 0.13 + +5 \times 0.1 + 7 \times 0.1] = 5.65$$

This means the weighted sum of the ratings for the Design P is 5.65

The lower triangle in the decision matrix contains the weighted rating for each criterion and the weighted sum is found by adding them together.

In the same way:

a. Design Q has the weighted sum of the rating is 6.48
b. Design R has the weighted sum of the rating is 7
c. Design S has the weighted sum of the rating is 5.48
d. Design T has the weighted sum of the rating is 6.45

9.4.2 Evaluation and decision

9.4.2.1 *Ensuring the validity of the sample from the design space*

Consider design R in the decision matrix shown in Figure 9.3. The sum of the weighted rating is $= [7 \times 0.12 + 7 \times 0.15 + 7 \times 0.2 + 7 \times 0.2 + 7 \times 0.13 + +7 \times 0.1 + 7 \times 0.1]$

$$= 7(0.12 + 0.15 + 0.2 + 0.2 + 0.13 + 0.1 + 0.1)$$

But $(0.12+0.15+0.2+0.2+0.13+0.1+0.1)=1$

This shows that *if the rating is the same for all criteria* the sum of the weighted rating is equal to the original rating which in this example is 7.

Now in general, if the design meets all criteria 100% the common rating will be 10. Therefore, the maximum weighted sum can only be 10.

Thus, an ideal solution will have a weighted rating sum of 10.

This follows that an optimal solution should be close to it (say around 9) and near optimal solutions can be close to the optimal (say around 8 or even 7).

The sample designs generated, can be assumed to have optimal or near optimal solutions in it, if and only if, some of the designs in the decision matrix have, Sum of Weighted Rating, in the range 8–10.

This observation follows that if the weighted sums in a decision matrix is below say 6, the design team should return to generate some more concepts. This is the decision where the design team ensures that the chosen design is of a high grade.

9.4.2.2 Need for revisiting concept generation

Consider design Q in the matrix in Figure 9.3. It got a rating of 10 out of 10 for criterion 'e'. The design with the highest sum (design R) has a score of only '7' for this criterion. It therefore would be a good idea to revisit conceptual design and see whether the technique in Q can be adapted to fit with R so that it can meet criterion 'e' in a better way. But this may not be compatible. Pugh's matrix will explain this more, later.

9.4.2.3 Choosing the design

After ensuring that the sample from the solution design space has better designs (optimal or near optimal designs) and no design in the sample space needs further enhancements, choosing a design can take place. The chosen design could be the design that has the highest 'Sum of Weighted Ratings'.

Example 9.5: Choice of kettles for a hotel

A large hotel management was in need of room kettles for their new hotel with 500 rooms. The hotel management has called for tenders from five companies for kettles that can boil up to five cups (1,250 mL) and their tenders are shown in Table 9.4. Available properties are given in the second row.

The kettle would be chosen according to the following criteria:

a. Speed
b. Preferable temperature setting option
c. Cost

Table 9.4 Tenders for kettles in hotel rooms

Company A	Company B	Company C	Company D	Company E
Temp setting Quick boiling Attractive noise (40 db) 1.2 kg 1,200 W 100 AED	Temp setting Quick boiling Coated outside low noise (40 db) 1 kg 1,500 W 75AED	Corded kettle Safety lock lid low noise (50 db) Water-gauge 1,000 W 0.5 kg 50 AED	Temp setting Quick boiling Stainless steel outside noise (70 db) 1 kg 1,200 W 40 AED	Corded kettle Safety lock lid Attractive coated Noise 60 db 1,000W 1 kg 60 AED

d. Appearance
e. Ease of use (cordless)
f. Less noise
g. Less weight

The hotel management has requested you to help them in the following:

i. Allocating weights for each criterion
ii. Formulating a decision matrix to establish that the Tenders are of High Value.
iii. Choosing the better design from the tenders

Answer

The team decided to get the order of preference of the individual members and found that two criteria d and f had similar maximum preference and another three, a, e and g had similar medium preferences. They used an allocation reflecting these and the weight for each of the criterion is given as shown in Table 9.5.

The weights determined by the team are used to construct the decision matrix shown in Figure 9.4.

The weighted sums are $\begin{bmatrix} \frac{231}{28} & \frac{244}{28} & \frac{184}{28} & \frac{219}{28} & \frac{175}{28} \end{bmatrix}$

Table 9.5 Weights of the criteria

No	*Criterion*	*Weight*
1	a	4/28
2	b	3/28
3	c	1/28
4	d	6/28
5	e	4/28
6	f	6/28
7	g	4/28
Sum		28/28

	Criterion a	Criterion b	Criterion c	Criterion d	Criterion e	Criterion f	Criterion g	Total
Weight	4/28	3/28	1/28	6/28	4/28	6/28	4/28	28/28
Company A	8 / 32/28	10 / 30/28	5 / 5/28	9 / 54/28	9 / 36/28	9 / 54/28	5 / 20/28	231/28
Company B	9 / 36/28	10 / 30/28	6 / 6/28	9 / 54/28	9 / 36/28	9 / 54/28	7 / 28/28	244/28
Company C	7 / 28/28	0 / 0/28	8 / 8/28	6 / 36/28	7 / 28/28	8 / 48/28	9 / 36/28	184/28
Company D	8 / 32/28	10 / 30/28	9 / 9/28	8 / 48/28	9 / 36/28	6 / 36/28	7 / 28/28	219/28
Company E	7 / 28/28	0 / 0/28	7 / 7/28	7 / 42/28	7 / 28/28	7 / 42/28	7 / 28/28	175/28

Figure 9.4 Decision matrix to evaluate the kettles.

The maximum out of the weighted sums is 8.7 out of 10 and the minimum is 6.25 out of 10. This means the sample contain rich kettles.

The kettle given by Company B is the best kettle out of the five tendered.

Example 9.6: Choice of a birthday cake for a five year old girl using Decision matrix

A birthday cake for a five-year-old girl has been designed by her several aunts and a choice has to be made. A set of criteria has to be formulated to make a choice. Design brief and customer requirements cannot help fully to make this formulation. An objective tree is more suitable to guide the formulation of the criteria.

The objective tree sets the targets the designer has to achieve in his product. The objective tree for the birthday cake is shown in Figure 9.5.

The Goal: The goal is to generate a 'Thrilling smile with glowing satisfaction' in the child's and in all participants' faces. To achieve this, six sub-objectives have to be achieved. They are:

a. Create a captivating appearance to the cake
b. Build-in mouth-watering taste
c. Include an eye-catching and tasty frosting
d. Employ a baking to result in a spongy and inviting formation to the cake
e. Blend a delightful flavour
f. The cake should have inserts containing emotionally loving messages

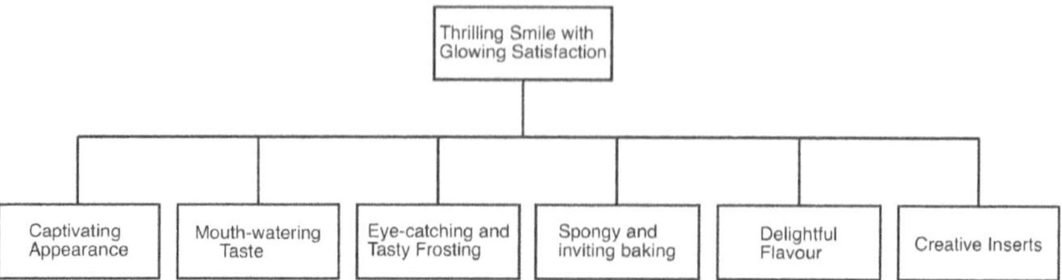

Figure 9.5 Objective tree for a birthday cake.

An additional consideration is the cost. These seven items can be considered as the criteria for choosing a cake design. Use the Tabulation method to estimate the weights for all the individual criterion.

Steps in the tabular method:

1. The tabular method starts with the creation of a table with $(n+4)$ columns and $(n+2)$ rows.
2. On the first column, enter the names of the criteria one in each cell starting from the second row.
3. On the first row, starting from the second column enter the names of the criteria one in each cell.
4. In the next column, add 1 to all cells.
5. The next column is entitled as score and the last column indicates the weight for each criterion in the respective row.
6. Fill the diagonal cells with a '-' to cancel them.
7. Now consider the criterion in the second row (the first criterion) with others designated in columns 3 onwards. Enter a '1' if the row criterion is more important than the column criterion. Otherwise enter a '0'.
8. Fill each of the subsequent rows in the same manner.
9. Now take a row sum for each criterion and enter them in the column 'Score'. This is the number of criteria which are less important than the one in row with an addition of an extra 1.
10. Find the total of the score and divide each row sum with the total. Enter the answers in the next 'weight' column.
11. The weights for individual criterion in the row are given in the weight column (Table 9.6).

To continue the narrative, the six aunts made sample cakes of their proposal and asked the grand mother to evaluate their cakes according to the above criteria. Grandmother's ratings of the cakes are given in Table 9.7. Formulate the decision matrix and choose the cake.

The decision matrix for choosing the cake is shown in Figure 9.6. The results show a keen competition between concepts B, D and F. Concept F by Aunt 6 was chosen as the winner.

Table 9.6 Tabulation method of weight calculation for cake design

	a	*b*	*c*	*d*	*e*	*f*	*g*	+1	*Score*	*Weight*
a	–	1	0	1	1	0	1	1	5	5/28
b	0	–	0	0	0	0	0	1	1	1/28
c	1	1	–	1	1	0	1	1	6	6/28
d	0	1		–	0	0	0	1	2	2/28
e	0	1	0	1	–	0	0	1	3	3/28
f	1	1	1	1	1	–	1	1	7	7/28
g	0	1	0	1	1	0	–	1	4	4/28
									28	1

Table 9.7 Grandmother's ratings of the cakes

	Criteria a	*Criteria b*	*Criteria c*	*Criteria d*	*Criteria e*	*Criteria f*	*Criteria g*
Concept A by Aunt 1	8	10	5	9	9	9	5
Concept B by Aunt 2	9	10	5	9	9	9	7
Concept C by Aunt 3	7	0	8	6	7	8	9
Concept D by Aunt 4	8	10	9	8	9	8	7
Concept E by Aunt 5	7	7	7	7	7	7	7
Concept F by Aunt 6	8	10	9	7	8	9	7

	Criterion a	Criterion b	Criterion c	Criterion d	Criterion e	Criterion f	Criterion g	Total
Weight	5/28	1/28	6/28	2/28	3/28	7/28	4/28	28/28
Concept A	8 / 40/28	10 / 10/28	5 / 30/28	9 / 18/28	9 / 27/28	9 / 63/28	5 / 20/28	208/28
Concept B	9 / 45/28	10 / 10/28	6 / 36/28	9 / 18/28	9 / 27/28	9 / 63/28	7 / 28/28	227/28
Concept C	7 / 35/28	0 / 0/28	8 / 48/28	6 / 12/28	7 / 21/28	8 / 56/28	9 / 36/28	208/28
Concept D	8 / 40/28	10 / 10/28	9 / 54/28	8 / 16/28	9 / 27/28	8 / 56/28	7 / 28/28	231/28
Concept E	7 / 35/28	7 / 7/28	7 / 42/28	7 / 14/28	7 / 21/28	7 / 49/28	7 / 28/28	196/28
Concept F	8 / 40/28	10 / 10/28	9 / 54/28	7 / 14/28	8 / 24/28	9 / 63/28	7 / 28/28	233/28

Figure 9.6 Decision matrix for choosing cake design.

Example 9.7: Choice of a table-top dessert server

Systematic design and manufacture of a table-top dessert server was given as a project to a class following product development course. The requirement was a compact, collapsible and easy to assemble dessert server which can carry a maximum of six different desserts. It is intended to be placed at the centre of the table and is rotatable permitting easy access to different desserts in the server. Six groups have produced the dessert servers and they are shown in Table 9.8. One of them has to be chosen for the class prize. The criteria for selection are given below. Use the ordering method to allocate the weights for each of the individual criterion. Use the weights and construct a decision matrix and choose the one for the prize.

The criteria are as follows:

a. Stability
b. Ability to be moved
c. Rotate ability
d. Durability
e. High class or richness
f. Cost
g. Compactness or size
h. Easy to assemble and dismantle

Table 9.8 Designs of table-top dessert server

The steps in the ordering method are

i. Arrange the criteria in the descending order of importance starting with the most important criterion at the top
ii. Allocate points to each of the criterion starting at the top and decreasing by 1 each time, while give the top one receives the number equal to the number of criteria.
iii. Now take the sum of the points, which is 36.
iv. The weights for each of the criterion is its points divided by the Sum

Table 9.9 shows the weight calculations and Figure 9.7 shows the decision matrix. Design R is the chosen one for the prize.

Table 9.9 Weights of the criteria

Criteria in descending order of importance	*Name*	*Points*	*Weight*
Stability	a	8	8/36
High class or richness	e	7	7/36
Compactness or size	g	6	6/36
Rotate ability	c	5	5/36
Durability	d	4	4/36
Cost	f	3	3/36
Ability to be moved	b	2	2/36
Easy to assemble and dismantle	h	1	1/36

	Criterion a	Criterion b	Criterion c	Criterion d	Criterion e	Criterion f	Criterion g	Criterion h	Total
Weight	8/36	2/36	5/36	4/36	7/36	3/36	6/36	1/36	
Design P	9 / 72/36	8 / 16/36	7 / 35/36	8 / 32/36	8 / 56/36	9 / 27/36	7 / 42/36	7 / 7/36	287/36
Design Q	7 / 56/36	8 / 16/36	7 / 35/36	9 / 36/36	9 / 63/36	8 / 24/36	10 / 60/36	9 / 9/36	299/36
Design R	8 / 64/36	10 / 20/36	9 / 45/36	9 / 36/36	9 / 63/36	8 / 24/36	8 / 48/36	8 / 8/36	308/36
Design S	10 / 80/36	8 / 16/36	7 / 35/36	8 / 32/36	8 / 56/36	9 / 27/36	7 / 42/36	8 / 8/36	296/36
Design T	7 / 56/36	10 / 20/36	10 / 50/36	10 / 40/36	9 / 63/36	7 / 21/36	7 / 42/36	7 / 7/36	299/36
Design U	9 / 72/36	7 / 14/36	9 / 45/36	8 / 32/36	8 / 56/36	9 / 27/36	8 / 48/36	8 / 8/36	302/36

Figure 9.7 Decision matrix for choosing the dessert server.

9.5 PUGH'S CONCEPT EVOLUTION – CONTROLLED CONVERGENCE METHOD

Controlled convergence method employs both synthesis evaluation and further synthesis as in gallery method until a good design is reached. It permits alternate convergent (analytic) and divergent (synthetic) thinking to occur. This is because as the reasoning proceeds a reduction in the number of concepts comes about for rational reasons. Also, this provides opportunities for thinking, and new concepts are generated. Thus, it is alternately a generative (creative) and a selection process. The method is illustrated in Figure 9.8. The process starts with developing several concepts based on product design specification (PDS), shown at the top of the figure. On evaluation some concepts prove weak and controlled convergence is applied to eliminate them. This is shown by the stage marked 'Initial number reduced'. The evaluation process gives better insight and few more concepts are added. Controlled convergence is applied again leading to further reduction. Now few more concepts are added before applying controlled convergence again. This process of adding concepts and evaluating and reducing continues until a really good design emerges.

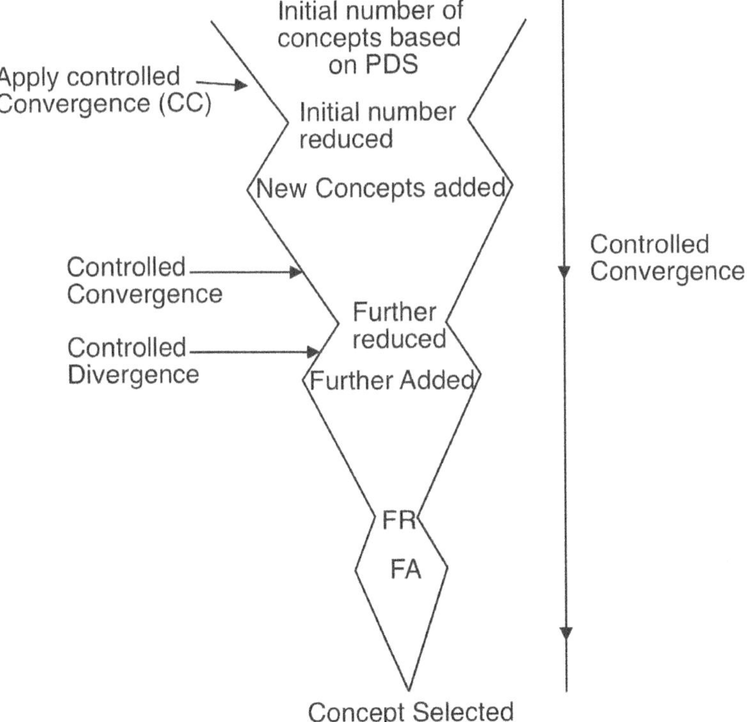

Figure 9.8 Controlled convergence.

9.5.1 Use of the matrix method - Phase I

A matrix aimed at comparing and evaluating design concepts against the criteria is proposed by Pugh. The rows represent the criteria one by one in the matrix starting from row 2. They are entered in column 1. The concepts are inserted in row 1, one by one in the columns starting from column 2. A peer concept, the one that is considered the best, that is, the one that is most likely to meet all the specifications in the most effective manner. This is considered to be the DATUM and other concepts compared with this to fill the matrix. The elements of the matrix are filled in the following way:

1. Consider each concept with the DATUM with respect to each criterion in the criteria.
2. If the concept is better than the Datum a+is entered in the element representing the criterion/concept combination
3. If the concept is similar to the Datum an S is entered in the element representing the criterion/concept combination
4. If the concept is worse than the Datum a - is entered in the element representing the criterion/concept combination
5. After all concepts are evaluated in this manner the total +, - and S for each design is entered in the next three rows to complete the comparison and evaluation matrix.

Figure 9.9 shows a typical comparison and evaluation matrix as given by Pugh.

Concept / Criteria	1	2	3	4	5	6	7	8	9	10	11
A	+	−	+	−	+	−	D	−	+	+	+
B	+	S	+		−	−		+	−	+	−
C	−	+	−	−	S	S	A	+	S	−	
D	−	+	+	−	S	+		S	−	−	S
E	+	−	+	−	S	−	T	S	+	+	−
F	−	−	S	+	+	+		+	−	+	S
+	3	2	4	1	2	2	U	3	2	4	2
−	3	3	1	4	1	3		1	3	2	2
S	0	1	1	1	3	1	M	2	1	0	2

Figure 9.9 Pugh's comparison and evaluation matrix.

9.5.1.1 *Evaluation and further action process*

1. Consider the negatives in the strong concepts – What has to be done to reverse the negatives? Is it possible at all? Can it be done without reversing any of the current positives? If it is possible, introduce it as an additional design into the matrix and rerun the matrix.
2. Attack the negatives in the weak concepts – See whether they can be improved with respect to the datum. Can this be done without reversing any existing positives? If it is possible, introduce it as an additional design into the matrix and rerun the matrix.

9.5.2 Phase 2

This phase proceeds if a decision is taken to develop the strongest concepts (more than one concept) that emerged from the phase 1. This entails further work on these concepts, to engineer them further and provide more detail than earlier. This will provide a better insight of the problem, specifications and solutions. Based on these, new set of criteria should be formulated and a new comparison and evaluation matrix should be compiled. The compiled matrix should be subjected to evaluation once again.

Example 9.8: Controlled convergence on the design of a memorabilia clock

A memorabilia wall clock is needed to celebrate the 15th year of the Master of Engineering Management program. The clock on the wall is expected to have the following characteristics. Propose concepts and subject the collection to controlled convergence method to choose a design.

Preferred characteristics

1. *Boasting the Strengths of the MEM program* – MEM mentioned. Providing colleges mentioned, the courses mentioned
2. *Reflection of strengths of Engineering* – Milestones on the development passage of engineering and technology
3. *Use of latest technologies* – Digital or analogue nature, showing a wide range of technologies
4. *Relates to local culture* – Religion or the philosophy, dress etc
5. *Relates to inheritance of UAE* – sand dunes, fishing boats, camels, palm trees
6. *Convenient size* – For laying out different themes or ideas, circular or polygon shaped
7. *Aesthetically pleasing appearance* – Shape of the clock, arrangement of the supporting items
8. *Technical Advancement of the country* – The biggest achievements like Hope explorer, Burj Khalifa etc
9. *Technological developments of the country* – Advancements of technology in the ordinary man's day to day life.

ANALYSIS OF THE FIRST ROUND – PHASE 1

The initial set of designs are shown in Table 9.10 and their Pugh's Matrix is shown in Table 9.11. Considering the scores of the first round reveals that designs B and D as emerging strong designs and designs A, C and E as the designs with many negatives.

Consider design B. Its negative is in reflection of strengths of engineering. In many respects it is similar to the datum design. The negative can easily be rectified by adding features of engineering and there is space for it. Consider design D. It received negatives in relating to (i) local culture and (ii) inheritance of UAE. A pentagon relates to the religious aspects in design C and this can be brought in and heritage can be harmoniously included in design D. These two designs can be improved as shown in designs G and H in Table 9.12.

Considering the designs with more negatives, designs A and C were discarded. Design E was considered good and was taken for further development and developed as design I. The developed designs are shown in Table 9.12. An additional design was proposed with more features and was added to the Table of improved designs as design J. From this collection, Design J was chosen to be the final design.

The designs were considered with the first set of criteria and a design that works well with all of them a second set of criteria and Phase 2 development was not needed.

Note: These are the considerations by the author; but the reader can have different opinions.

Table 9.10 Initial set of conceptual designs

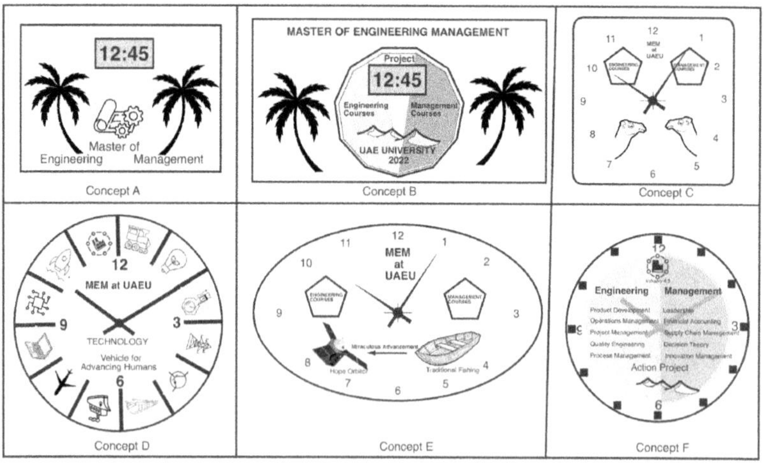

12:45 Master of Engineering Management Concept A	MASTER OF ENGINEERING MANAGEMENT Project 12:45 Engineering Courses Management Courses UAE UNIVERSITY 2022 Concept B	MEM at UAEU Concept C
12 MEM at UAEU 9 3 TECHNOLOGY Vehicle for Advancing Humans 6 Concept D	MEM at UAEU Concept E	Engineering Management Action Project Concept F

Table 9.11 Pugh's matrix

Criteria	Initial Designs						Variations to Initial Designs			
	Design A	Design B	Design C	Design D	Design E	Design F	Design G	Design H	Design I	Design J (New)
Strengths of the MEM	-	+	-	S	-	DATUM	+	S	S	S
strengths of Engineering	-	-	-	+	-		S	+	+	+
Use of latest technologies	+	+	-	S	-		+	S	S	+
Relates to local culture	S	S	+	-	+		S	S	+	+
inheritance of UAE	S	+	S	-	S		+	S	S	+
Convenient size	-	S	-	+	-		S	+	-	+
Aesthetically pleasing	S	+	-	+	-		+	+	-	+
Country's Technical Advancement	S	S	-	+	-		S	+	+	+
Country's Technological developments	-	S	S	S	S		S	S	S	S
$\sum -$	4	1	6	2	6		0	0	2	0
$\sum S$	4	4	2	2	2		5	4	4	1
$\sum +$	1	4	1	5	1		4	5	3	8

Table 9.12 Improved designs

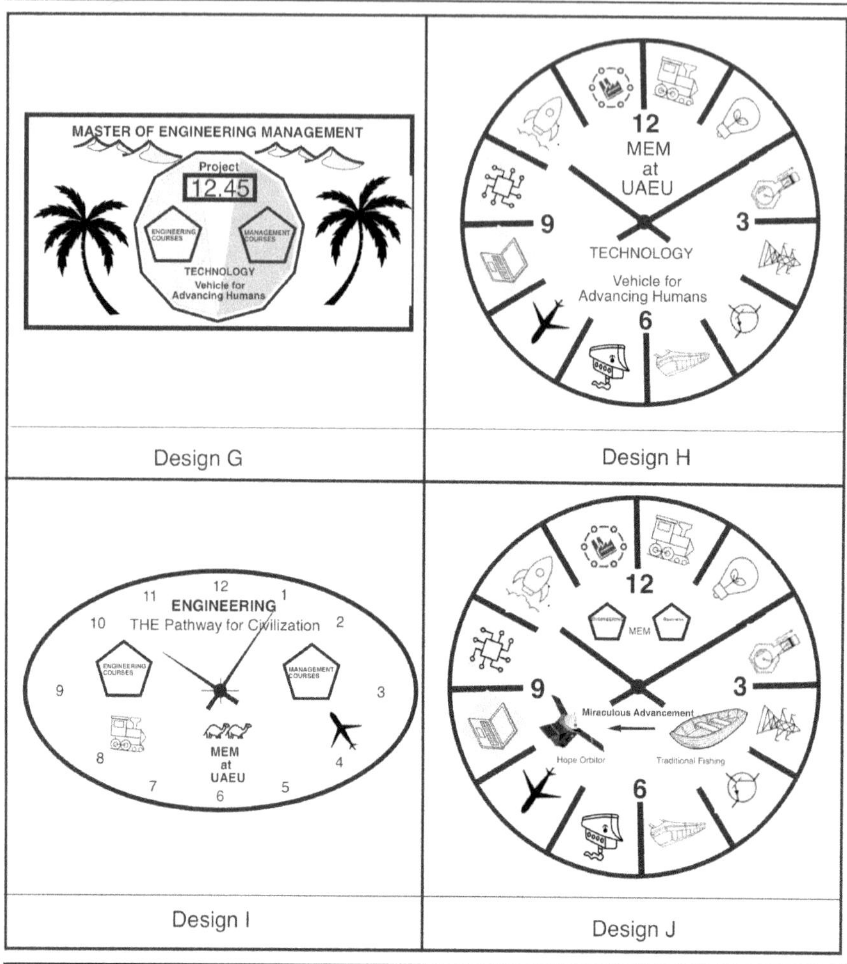

Example 9.9: Controlled convergence on the design of a leaflet display stand

Leaflet stands are designed and used to hold and display leaflets. They are commonly used in exhibitions, conferences, and education sectors like universities and institutes. Universities use the stand to display the program leaflets which detail programs offered by the colleges and departments, their entry requirements, content, duration, study plan, tuition fees and other details. Stands are usually placed in front of college offices or university registration offices to make it accessible to students as well as their parents. The task here is to design such a leaflet holder to hold at least 12 leaflets. The product designed to hold at least 12 leaflet holders and consists of three main parts that are base with wheels, leaflet holders, and storage area. It was expected that the product would clearly display leaflets and be attractive to users. Target market is education sectors like universities, institutes, and technical colleges. The criteria for choosing a concept are as follows:

1. Easy visibility of all programs
2. Storage for each program details
3. Attractive and high-class
4. Compact storage
5. Lightweight
6. Portable and easy to move
7. Easy to assemble
8. Low cost
9. Minimum space

A team of four has worked on this design. They produced their initial set of designs and they are shown as designs A, B, C, D and E in Table 9.13. Design C was considered as the datum and evaluation was carried out for every design for every criteria. The evaluation shown in Table 9.14 resulted in more ideas towards improving the designs eliminating the negatives. As can be seen, designs F, G and H were developed to eliminate negatives in the previous designs.

A joint discussion at the end led to the design that was manufactured. Figure 9.10 shows the display stand built by the students.

Table 9.13 Conceptual designs

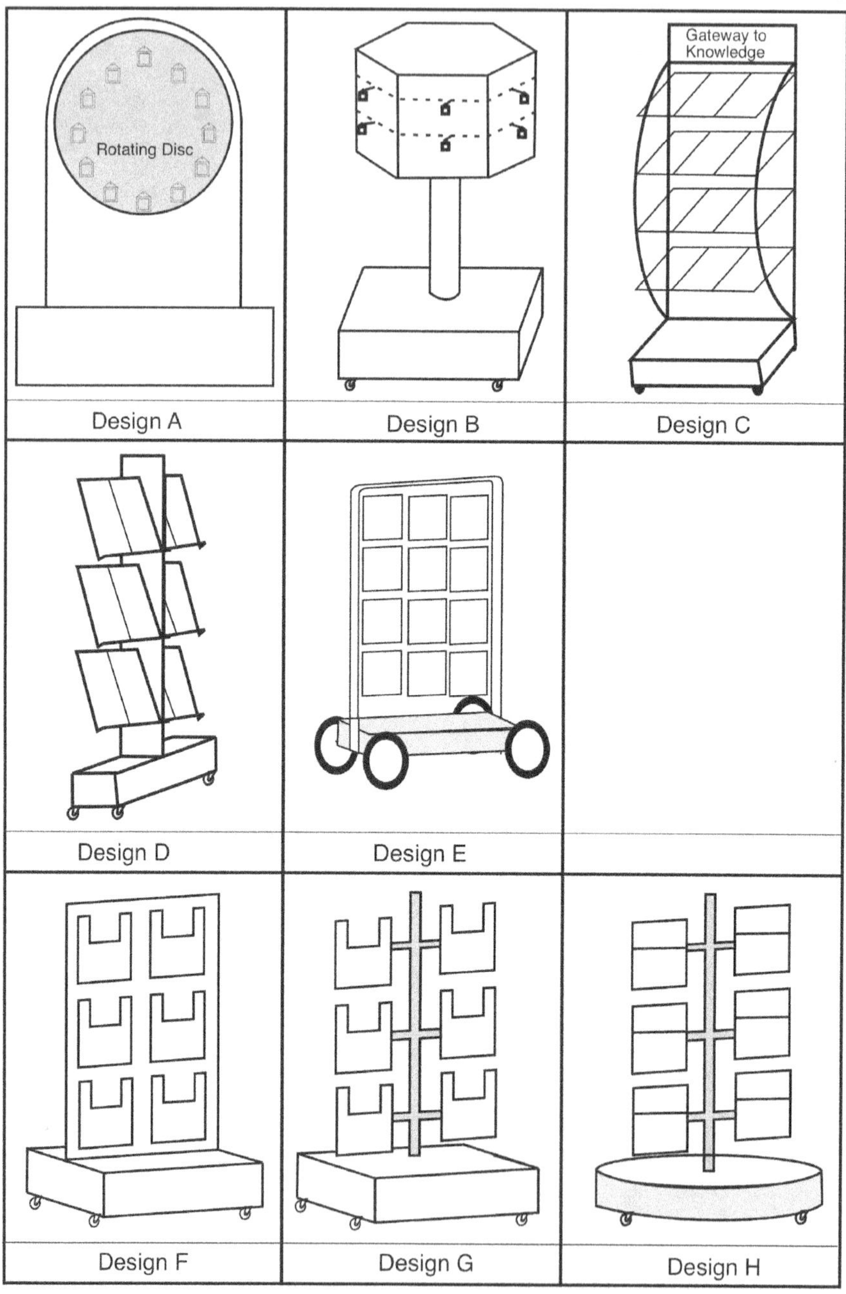

Table 9.14 Pugh's matrix

Criteria	Initial Designs					Variations to Initial Designs		
	Design A	Design B	Design C	Design D	Design E	Design F	Design G	Design H
Easy visibility of all programs	+	-	DATUM	S	+	+	+	+
Storage for each program details	S	S		S	S	S	+	+
Attractive and high-class	S	+		-	S	S	S	S
Compact storage	-	-		S	S	+	+	+
Light weight	-	S		S	-	+	S	S
Portable and easy to move	-	-		-	+	+	S	S
Easy to assemble	+	S		-	-	S	+	+
Low cost	-	-		+	-	S	+	+
Minimum space	-	-		+	S	S	+	+
$\sum -$	5	5		3	3	0	0	0
$\sum S$	2	3		4	4	5	3	3
$\sum +$	2	1		2	2	4	6	6

Figure 9.10 Leaflet display stand built by the students.

REFERENCES

1. Haik Y., Sivaloganathan S. and Shahin T.M., *Engineering Design Process*, 3rd edition, Cengage, Noida, India, 2017.
2. Pugh S., *Creating Innovative Products Using Total Design*, edited by Don Clausing and Ron Andrade, Addison Wesley Publishing Company, Reading, MA, 1996.

Chapter 10

Embodiment design

10.1 INTRODUCTION

Embodiment design defines the physical form to the scheme established in the conceptual design. Several methods for embodiment design have been proposed by researchers. One of them proposes to achieve this in three major steps: (1) product architecture, (2) configuration design, and (3) parametric design. The method proposed Pahl and Beitz achieves this in 13 steps and this method is adopted in this book.

The design of a device starts with the goals or the purpose for the device, as explained in chapter 5, and the actions by the user ensure the setting of the right environment for the product to produce the intended functions. The functions are the desired roles. They are the results of the behaviour of the structures or structural components. The functions are defined during the 'Product Concept' stage of the design model. Howard, Culley and Dekoninck [1] state that during conceptual design phase the designers will produce concepts which have behaviours that will satisfy the Functional Requirements and during the Embodiment Design Phase the physical structures of these concepts are realized.

Durward K Sobek [2] states that the transition between concept level design and detailed level design is characterized by exploration of and decisions about

1. what the components and subsystems are, and what their function will be
2. how the different pieces will be arranged, including location, orientation, and grouping and
3. how the pieces will connect or interface.

On a different reference Sobek [3] states that "one of the challenges we see is that system-level work does seem to involve detailed investigation, but not on everything. Good system-level work seems to entail knowing which details (the 'vital few') are important at that stage of design and investigating those thoroughly, leaving the many, less significant details until later".

DOI: 10.1201/9781003484950-12

This book refers to embodiment design stage as the bridge between conceptual design and detail design in that the functions explored and chosen in the conceptual design are translated into physical structures by the appropriate choice of form or geometry, engineering materials, and the manufacturing processes. The designed concept, as described earlier, will have a scheme built into it. In the embodiment design phase physical function providers to achieve the function sequence, are integrated together to achieve the desired goal. As Otto and Wood [4] put it, embodiment design becomes challenging because of the highly coupled nature of the parameters, which results in a sequence of changes in several parameters due to a single change in one parameter. Because of this reason several repeated calculations are carried out during the embodiment design phase. The dimensions for the structural parts are decided tentatively during this phase.

10.2 EMBODIMENT DESIGN METHOD BY PAHL AND BEITZ

Pahl and Beitz [5] proposed a 13-step method for embodiment design. The process from concept to preliminary layout takes ten steps, and the process from preliminary layout to definitive layout takes three steps. The steps suggested here are progressively refining steps and are very general. They help the designer to plan his or her work and, as such are very useful. The schematic diagram for the method is shown in Figure 10.1.

To understand the process, consider example 10.1, the design of a barrier entrance in a road as shown in Figure 10.2.

Example 10.1: Design of a barrier gate

Step 1. Using the requirements and specifications, identify those requirements that have a crucial bearing on the embodiment design – embodiment determining requirements:

a. size-determining requirements such as road width, clearance required, available backspace, etc.
    i. Road width: known (say 6 m).
    ii. Clearance required: known (say 1 m).
    iii. Backspace available: known (say 2 m)
b. arrangement-determining requirements such as direction of flow, pivot position, motion, position of barrier, etc.
    i. Pivot position: Left-hand side to the traffic flow
    ii. Motion: Rotational
    iii. Position: Horizontal.
c. Material-determining requirements such as resistance to corrosion, service life, specified materials, etc.

Step 2. A scale drawing of spatial constraints determining or restricting embodiment design must be produced at this stage: space available: known (say more than 15 m)

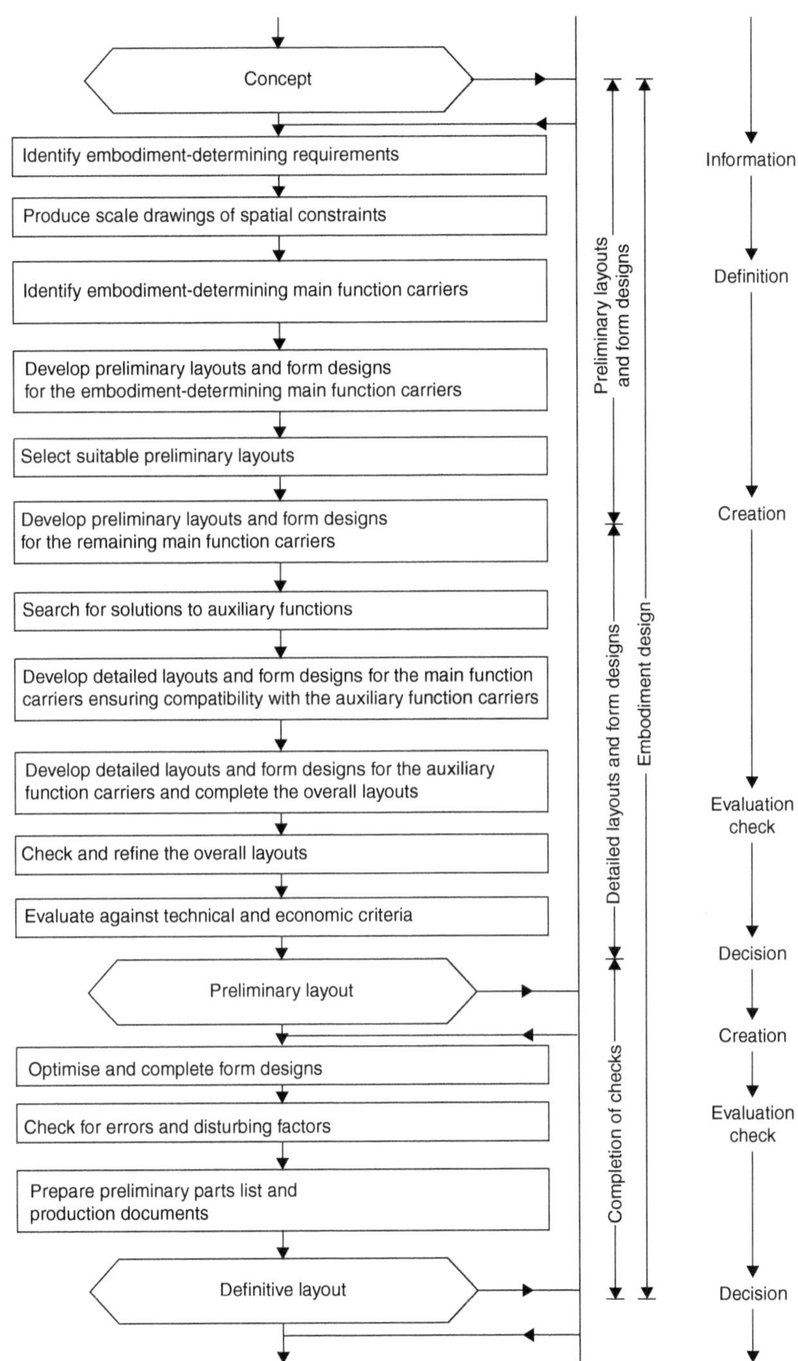

Figure 10.1 Systematic embodiment design process by Pahl and Beitz [5].

Figure 10.2 A road barrier.

Step 3. Once the embodiment determining requirements and spatial constraints have been established, the concept is used to identify the main function carriers, ie, the assemblies and components fulfilling the main functions.

a. Which main functions and function carriers determine the size, arrangement and component shapes of the overall layout.
   Lever arm, lever type (class 1)
   Maximum height at middle of the road
   Height of the pivot point
   Front length of the barrier l_1
   Back length of the barrier l_2
b. What main functions must be fulfilled, by which function carriers jointly or separately?
   Amplifying force by specified amount by lever mechanism.
   Dead weight magnitude
   Rope position

Steps 4 and 5. Preliminary layouts and form designs for the embodiment-determining main function carriers are developed at this stage. This means the general arrangement, component shapes

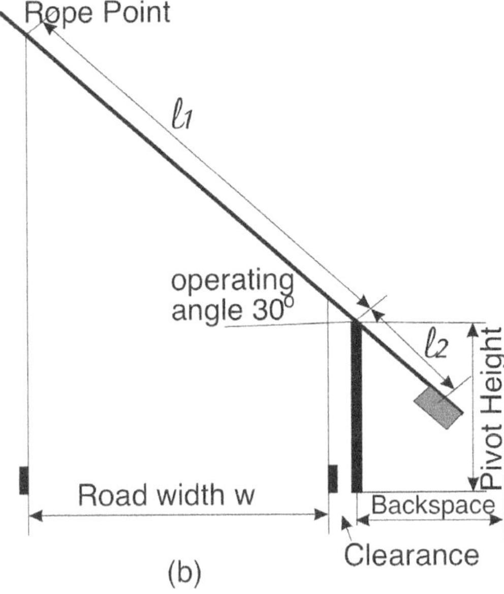

Figure 10.3 Concept of the barrier as a class I lever.

and materials are determined provisionally at this stage. The result must meet the overall spatial constraints. After developing a few ideas, a suitable one is chosen. Considering all possible options is recommended here.

a. The design starts with the initial arrangement shown in Figure 10.3. The front length of the barrier and the back length of the barrier are the first ones to be developed. With an assumption of lengths based on the following two constraints:
   i. Front barrier length/back barrier length = amplification factor
   ii. space constraint: space available, and the positions of the rope, dead weight and forces
b. The force analysis is carried out in order to find types and the range of loading. In this case, a bending moment diagram as in Figure 10.4 is found suitable.
   The design of the barrier bar is carried out by solving for the strength criteria:
   i. strength greater than or equal to the permissible stress.
c. In order that it can be solved, a set of assumptions are needed. The choice of material is one of these. By looking up the table for suitable/available section shapes, a section shape is chosen (steel pipe with outer diameter 3.5 inches and linear density 7.576 lbs/linear foot). Now the bending stress criterion is applied to find the dimensions.

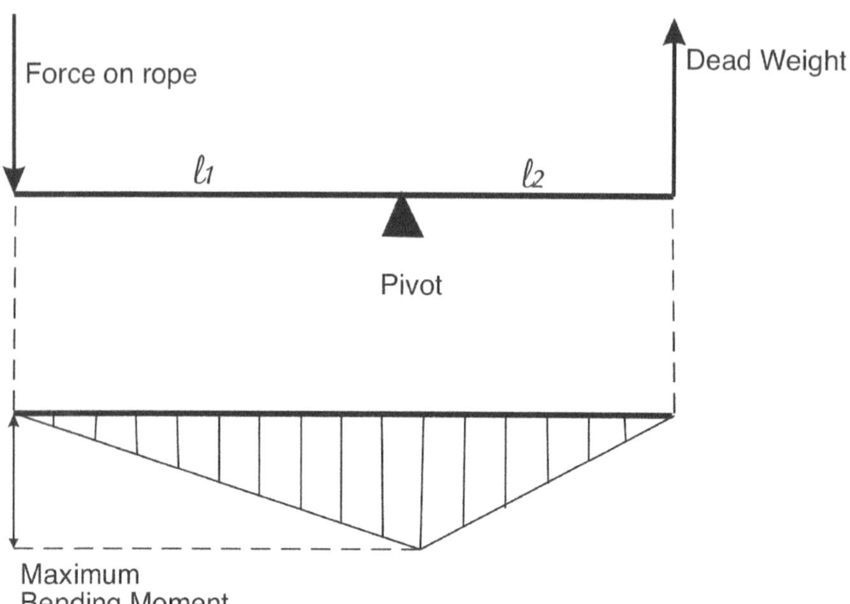

Figure 10.4 Bending moment diagram.

Step 6. The same procedure is repeated for the remaining function carriers. In this case, other main function carriers are the axle in the pivot and the pivot column.

Assuming the support column is made up of two channel irons of dimensions $126 \times 53 \times 5.5$ mm separated by 300 mm apart the layout can be fixed in the way shown in Figure 10.5. For the force calculations, the barrier bar's weight is considered.

If the length of the front barrier bar is assumed as 20 feet and the back length is assumed as 5 feet total weight is $25 \times 7.576 = 174.2$ lbs $= 86$ kg.

This is a small load, and hence further calculations were not carried out.

Step 7. Next determine what essential auxiliary functions (such as locking the axle to keep the barrier bar at centre, rest for the dead weight on ground, etc.) are needed.

Step 8. Now detailed layouts and form designs are developed for the main function carriers ensuring compatibility with the auxiliary function carriers. The designer proceeds to develop the detailed layouts and form designs for the auxiliary function carriers, adding standard and bought-out parts.

Step 9. Now check overall layouts for mistakes in function, spatial compatibility, performance, durability, safety, aesthetics, etc.

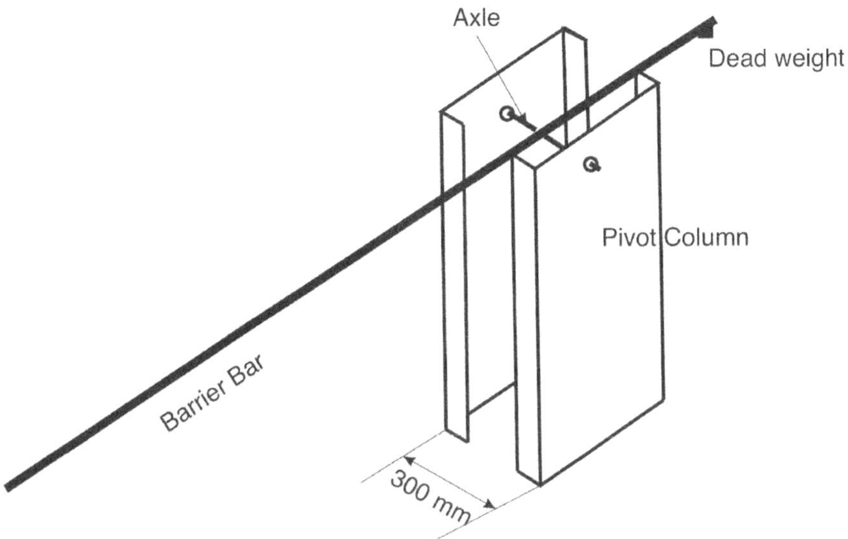

Figure 10.5 Schematic of the layout of the components.

The length constraint is re-checked against the values of restrictor pin hole diameter and pivot thickness. If it is violated, step 8 is repeated with the new value of length.

Step 10. Now check whether the design is economically viable. If not, go back to the beginning.

10.3 APPROACH OF OTHER AUTHORS FOR EMBODIMENT DESIGN

Dieter [6] explains a method incorporating three steps or components, namely (a) product architecture, (b) configuration and (c) parameter design. The architecture establishes the building blocks made up of a collection of components meant to carry out a specified function. The building blocks are connected by interfaces. The general dimensions and form of components are realized in the next configuration design. It provides modelling or sizing of parts, which strongly depends on the preliminary selection of materials, available forms of the materials and production techniques that would be used to create the components. Succeeding parametric design aims to assign values for the design variables in the configuration and are within the control of the designer, to produce the optimal design.

Sobek [3] calls a stage in between conceptual design and detail design as the system-level design. He states that it involves the exploration of and decisions about what the components and subsystems are, and what their function will be; the basic geometry of the different pieces and how they will be arranged, including location, orientation and grouping; and how the pieces will connect or interface together and with the environment.

In general, there is a stage that translates the functional scheme given by the conceptual design into defined structures that can be manufactured. In this book, the method proposed by Pahl and Beitz [5] is followed.

10.4 MODEL QUESTIONS AND SOLUTIONS

Example 10.2: Embodiment design of a step ladder

A step ladder to reach heights of about 3.5 m is needed. The customer requirements are given in Table 10.1. A conceptual design given in Figure 10.6 has been chosen as the preferred design. Develop the embodiment design.

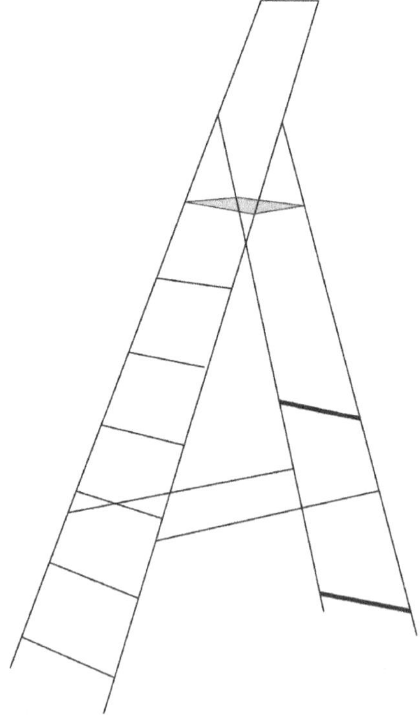

Figure 10.6 Conceptual design – sketch of the ladder.

Table 10.1 Prioritized customer needs for step ladder

Verbatim	*Need*	*Importance*
Should be able to reach the Ceiling of a house, using it	Sufficiently long	9
Should follow the guidelines on the use of step ladders	Leave the top step and platform free	9
Should have broader steps	Should have broader steps	8
Should have sufficient number of steps	Number of steps to meet the required height	8
Should have a guide rail at the top	Should have guide rail extension at the top	7
The support rail should be placed to ensure stability	Support rail should be hinged above platform	7
The rails should have shoes with sufficient coefficient of friction	Rails should have shoes with sufficient coefficient of friction	7
The ladder should be light-weighted	The ladder should be light-weighted	8

Answer

The distance between rungs is about 0.3 m, and therefore, the usable steps should be at least five for a person of about 1.5 m tall. From the list of requirements, embodiment determining requirements are the recommended practice to leave the platform and top step without use. This means the bottom steps should ascend a height of 1.5 m. The distance between rungs is about 0.3 m and therefore usable steps should be at least five. If the guard rail is about 0.6 m these together form the first main function provider, the front rail. The second main function provider is the back rail because it provides the stability to the ladder. The angles of inclination are kept as 75° and 80° as shown in Table 10.2, which shows the entire embodiment design process.

The length of the front rail $= [0.2 + 1.8 + 0.3 + 0.3] = 2.6\,\text{m}$

Vertical height from the hinge $= 2.3 \times \sin 75 = 2.2216\,\text{m}$

Therefore, the length of the rear rail $= \dfrac{2.2216}{\sin 80} = 2.2256\,\text{m}$

Other main function carrier is the step. The length is the parameter that is needed at this stage. This can be assumed to be about 450 mm to provide more stability on parallel loading (or working).

The steps followed are given in Table 10.2.

While the dimensions are considered, a quick check on the availability of the raw material in the market is necessary to ensure that manufacturing is feasible with minimum effort. Rectangular Aluminium sections for the rails are available in sizes $(35 \times 16 \times 1.4,\ 38 \times 19 \times 1.6\ \text{and}\ 38 \times 19 \times 3.3)$. Aluminium sheet metal that can be used for the steps is also available in thicknesses 3 and 4 mm.

Table 10.2 Steps followed for embodiment design

Embodiment determining requirements	Reach heights of 3 m by a 1.5 m tall person. Follow the recommended practice not to use top step and platform. Should have a guard rail.
Scaled sketch initiation	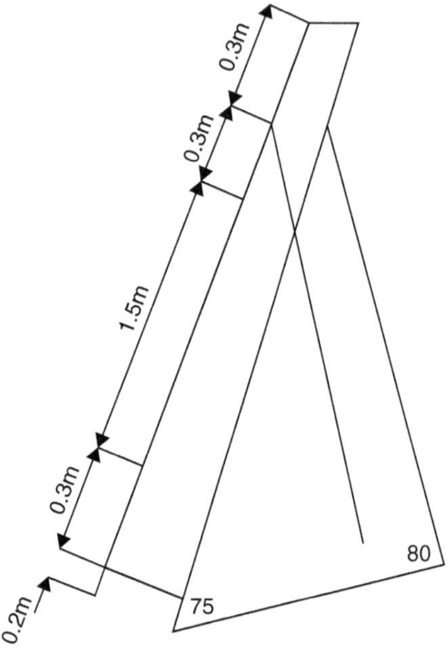
Embodiment determining main function carriers	Front rail including the guard rail (0.3 m from the hinge) Rear rail
Develop Preliminary layouts for the above main function carriers	Front rail has options to change the spacing of rungs to 0.25, 0.3 and 0.35 m. Also, it can have an extended guard rail. Other very seldom-used option is to have foldable side supports. Rear rail takes suitable dimensions for the angles specified. The rails can be made out of rectangular sections from the market.
Choose preliminary layout for the above function carriers.	Chosen layout is shown in the scaled drawing in (ii) above.
Develop preliminary layouts for main function carriers remaining	This includes the rungs and ties in the rear rail. The form for these can be rectangular sections with suitable supports at the rails. Another main function carrier is the ties that are tied between the front and rear rails.
Auxiliary function carriers	The top platform, the shoes at the bottom of the rails and the interfaces between, the rungs and the rails, fall in this category.

(*Continued*)

Table 10.2 (Continued) Steps followed for embodiment design

Evaluate for technical and economic criteria	At this point, the use of standard material sections available in the market is the main technical concern. Hand calculations on stresses at the interfaces and overall stability calculations are also carried out at this point. Design for manufacture is the main focus here.
Complete preliminary Layout.	At this point a complete physical structure should become available (Figure 10.7) and checking that, is the main target here.

The embodiment design for the step ladder is shown in Figure 10.7. No additional optimization was needed in this design and the parts list are shown in Table 10.3.

The product can now be modelled with all details needed. However, proving of the product needs some engineering analyses and thus this is an Engineering Design problem. Stability, strength of the front rail, rear rail, step and ties have to be calculated at the worst loading conditions and the stresses at critical points should be proven to be within acceptable range. These calculations will be carried out at the detailed design stage, which is part of the detail design stage.

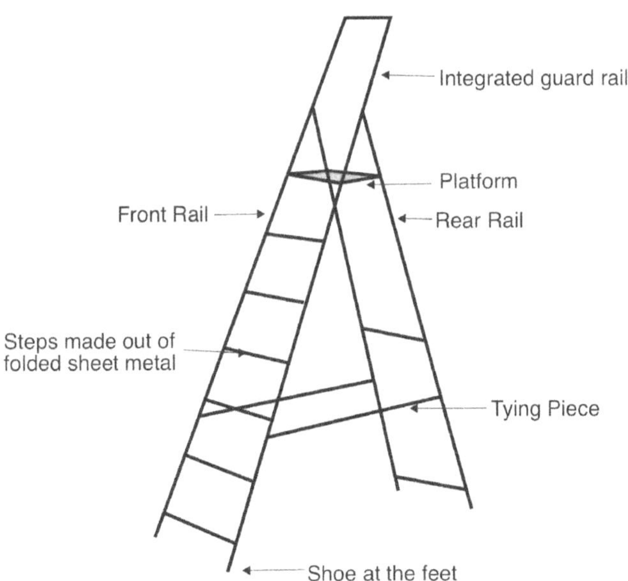

Figure 10.7 Labelled sketch – embodiment of the ladder.

Table 10.3 Parts in the step ladder

Part name	*Details*
Front rail	Required length is around 2.6 m.
Steps	Six steps plus the platform to meet the recommendation
Shoe at the feet	Suit the cross section
Guard rail	This may be integrated with the front rail pair
Platform	Correct dimension can be found during modelling
Rear rail	Required length is around 2.25 m
Tying piece	Depends on the fixing height

Example 10.3: Paperweight for a company

A company wishes to design and make a few paperweights for their use. They prefer it to be made of aluminium using CNC machining. They have collected the requirements and developed several concepts. The chosen conceptual design is shown in Figure 10.8. The prioritized needs are given in Table 10.4. Develop the embodiment design.

Figure 10.8 Conceptual design of a paperweight.

Table 10.4 Prioritized customer needs for paper weight

Verbatim	*Need*	*Importance*
Should be heavy to keep papers	Reasonable weight	9
Well thought-out quotations	Text should state admirable behaviour or character	9
Flat shape holdable by hand to accommodate paper	Hold paper through a wide area easily pickable by hand	8
Should be attractive	Should be bright and attractive	8
Should reflect the character of the company	Reflect the character of the company	8
Should be unique	Should be rare in the class	8
Should have illustrations (Figures etc)	Should have inspiring illustrations	7
Made up of attractive material	Attractive material	7

Answer

The first step in embodiment is to identify the embodiment determining requirements. In this example, they are (a) flat shape holdable by hand to accommodate the paper and (b) the quoted text to be accommodated within the space available (Figure 10.9, Table 10.5).

The stork is modelled in the following way. The sketch was made on a graph paper as shown in Figure 10.10. The body of the stork was divided into curve segments by introducing node points A, B, C, - - -J to represent the body. A composite curve KLMNOPQ represents the leg of the stork (Table 10.6).

Composite curve KLMNOPQ passes through the vertices K, L, M, N, O, P and Q. Their coordinates are K(29, 24), L(29, 18), M(20, 15.5), N(26, 14), O(32, 24), P(30,27.5) and Q(30, 23.5). This can be modelled using Bezier curves or B spline curves.

The above details are sufficient to produce the finished embodiment design of this product. Since this is a product design problem no engineering analyses are needed and production drawings can be produced straight away. Table 10.5 summarizes the steps followed in the embodiment process.

Figure 10.9 The completed design.

Table 10.5 Steps followed for embodiment design

Embodiment determining requirements	Flat shape holdable by hand to accommodate paper Quoted text should state admirable behaviour or character
Scaled sketch initiation	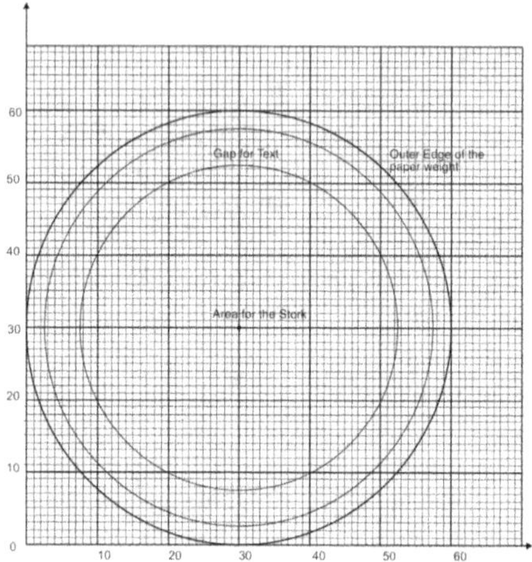
Embodiment determining main function carriers	The main function carrier is the cylindrical block with large diameter to (i) accommodate the paper and (ii) Sufficient area to accommodate the text.
Develop Preliminary layouts for the above main function carriers	A cylindrical block with about 30mm radius and about 15mm height would be easily holdable by hand. The area for the text can have height for text between 5 and 10mm.
Choose preliminary layout for the above function carriers	Chosen layout is shown in the scaled drawing in (ii) above. The text chosen is in two parts: (i) Wait like a stork for opportunities and (ii) strike like a stork at opportunities.
Develop preliminary layouts for main function carriers remaining	This is the area for the illustration of the stork. The stork was drawn as a collection of composite curves as explained later.
Auxiliary function carriers	In this design the area between the outer edge and the outer guideline for text can have a small dip say about 2–3 mm with different colour to make the text prominent. In the same way, background colouring for other areas also can be considered.
Evaluate for technical and economic criteria	At this point material and associated manufacturing process have to be considered. Brass block in CNC machining is chosen for this product.
Complete preliminary layout	At this point a complete physical structure should become available and checking that, is the main target here.

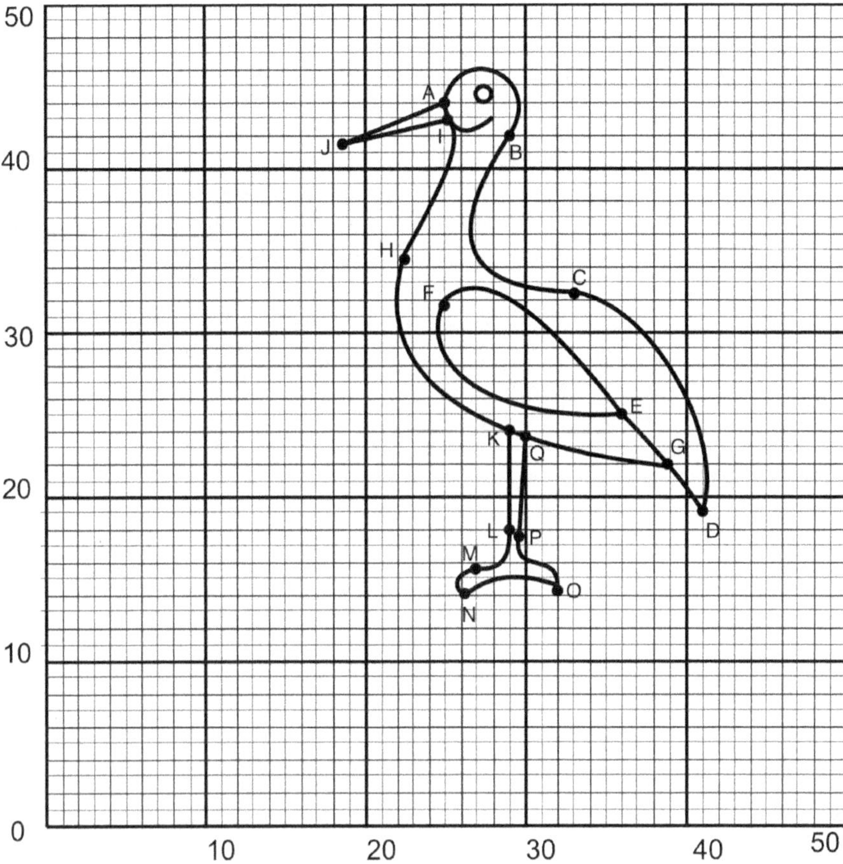

Figure 10.10 Processing the stork.

Table 10.6 Curve segments and chosen points

Segment	*Starting and finishing points*	*Intermediate points*
AB	A(25, 44) and B(29, 42)	(27, 46) and (29.5, 44)
BC	B(29, 42) and C(33, 32.5)	(27,38) and (29, 33)
CD	C(33, 32.5) and D(41, 19)	(38, 29) and (40.5, 35)
DE	D((41, 19) and E(36, 25)	Straight line
EF	E(36, 25) and F(25,32)	(29,32) and (33.5, 28)
FE	F(25, 32) and E(36, 25)	(26.5, 27) and (30, 25.5)
GH	G(39, 22) and H(22.5, 34.5)	(23, 27) and (32, 23)
HI	H(22.5, 34.5) and I(25, 43)	(24, 38) and (25, 40)
IJ	I(25, 43) and (18.5, 41.5)	Straight Line
JA	J(18.5, 41.5) and A(25, 44)	Straight line

Example 10.4: Embodiment design of a single-stage scissor lift

Figure 10.11 shows the sketch of the structure of a single-stage scissor lift. One kind of powering the lift is to drive the power link in the to and fro motion. Two equal links are connected by a midpoint pin to form a scissor. Two such scissors carry the platform on which a person goes up and down. One end of one link of the scissor is connected to the base frame at the pivot point about which it can rotate. This link is called the fixed link. The other link is called the free link and it is connected to the platform through a pin about which it can pivot. The free end of the free link slides over the frame when the power link moves to and fro. Labelled members in Figure 10.11 explains the concept of the scissor lift.

When the free ends of the free links of the two scissors are at the right-hand end, the platform is at the lowest position. When the power link moves towards the left, the platform will start going up until the power link reaches its limit.

Consider Figure 10.12 when the fixed link or lever makes the minimum angle with the base frame. The platform is at its lowest position in position 1. When the angle increases, the lever takes intermediate positions and the platform moves up to higher position like position 2.

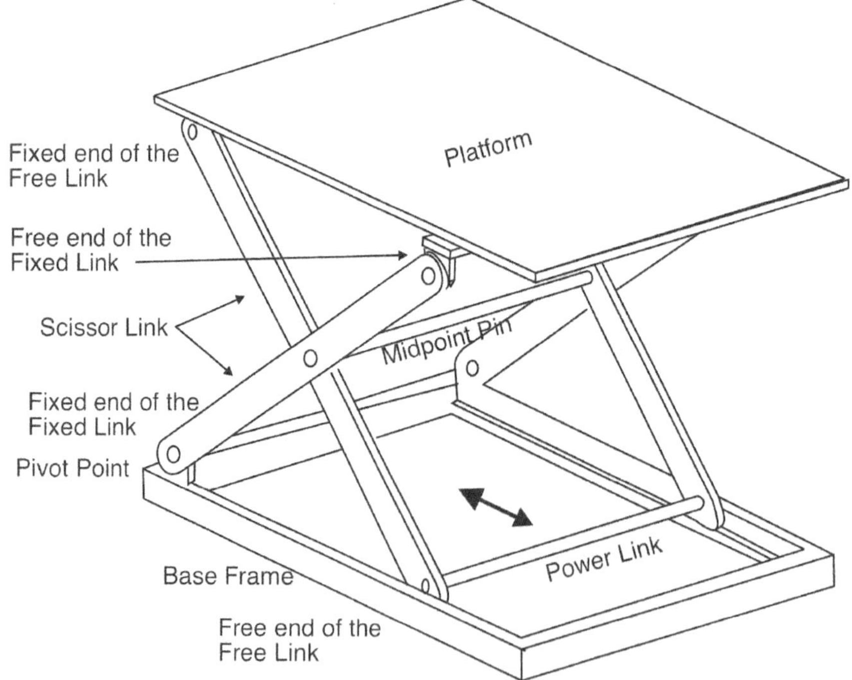

Figure 10.11 Single-stage scissor lift.

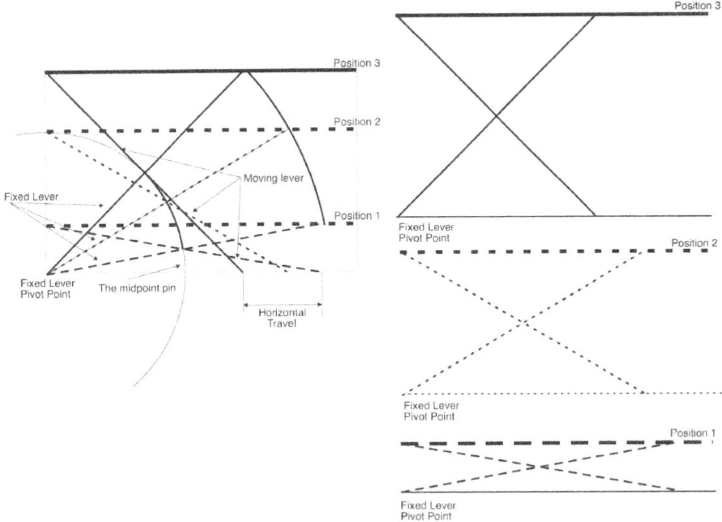

Figure 10.12 Platform position at various rotations.

When the angle increases to the maximum, the platform takes the highest position, position 3. Also, the free end of the free link travels towards the left. The figure shows the horizontal travel. The figure also shows the locus of the midpoint pin and the end point of the fixed lever as two arcs.

The design of such a lift is the task in hand. The lift should give height rise of 1.0 m from its stowed position. The prioritized customer needs are given in Table 10.7 shows the prioritized needs.

Table 10.7 Prioritized customer needs for a single-stage scissor lift

Verbatim	*Need*	*Importance*
The stowed height of the lift should be reasonable	Low stowed height of the platform	9
Should be stoppable at intermediate positions	The drive should hold at intermediate positions	9
A 1.5 m person should be able to reach ceiling of a house	Net travel should be around 1 m	8
Should be able to take in lifts in domestic houses	Should be able to get into lifts of 1.5 m × 1.5 m	8
Should have no possibility of oil spills	Should have no hydraulic mechanisms	8
Capacity of the lift should be around 200 kg	Capacity of the lift should be around 200 kg	8
The platform should be non-slippery	Non-slippery platform surface	7
Power driven from national electricity grid	Use single-phase electricity to power the lift	7

Answer

The input for embodiment design is the conceptual design. One of the requirements is 'Should have no possibility of oil spills' and hence powering should exclude hydraulic drives. Accordingly, a power screw is chosen to drive the power link.

Embodiment determining requirements are

i. Net travel should be around 1 m
ii. Should be able to get into lifts of 1.5 m×1.5 m
iii. Low stowed height

The main function provider for these requirements is the length of the scissor link.

Consider the link in position 1 as shown in Figure 10.13.

Stowed height = (L Sin θ + Height of wheel and accessories) where θ takes the minimum value of say 10°.

Similarly, the position of the free end of the free link = $L \cos 10$

Now consider the highest position of the platform, Position 3 shown in Figure 10.14.

Maximum height = (L Sin θ + Height of wheel and accessories) where θ takes the maximum value of say 50°.

Similarly, the position of the free end of the free link = $L \cos 50$ (Table 10.8)

No additional optimization was needed in this design. Detailed design involving engineering calculations is needed for this Engineering Product design.

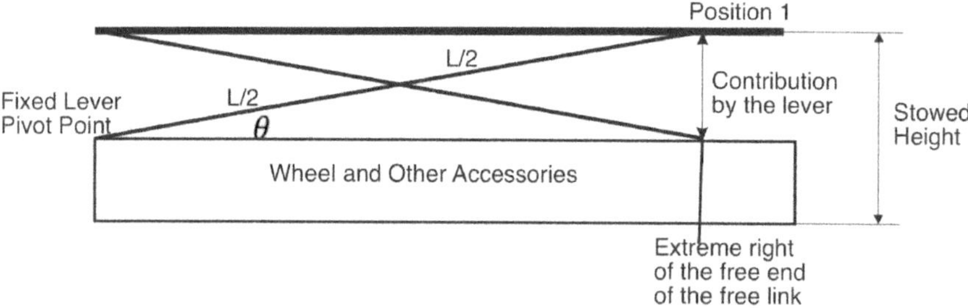

Figure 10.13 Stowed position of the lift.

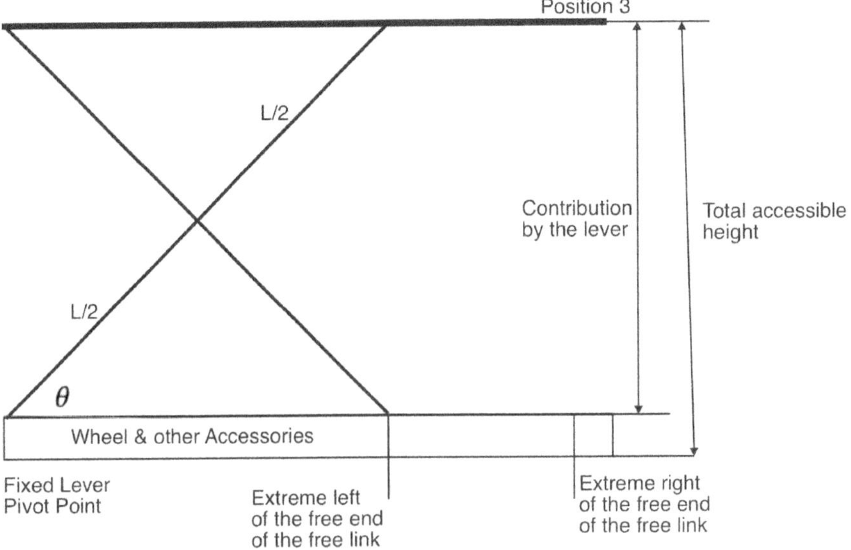

Figure 10.14 Maximum position of the lift.

Table 10.8 Steps followed for embodiment design

Embodiment determining requirements	i. Net travel should be around 1 m ii. Should be able to get into lifts of 1.5 m × 1.5 m iii. Low stowed height
Scaled sketch initiation	Refer Figures 10.12, 10.13 and 10.14.
Embodiment determining main function carriers	The embodiment determining function carriers are i. The scissor lever. ii. Provision for frame and wheel
Develop Preliminary layouts for the above main function carriers	From the calculations above Total height = Contribution by scissor link + Stow height $= L\,\sin 50 +$ wheel and accessories Assuming wheel and accessories = 100 mm and $L = 1{,}000$ mm gives total height as 866 mm. If $L = 1{,}200$ mm total height achievable = (919 + 100) mm The length of the platform can be assumed to be 1,200 mm the length of the lever.

(Continued)

Table 10.8 (Continued) Steps followed for embodiment design

Choose preliminary layout for the above function carriers.	From the above calculations let us choose $L = 1{,}200$ mm and the width of the frame as 800 mm. The platform should be greater than $1{,}200 \times \cos 10 = 1{,}118$ and let it be 1,200 mm as well.
Develop preliminary layouts for main function carriers remaining	The next main function carrier is the powering mechanism, the power screw. The length of the travel is $1{,}200(\cos 10 - \cos 50) = 410$ mm. Let it be 450 mm. The motor gearbox, bearing housings should be accommodated within the remaining 750 mm of the base.
Auxiliary function carriers	The wheels, frame, etc are the auxiliary function carriers.
Evaluate for technical and economic criteria	At this point, material and associated manufacturing process have to be considered. Aluminium extrusions can be the material and welding is the main manufacturing process.
Complete preliminary layout	At this point a complete physical structure should become available and can be modelled in the detail design part. Checking the details is the requirement to complete the preliminary design part of the embodiment design.

REFERENCES

1. Howard C. and Dekoninck H., Describing the creative design process by the integration of engineering design and cognitive psychology literature, *Design Studies*, 29(2), 2008, 160–180
2. Sobek II D.K., Transitions: From Conceptual Ideas to Detail Design. In: *Proceedings of the 2005 American Society for Engineering Education Annual Conference & Exposition, 2005,* https://www.montana.edu/dsobek/career/documents/ASEE05_final.pdf
3. Sobek II D.K., System level design: a missing link? *International Journal of Engineering Education*, 22(3), 533–539, 2006.
4. Otto and Wood, *Product Design: Techniques in Reverse Engineering and New Product Development*, Pearson, 1991.
5. Pahl G. and Beitz W., *Engineering Design*, The Design Council, London 1984.
6. Dieter G., *Engineering Design*, McGraw Hill, New York, 1983.

CONCLUDING REMARK FOR PART 2

Part 1 of this book has explained how to define the design problem well through the specifications. To develop the specifications, it started with the design brief and established the types of needed requirements, customer verbatim, prioritized needs, need-metric matrix and the functional representation.

In Part 2, various conceptual design methods have been explained with examples on developing conceptual designs. The objective here is to find an optimal or near-optimal solution from the unknown solution space and develop it as a complete design. The strategy adopted is to generate a sample consisting few sample designs, evaluate and establish that they are high-grade designs and then choose one of them for further development. Following this embodying the chosen concept was explained in detail. Parts 1 and 2 together provide adequate exposure and experience in handling activities at different stages of the systematic design process. The remaining part is to experience how to plan and carry out a sequence of activities to start and complete a design project. An English proverb reads 'A good example is the best sermon'. Part 3 provides complete examples to experience the full weight of some simple but complete projects. Move in and experience them.

Part 3

Complete design projects

Chapter 11

Complete design projects

11.1 INTRODUCTION

This chapter describes the integrated experiences in product development and engineering product development. These projects were done as part of the Product Development course in the Master of Engineering Management program and in the Design and Manufacturing Lab course in the Mechanical Engineering undergraduate program. The Chapters in Part 1 covered the steps to define the problem well. In part 2, generation of a sample of designs from the solution space for that problem, evaluation of the proposed solutions, choice of one of those solutions, and development of it further with the embodiment design, have been described. As mentioned in the concluding paragraph of part 2, product design needs geometry definition to prepare the CAD model and the production drawings, and engineering design needs engineering analyses from various subject domains to develop and prove the design. The breadth and depth of domains for engineering products are very wide and are problem dependent. Also, the quantity of details that can be considered for detailed design of engineering products is huge. Accordingly, the objective of the examples here is to explain the methodology only, and the reader is advised to refer subject materials for in-depth details of the analyses.

11.2 READING ASSISTANT EXAMPLE

Reading good books is like having conversation with the finest persons of the past. In the journey of discovering something no one knew before, research and reading, guide the formulation of the destination, and reading continues to be widely used in the research-oriented world. In reviewing a topic, the researcher often requires access to several books almost at the same time. If such access can be made without the necessity of getting up from the seat, it will greatly enhance the review process. A 'Reading Assistant' can keep several books in open condition and allow instant and easy access to any of the books, without getting up from the seat by the reader. Developing such a 'Reading Assistant' that can handle up to six open books is the task in this product design example and its design brief is given in Table 11.1.

DOI: 10.1201/9781003484950-14

Table 11.1 Design brief for a 'reading assistant'

Reading assistant	
Product description	The product is a simple, lightweight, and portable mechanism that enables users to easily switch between several books. It keeps all the current set of books in open position which facilitates note taking and eliminates the need for frequent opening and closing of books and walking towards the shelf to pick a different book.
Product concept	It should be an object that has several planar faces each of which can support a single open book. Ideally it is an object standing next to the seated reader.
Benefits to be delivered	Holding up to six books Books at convenient height Attractive and unique Portable Should be able to have other uses
Positioning & target price	Upper quartile of the market with a price tag of $250–$500
Target market	University academic staff Research students Reference libraries
Assumptions and constraints	CAD/CAM and CNC facilities are available for the design and manufacture.
Stakeholders	University academic staff Research students Reference libraries
Possible features and attributes	No restriction on the foot space on floor
Possible area for innovation	Space providers for keeping several single books

Drafted by Dr Sangarappillai Sivaloganathan

11.2.1 Design process plan

A widely circulating saying goes 'A Good Carpenter always cuts twice'. The saying says that he is a good carpenter and continues to say that he always cuts twice. The first cut is made in his head, i.e., he plans every cut in detail. If the saying is accepted for design, it would read as 'a good designer would first plan and then executes the plan to execute the design'. Figure 11.1 shows the design plan for the design of a reading assistant.

11.2.2 Establishing the needed types of requirements

After establishing the design process plan, the first step was to identify the types of requirements that would enhance the design team to achieve a

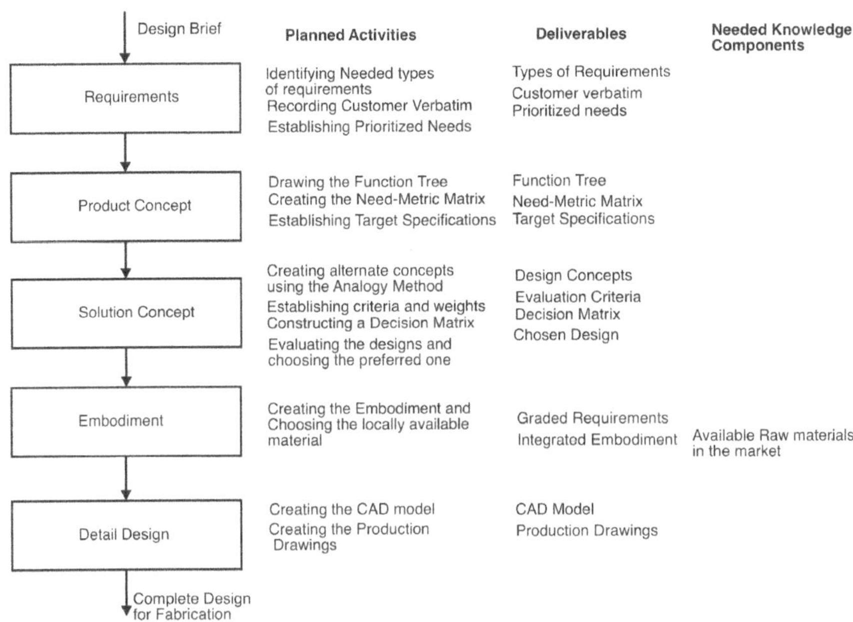

Figure 11.1 Design process plan for the reading assistant.

product that would attract customers. The types of relevant requirements identified by referring to the checklist by Pahl and Beitz [1] are given in Table 11.2.

Table 11.2 Types of needed requirements

Type	*Description*
Geometry	Size, height, breadth, length, diameter, space requirement, number, arrangement, connection, and extension
Kinematics	Type of motion, direction of motion, velocity, acceleration
Forces	Direction of force, magnitude of force, frequency, weight, load, deformation, stiffness, elasticity, stability, resonance
Ergonomics	The man–machine relationship, type of operation, clearness of layout, lighting
Assembly	Assembly Special regulations, installation, siting, foundation, transport limitations due to lifting gear, clearance, means of transport (height and weight), nature and conditions of dispatch
Materials	Materials Physical and chemical properties of the initial and final product, auxiliary materials, prescribed materials (food regulations, etc.)
Maintenance	Servicing intervals (if any), inspection, exchange and repair, painting, cleaning

Table 11.3 Stakeholder verbatim

Geometry	
	Book holder should be able to accommodate an A4 book
	Base should be sufficiently large
	The book should be accessible by a seated person
Kinematics	The books should be rotatable for easy access
Forces	Holding the book on the presenter should be easy
Ergonomics	The RA should be a stand-alone unit to stand by the reader
	Easy to rotate and access the book
	Should be able to hold up to six books
Assembly	The RA should be easy to assemble
	Should be moveable
Materials	Material should be durable
Maintenance	Minimum maintenance or maintenance free

11.2.3 Stakeholder interviews and verbatim

Stakeholders' requirements can be obtained by questionnaires, individual interviews, focus group meetings and observing an object in operation. In this project, individual stakeholders were interviewed one by one and their verbatim were recorded. Table 11.3 contains the identified types and the stakeholders' verbatim.

11.2.4 Establishing prioritized needs

Normally stakeholder verbatim have to be translated into needs by the design team so that the information content of all needs is somewhat similar and technically clear. But the stakeholders' verbatim in this case are simple and uniform and they can be treated as needs. The interviewing team has assigned the importance ratings for the needs as shown in Table 11.4.

Table 11.4 Prioritised needs

No	*Needs*	*Importance*
1	Book holder should be able to hold an A4 book	9
2	Base should be sufficiently large	9
3	The book should be accessible by a seated person	7
4	The books should be rotatable for easy access	7
5	Holding the book on the presenter should be easy	6
6	The RA should be a stand-alone unit standing nearby.	8
7	Easy to rotate and access the book	9
8	Should be able to hold up to six books	8
9	The RA should be easy to assemble	9
10	Should be moveable	9
11	Material should be durable	7
12	Should be maintenance free	7

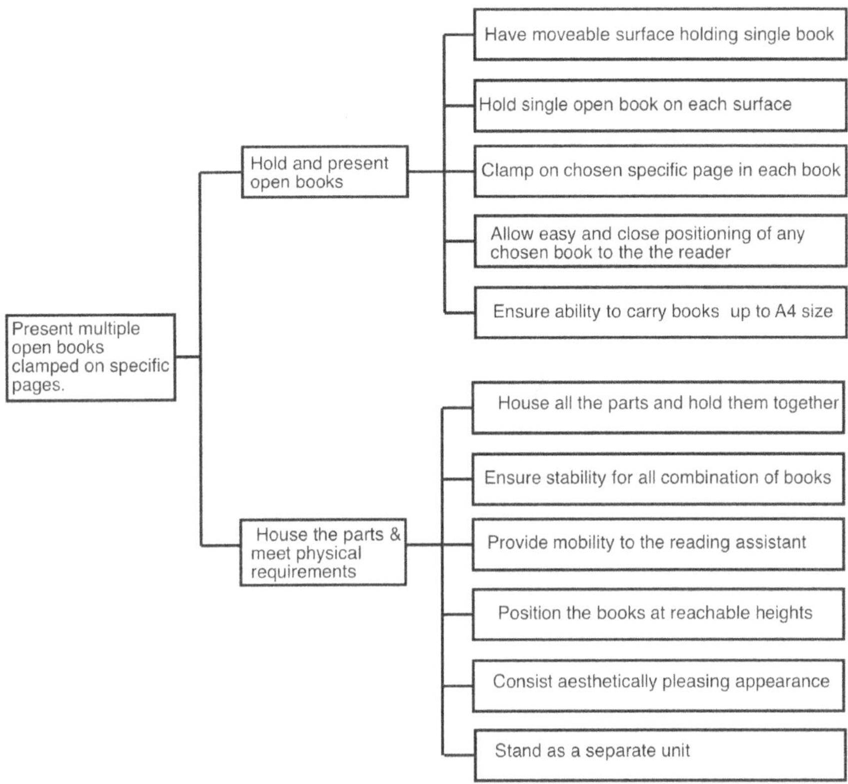

Figure 11.2 Function tree of a reading assistant.

11.2.5 Function tree

The objective or goal of the reading assistant is to present multiple open books clamped on specific pages. To achieve this goal two purpose functions are needed: (i) hold and present the open books by several presenters and (ii) house all the parts and dimensional or physical requirements. Each purpose function is realized by a set of action functions. The function tree is shown in Figure 11.2.

11.2.6 Need-metric matrix

Twelve needs are identified and given in Table 11.4. In order to draw specifications deploying them they should be converted into measurable quantities or metrics. Need-metric matrix is the mechanism used to identify the metrics and their units. Table 11.5 shows the need-metric matrix developed for this purpose.

Table 11.5 Need-metric matrix

		Metric	Presenter Length & width	Base Length & width	Distance to pivot point	Rotatable presenter	Clamping Clips	Separable and stand alone	Easy Rotating ability	Number of book holders	Number of Operations	Easy to move	Chosen Material	MTBF
		Units	mm	mm	mm	Yes/No	Yes/No	Yes/No	Yes/No	Number	Number	Yes/No	List	Number hrs
1	Should be able to accommodate A4 books		■											
2	Base should be sufficiently large			■										
3	Should be accessible by a seated person				■									
4	The books should be rotatable for easy access					■								
5	Holding the book on the presenter should be easy						■							
6	Should be a stand-alone unit to stand by the reader.							■						
7	Easy to rotate and access the book								■					
8	Should be able to hold up to 6 books									■				
9	The RA should be easy to assemble										■			
10	The RA should be moveable											■		
11	Material should be durable												■	
12	Should be maintenance free													■

Table 11.6 Target specifications of the reading assistant

No	*Specification*	*Value*
1	Presenter dimensions	>210 × 297
2	Base Dimensions	<500 × 500
3	Pivot point of book within base	Yes
4	Rotatable presenter	Yes
5	Clamping clips in presenter	Yes
6	Stand-alone unit	Yes
7	Easy rotation to access book	Yes
8	Number of book presenters	6
9	Steps to assemble	<5
10	Easy moving	Yes
11	Chosen Material	List
12	Maintenance (high MTBF)	>4,000 hours

11.2.7 Target specifications

Target specifications are formed by considering the design brief, function tree and mainly the stakeholder needs and the need-metric matrix. The target specifications for the reading assistant are given in Table 11.6.

11.2.8 Conceptual designs

The concepts developed for example 8.23 can be embodied to make a stand-alone reading assistant. The designs are shown again in Table 11.7.

11.2.9 Evaluation criteria and weights

To assign weights for the criteria they are arranged in the descending order of importance. The most important criterion is assigned a value equal to the number of criteria and each criterion is assigned a value in the descending order as shown in Table 11.8. The assigned values are divided by the sum of the assigned values to get the weights for each criterion. The criteria from the design brief and important requirements are as follows:

a. Holding up to six books
b. Books at convenient height
c. Attractive and unique
d. Portable (lightweight)
e. Should be able to have other uses
f. The RA should be easy to assemble
g. Should be moveable

Table 11.7 Conceptual designs

Base Product	Conceptual Design

Table 11.8 Calculation of weights using ordering method

No	*Criterion*	*Assigned value*	*Weight*
1	a	7	7/28
2	b	3	3/28
3	c	5	5/28
4	d	4	4/28
5	e	2	2/28
6	f	1	1/28
7	g	6	6/28
Sum		28	28/28

11.2.10 Constructing the decision matrix

The designs were considered one by one for fulfilment of each criterion and a score between 1 and 10 is given. The given scores are recorded in the upper triangles in the row allocated for the design. For example, the lamp design has $\begin{bmatrix} 10 & 8 & 7 & 6 & 0 & 7 & 6 \end{bmatrix}$ as the scores. The weighted score is obtained by multiplying the score by the weight. This is stored in the bottom triangle for each criterion. The last column shows the sum of all weighted scores in a row belonging to each of the designs. The Decision matrix is shown in Figure 11.3.

11.2.11 Evaluation and concept selection

The total weighted scores are 196/28, 240/28, 148/28 and 226/28 for the designs. The maximum possible score is 10, and thus two designs are in the very good category (greater than 7) while two designs are moderate.

	a	b	c	d	e	f	g	
	7/28	3/28	5/28	4/28	2/28	1/28	6/28	
Lamp	10 / 70/28	8 / 24/28	7 / 35/28	6 / 24/28	0 / 0/28	7 / 7/28	6 / 36/28	196/28
Rubber Plant	10 / 70/28	7 / 21/28	8 / 40/28	8 / 32/28	10 / 20/28	9 / 9/28	8 / 48/28	240/28
Pyramid	0 / 0/28	8 / 24/28	9 / 45/28	7 / 28/28	0 / 0/28	8 / 8/28	7 / 42/28	148/28
Drawyer	10 / 70/28	6 / 18/28	7 / 35/28	7 / 28/28	10 / 20/28	7 / 7/28	8 / 48/28	226/28

Figure 11.3 Decision matrix for reading assistant.

The pyramid and lamp have the criterion b, books at convenient height, better than the rubber plant design. This relates to the nature of the rubber plant and therefore cannot be improved as well. In a similar fashion, attractiveness and uniqueness cannot be improved. Hence, the chosen design is the rubber plant design.

11.2.12 Embodiment design

A column with journal bearings carrying plates for holding the book is the concept. The embodiment determining requirement for the plate is 'Ability to accommodate A4 books.' A size of 350 mm × 250 mm is suitable for the plate (Figure 11.4).

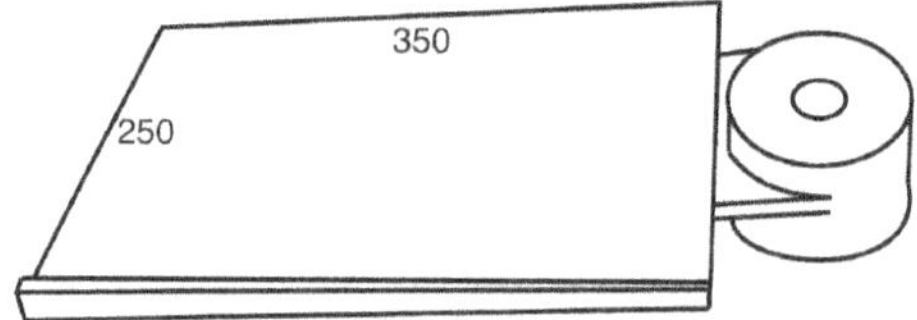

Figure 11.4 Book holder.

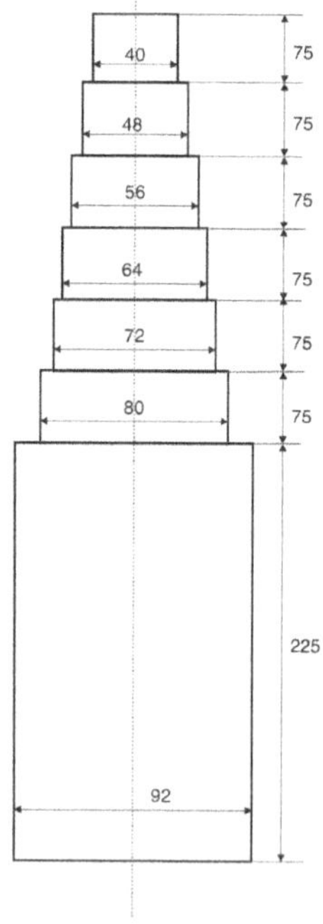

Figure 11.5 The column or shaft of the reading assistant.

'Should be able to hold up to six books' is the embodiment determining requirement for the column. It has six steps each having a height of 75 mm. Total height of the column is 675 mm to keep the height acceptable to a seated user. Figure 11.5 shows the embodiment of the column or shaft.

Stability when having any number of books is the embodiment determining requirement for the base. The holding plate is 350 mm × 250 mm in size. The plate will be just touching the bearing and this means the centre of gravity of the plate will be (175 + 50) mm away from the axis of the column. A square base with 500 mm sides will ensure stability in holding 1–6 books.

11.2.13 Manufacturing and finished product

Manufacturing the product involves machining the column, bushings and the arms and cutting the wooden base and presenters. The finished product is shown in Figure 11.6.

Figure 11.6 Finished reading assistant.

11.3 DISPLAY BOARD/FLOWER VASE STAND – EXAMPLE

Halls in big cities and towns host big conferences and, other smaller leisure activities and events. Normally they have several rooms and during a conference, parallel sessions of presentations take place in these rooms. The participants may cross between them to attend specific papers from more than one session. To facilitate their quick access, a display board in front of each room is needed. When small functions take place in these centres, only a small number of rooms will be in use. The display stands should be convertible as a flower vase stands in these circumstances. This project is concerned with the design of a convertible flower vase/display stand using an adjacent design as the starting point for conceptual design. The design brief is given in Table 11.9.

Method starting with the adjacent design: In this method, a product is visualized as a point in an n-dimensional hyperspace where each dimension is a feature. Then it is possible to identify the n number of desired features and a corresponding n-dimensional hyperspace of the desired or acceptable design. If this hyperspace and the feature vector of the unacceptable adjacent solution are compared, two things, (i) the unacceptable features present in the solution and (ii) desired features that are absent in the solution, can be identified. Now it is possible to adopt and produce concepts that have (i) none of the unacceptable features and (ii) all or most of the desired features that were absent in the original unacceptable solution with which the search was started.

The steps for this method for conceptual design are as follows:

i. Establish the n-dimensional hyperspace for the proposed product
ii. Identify the nearest product as an unacceptable solution and establish its features
iii. Compare the two and identify the unwanted features and remove them from the design of the unacceptable product
iv. Identify the mediocre features and improve them
v. Identify the missing features and add to enhance the product

Having thus seen the method the steps can be applied in the conceptual Design.

11.3.1 Design process plan

To achieve the design following the systematic method a design process plan was developed. It is shown in Figure 11.7.

Table 11.9 Design brief for a display board/flower vase stand

Display board/flower vase stand	
Drafted by	*Dr Sangarappillai Sivaloganathan*
Product description	A simple, lightweight, and portable appliance that enables the use of it as a display board for A3 size posters or the stand for a flower vase. Such an appliance would be useful in big conference halls where there can be several parallel sessions.
Product concept	An attractive artefact, that reflects features of the local surroundings and culture, that can be used as a display board mount or a flower vase stand as required.
Benefits to be delivered	Enable easy setting of it as a display board mount Enable easy setting it as a flower vase stand Provide storage for accessories Easily portable
Positioning & target price	Upper quartile of the market with a price tag of $250–$500
Target market	Conference halls in cities
Assumptions and constraints	CAD/CAM and CNC facilities are available for the design and manufacture.
Stakeholders	Conference hall management Conference hall maintenance staff Conference attendees Welfare officers in Councils
Possible features and attributes	Detachable mounting units for display board Permanent feature to function as flower vase stand
Possible area for innovation	Innovative manufacturing methods

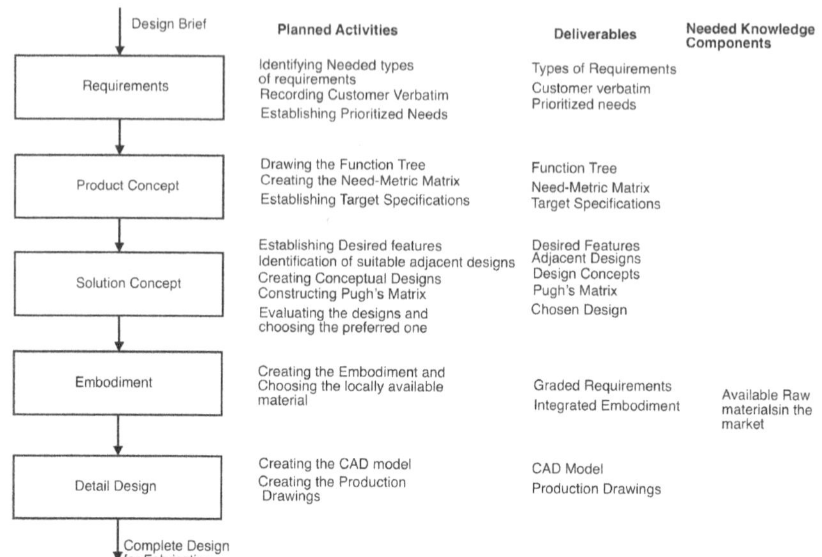

Figure 11.7 Design process plan for the display board/flower vase stand.

From the process plan the activities to be carried out are

1. Identifying needed types of requirements
2. Recording customer verbatim
3. Establishing prioritized needs
4. Drawing the function tree
5. Creating the need-metric matrix
6. Establishing target specifications
7. Establishing the desired features/criteria
8. Identification of suitable adjacent designs
9. Creating concepts using the method 'Starting with an adjacent unacceptable design'
10. Constructing Pugh's matrix
11. Evaluating, improving and choosing
12. Creating embodiment
13. Creating the CAD model
14. Manufacturing the finished product

11.3.2 Identifying the types needed of requirements

To identify the types of requirements that would enhance the design team to achieve a product that would attract customers, the types of needed requirements were identified by referring to the checklist by Pahl and Beitz [1]. The identified types are shown in Table 11.10.

Table 11.10 Types of needed customer requirements

Main group	*Sub-groups*
Geometry	height, breadth, length, space requirement, arrangement, extension
Forces	weight, load, and stability
Materials	Materials Physical properties of the material, properties of the final product
Safety	operational, operator and environmental safety
Ergonomics	Type of operation, clearness of layout
Production	Preferred production methods, means of production
Operation	special uses
Maintenance	Servicing intervals (if any), painting, cleaning
Cost	Costs maximum permissible manufacturing costs

Table 11.11 Types of needed customer requirements

Main group	*Customer verbatim*
Geometry	Should be able to display A3 posters Should be able to hold a vase of height of 400 mm The height of the poster should be around eye level Breadth and length could be around 500–700 mm There should be sufficient space to store needed auxiliaries Display and flower vase stand arrangement should be easy
Forces	The appliance should have lightweight Appliance should be stable in both arrangements
Materials	The material should be beautifully grainy or stranded. The final product should have an attractive surface
Safety	Should be easy to convert as display or vase stand Should be easily moveable from place to place
Ergonomics	The layout should be having sufficient handles
Production	The production methods and means should be easy
Operation	The appliance should be easily convertible
Maintenance	Minimal cleaning and maintenance
Cost	Target cost is around \$250–\$500

11.3.3 Recording customer verbatim

Stakeholders were chosen and interviewed by the design team members, and statements by the stakeholders were recorded. Table 11.11 contains the verbatim made by the customers.

11.3.4 Establishing prioritized needs

The verbatim was simple, and they are translated to needs in the following way:

1. Should be able to display A3 posters
2. Should be able to hold a vase of height of 400 mm
3. The height of the poster should be around eye level
4. Breadth and length could be around 500–700 mm
5. There should be sufficient space to store needed auxiliaries
6. Display and flower vase stand arrangement should be easy
7. The appliance should have lightweight
8. Appliance should be stable in both arrangements
9. The material should be beautifully grainy or stranded
10. The final product should have an attractive surface
11. Should be easy to convert as display or vase stand
12. Should be easily moveable from place to place
13. The layout should be having sufficient handles
14. The production methods and means should be easy
15. The appliance should be easily convertible
16. Minimal cleaning and maintenance
17. Target cost is around \$250–\$500

11.3.5 Establishing the function tree

Three purpose functions were identified, and the function tree shown in Figure 11.8 was developed.

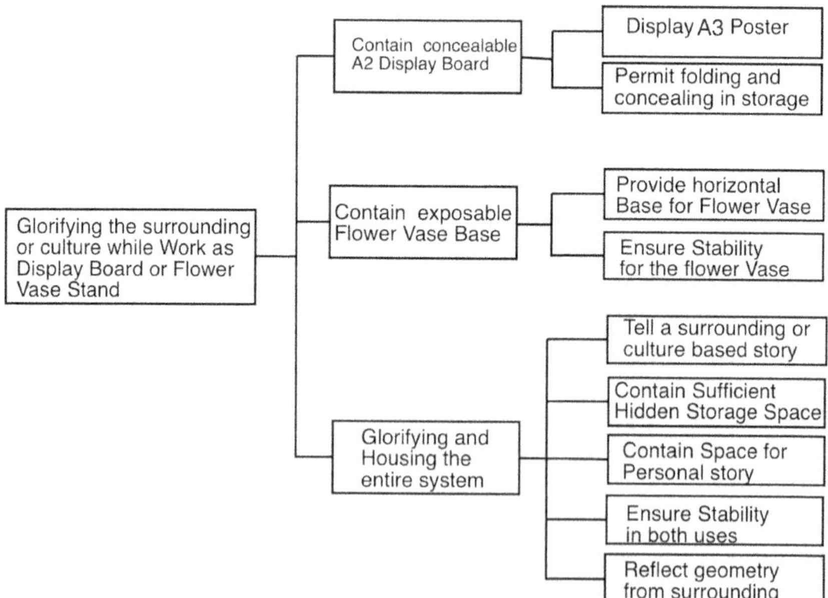

Figure 11.8 Function tree for a convertible display/flower vase stand.

11.3.6 Establishing the need-metric matrix

Table 11.12 Need-metric matrix

Units	mm	mm	mm	mm	mm	Number	kg	Yes/No	Yes/No	Yes/No	Number	Yes/No	Yes/No	Number	Dollars
Metric	Dimensions of Board	Vase height	Height of poster top	Dimensions of foot	Dimensions of space	Number of steps	Weight of appliance	C.G. inside the foot	Membership in list	Membership in List	Number of steps	Wheels provision	Material list	Number of Operations	Cost
Should be able to display A2 posters															
Should be able to hold a Vase of height of 400 mm															
The height of the poster should be around eye level															
Breadth and length could be around 500 – 700 mm															
Sufficient space to store needed auxiliaries															
Easy display and flower vase stand arrangement															
The appliance should have light weight															
Appliance should be stable in both arrangements															
The material should be beautifully grainy or stranded.															
The final product should have an attractive surface.															
Should be easy to convert as display or Vase stand															
Should be easily moveable from place to place															
The production methods and means should be easy															
Target cost is around $250 to $500															

11.3.7 Establishing the specifications

Table 11.13 Specifications for the convertible flower vase/display stand

Specification	*Importance*	*Values*
Length and width	9	A3 + 50 mm
Height of Vase	8	650–750
Storage chamber	7	Store vase & board
Retract/remove ability for display board	6	Yes
Portable – Has wheels	9	Yes
Square foot side	8	400–450 mm
Assembled height	9	1,500–1,700 mm
High-class finish	8	Varnished
Culturally Rich resembling historic things	9	Yes
Cleaning time	6	15–25 minutes
Lifetime	6	10 years

11.3.8 Establishing preferred feature vector

The stakeholder requirements and the derived specifications help to establish the desired features of a display/flower vase stand. They are put in a column vector in the following way:

$$\begin{bmatrix} \textit{Functioning as an attractive A3 display board} \\ \textit{Functioning as an attractive flower vase stand} \\ \textit{Easy conversion between the two modes} \\ \textit{Storage provision for the vase when not in use} \\ \textit{Display different features of the local area} \\ \textit{Light weight} \\ \textit{Easy to handle and move} \\ \textit{Easy to clean and maintain} \\ \textit{Easy to Manufacture} \end{bmatrix}$$

The preferred characteristics help to establish or identify objects that have some of these desired features. Such objects are called adjacent objects and new conceptual solutions can be derived from them. Section 11.3.9 identifies some suitable adjacent designs.

11.3.9 Identification of suitable adjacent concepts

Five adjacent designs were identified for consideration as starting point for the conceptual design process. They are:

1. Lectern – A lectern is a reading desk with a slanted top. In this high sloping-top someone can put his notes when he is standing up and giving a lecture. The slope, size and facilities to add other audio and video gadgets are some of the features in the modern lecterns.
2. Pulpit – A pulpit is a raised stand for preachers in a Christian church. The traditional pulpit is raised well above the surrounding floor for audibility and visibility, accessed by steps, with sides coming to about waist height.
3. Display Board – a board-shaped material that is supported by poles, rigid and strong. Generally, paper or other materials are affixed to it for people to see.
4. 'A' Board – An A board display adds to anyone like restaurants, cafes, and retail stores, to advertise to passers-by with ease. It has dual sides to place the poster or advertisement. Its cross section looks like a capital alphabet A and hence gets the name.
5. Column – A proper column has a flat surface at the top which can easily carry anything like a lamp or flower vase.

11.3.10 Creating conceptual solutions

Table 11.14 Conceptual designs derived from adjacent designs

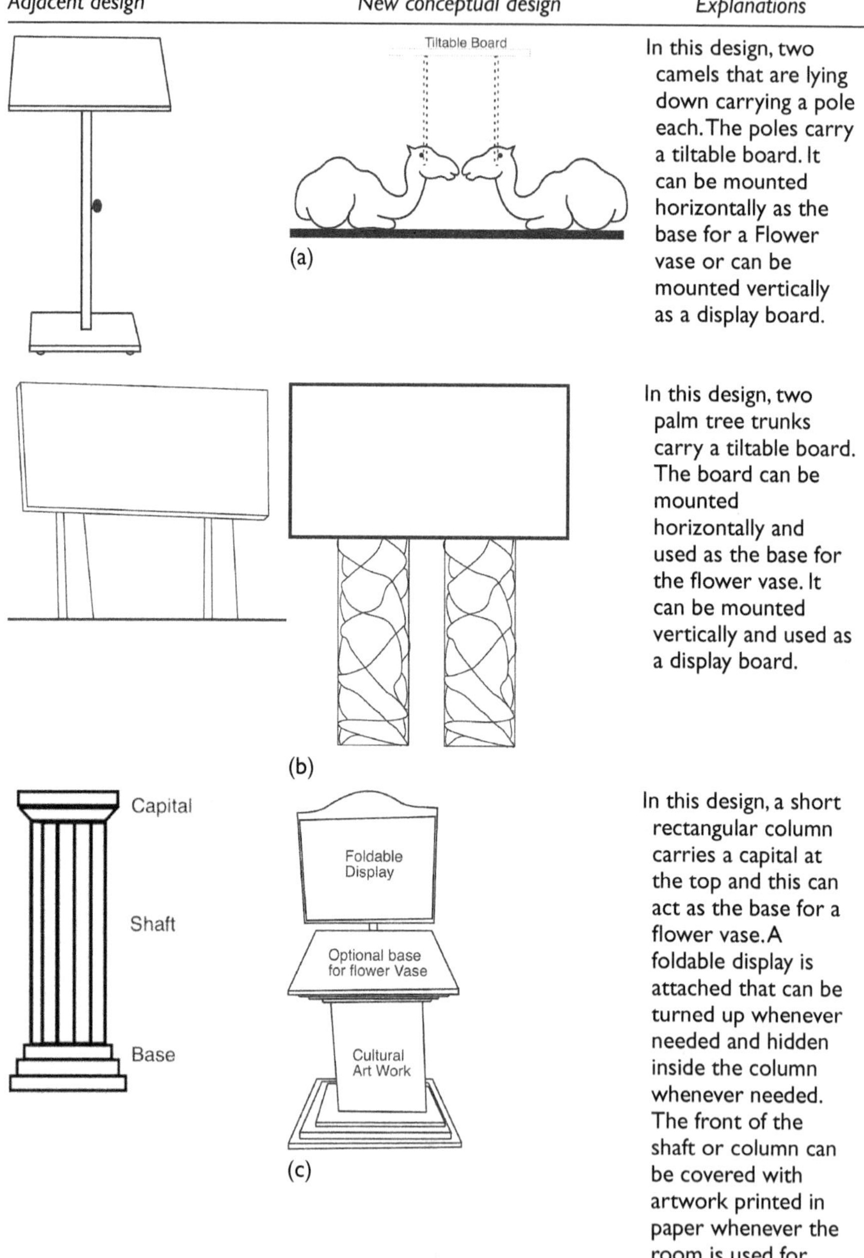

Adjacent design	*New conceptual design*	*Explanations*
	(a)	In this design, two camels that are lying down carrying a pole each. The poles carry a tiltable board. It can be mounted horizontally as the base for a Flower vase or can be mounted vertically as a display board.
	(b)	In this design, two palm tree trunks carry a tiltable board. The board can be mounted horizontally and used as the base for the flower vase. It can be mounted vertically and used as a display board.
	(c)	In this design, a short rectangular column carries a capital at the top and this can act as the base for a flower vase. A foldable display is attached that can be turned up whenever needed and hidden inside the column whenever needed. The front of the shaft or column can be covered with artwork printed in paper whenever the room is used for special functions.

(Continued)

Table 11.14 (Continued) Conceptual designs derived from adjacent designs

Adjacent design	*New conceptual design*	*Explanations*
	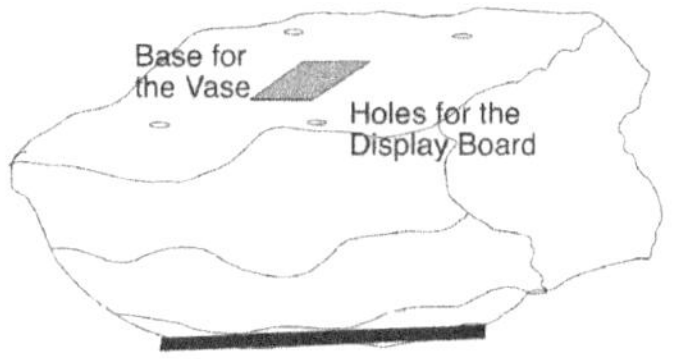 (d)	In this design, the display board is mountable on a hollow rock base whenever it is needed. It can be stored inside the hollow rock whenever it is not needed. On the top of the rock, a horizontal planar surface to seat the flower vase is made available. The vase can be stored inside the rock when not needed.
Pulpit in the church	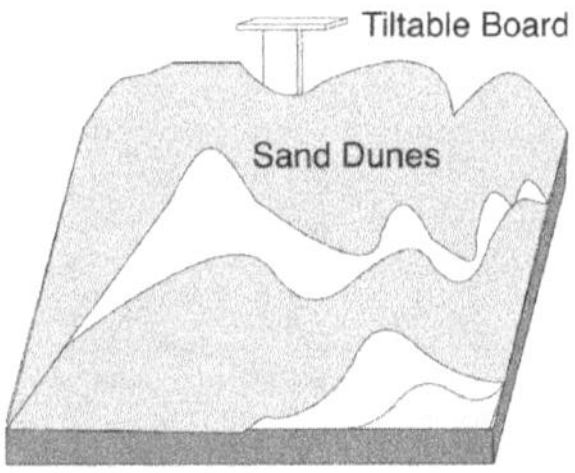 (e)	In this design, a tray filled with sand that can be arranged to mimic sand dunes in the desert is used as the base. A rectangular column located in the sand carries a tiltable board that can be fixed horizontally to carry the vase and fixed vertically to display the poster.
Pulpit in the Church	 (f)	In this design, a rectangular column fixed at the centre of a small platform carries a tiltable board. As in design (iii) the cultural artwork is placed at the front of the column.

By considering elements in the preferred feature vector and the adjacent design identified, the features that have to be (i) eliminated and (ii) added can be established. Once they are established, they can be added or eliminated to the features that have to be retained and new conceptual designs can be established. Consider Table 11.14.

11.3.11 Pugh's matrix

Table 11.15 Pugh's matrix

Preferred Feature	Design a	Design b	Design c	Design d	Design e	Design f
Functioning as an attractive A3 display board	-	-	+		S	S
Functioning as an attractive flower vase stand	-	-	+		S	S
Easy conversion between the two modes	S	S	+		S	S
Storage provision for the vase when not in use	-	-	S		-	S
Display different features of the local area	S	S	-		S	-
Light weight	S	S	S	DATUM	-	S
Easy to handle and move	S	-	+		-	S
Easy to clean and maintain	+	+	+		S	+
Easy to Manufacture	S	S	+		+	S
$\sum +$	1	1	6		1	1
$\sum -$	3	4	1		3	1
$\sum S$	5	4	2		5	7

11.3.12 Evaluation and selection of conceptual design

Analysis of Pugh's matrix shows that the majority of the designs have similar features excepting design c above. It has six pluses and two similar features. It has one negative. Storage and weight due to material can be considered during detail design to make sure that they are made better as far as the circumstances permit. The negative in 'Display different features of the local area' all other designs have at least one feature incorporated in their design. The conceptual design was revisited to see whether any improvement can be made to rectify the shortcoming. It was decided to place adhesive artwork that can be changed routinely to make the artefact fresh and rejuvenated.

11.3.13 Manufacturing and finished product

The product was mainly made out of timber. The making involved boards to build the column, the top face and the base. Inside the column a horizontal shaft was mounted on which the display arm rotated to bring the display out in position to carry and display the poster, and rotate in, to keep the display in storage. The finished product is shown in Figure 11.9.

Figure 11.9 Designed and built display/flower vase stand.

11.4 TABLE-TOP MEMORABILIA FOR MEM PROGRAM EXAMPLE

This example is adopted from a work carried out by a group of students in the same class as those in example 8.14. The task is to develop table-top memorabilia for the MEM program. The design brief for the task is given in Table 11.16.

11.4.1 Planning the design process

Figure 11.10 shows the design process plan in a systematic framework.

Table 11.16 Design brief of the memorabilia item for the MEM program

Design brief for a 'table-top memorabilia for the mem program'	
Drafted by	Dr Sangarappillai Sivaloganathan
Product description	A high-class and attractive memorabilia item boasting the strengths and features of the MEM program at UAEU and the heritage of UAE.
Product concept	A prismatic, or free-form solid with a control volume of 150mm × 150 mm × 150 mm.having 'MEM at UAEU' engraved. It should incorporate striking features of the MEM program and heritage of UAE.
Benefits to be delivered	Strengths of the MEM program Clear display of the strengths High class Relates to the heritage of the UAE
Positioning & target price	Target cost is around 60–100 dollars for the prototype.The product is not for sale in the market.
Target market	N/A
Assumptions, constraints and standards	Concepts have to be generated using the Gallery Method
Stakeholders	MEM Program coordinator Master of Engineering Management Students Alumni of the MEM Program
Possible features and attributes	–
Possible area for innovation	core strengths of MEM to be portrayed. Environment and Achievements of UAE to reflect

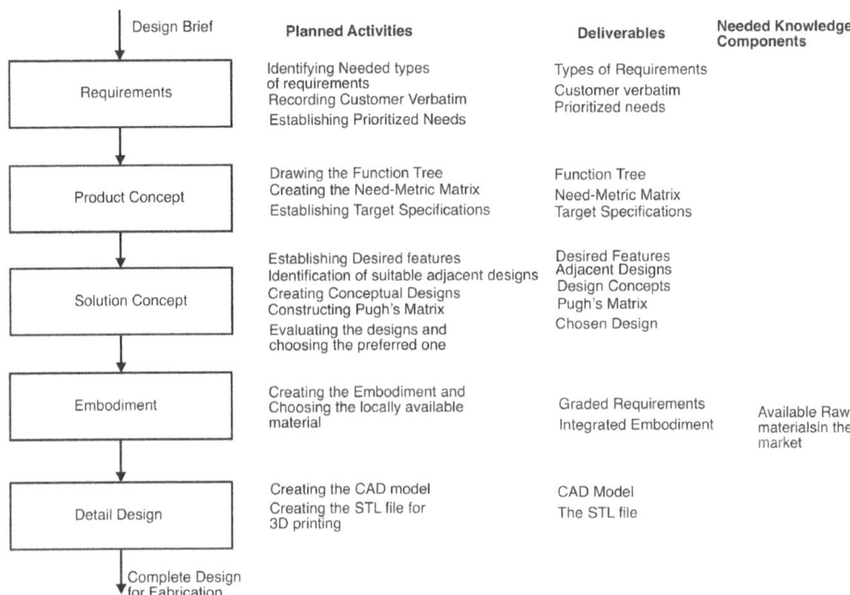

Figure 11.10 Design process plan for designing some table-top memorabilia for the MEM program.

11.4.2 Identifying the types needed of requirements

Checklist by Pahl and Beitz was referred to identify the types of needed requirements for the memorabilia. Table 11.17 shows the identified relevent types from the list by Pahl and Beitz [1]. The unrelated topics have been removed.

Table 11.17 Needed types of stakeholder requirements [1]

Main group	*Sub-groups*
Geometry	Size, height, breadth, length, diameter, space requirement, number, arrangement, connection, and extension
Materials	Materials physical and chemical properties of the initial and final product, auxiliary materials, prescribed materials (food regulations, etc.)
Ergonomics	The man–machine relationship, type of operation, clearness of layout, lighting, aesthetics
Production	Factory limitations, maximum possible dimensions, preferred production methods, means of production, achievable quality and tolerances
Assembly	Assembly Special regulations, installation, siting, foundation, transport limitations due to lifting gear, clearance, means of transport (height and weight), nature and conditions of dispatch
Operation	Quietness, wear, special uses, marketing area, destination (e.g., sulfurous atmosphere, tropical conditions)
Maintenance	Servicing intervals (if any), inspection, exchange and repair, painting, cleaning
Cost	Costs maximum permissible manufacturing costs; cost of tooling, investment, and depreciation

11.4.3 Recording customer verbatim

Interviewing the customers was the method adopted in this project. The design team interviewed identified stakeholders and recorded their verbatim. Their recordings are shown in Table 11.18.

Table 11.18 Requirement type and customer verbatim

Type	*Customer verbatim*
Geometry	Should include prismatic shapes
	Should include natural shapes like those in animals and plants
	Can reflect the local scenes
Material	Should last for relatively long periods (say 4–5 years)
Ergonomics	Could include features of the landscape of UAE
	Could include UAE culture (Dress, religion, cordiality, etc.)
	Could include a brief description of functions of MEM
	Could include the providers of the program
Cost	Should not cost more than 30–50 US dollars
Assembly	Should have an integrated appearance
	Should be harmonious
Operation	Should be light for easy to move around
Maintenance	Should be easy to clean
	Should not be fragile

11.4.4 Establishing prioritized needs

It is essential and customary that the recorded verbatim is analyzed by the design team for clarity and to ensure similarity in content between requirements. Since this is a small project bulk of the verbatim can be similar enough to be translated as need without any change. After the preparation of the needs, ideally they should be referred to the stakeholders for giving importance ratings. In this project, the importance ratings were given by the interviewing team, reflecting on the interviews. The translation and the importance ratings assigned are shown in Table 11.19.

Table 11.19 Verbatim need translation and assigned importance ratings

	Customer verbatim	*Need*	*Importance*
1	Should include prismatic shapes	Include prismatic shapes	9
2	Should include natural shapes like those in animals and plants	Include natural shapes	8
3	Can reflect the local scenes	Can reflect the local scenes	9
4	Should last for relatively long periods (say 4–5 years)	Should last for long periods	7
5	Could include features of the landscape of the UAE	Include features of the landscape of the UAE	9
6	Could include UAE culture (Dress, religion, cordiality, etc.)	Include UAE culture	8
7	Could include a brief description of functions of MEM	Add descriptions of functions of MEM	9
8	Could include the providers (the colleges) of the program	Include the providers of the program	8
9	Should not cost more than 30–50 US dollars	Should cost 30–50 US dollars	6
10	Should have an integrated appearance	Should have an integrated appearance	9
11	Should be harmonious	Should be harmonious	8
12	Should be easy to move around	Light for easy to move around	6
13	Should be easy to clean	Should be easy to clean	7
14	Should not be fragile	Should not be fragile	7

11.4.5 Function tree

The goal of the memorabilia item is to display the MEM program and UAE as a country to the world. To achieve this goal, three main functions are identified as essential. These purpose functions are:

1. Reflect contributions by the College of Business
2. Reflect contributions by the College of Engineering
3. House all features and glorify UAE

The first two are the specialties of the program while the third one is the nature and landscape of the country in which the university and the program are located. The function tree showing the goal, purpose functions and action functions is given in Figure 11.11.

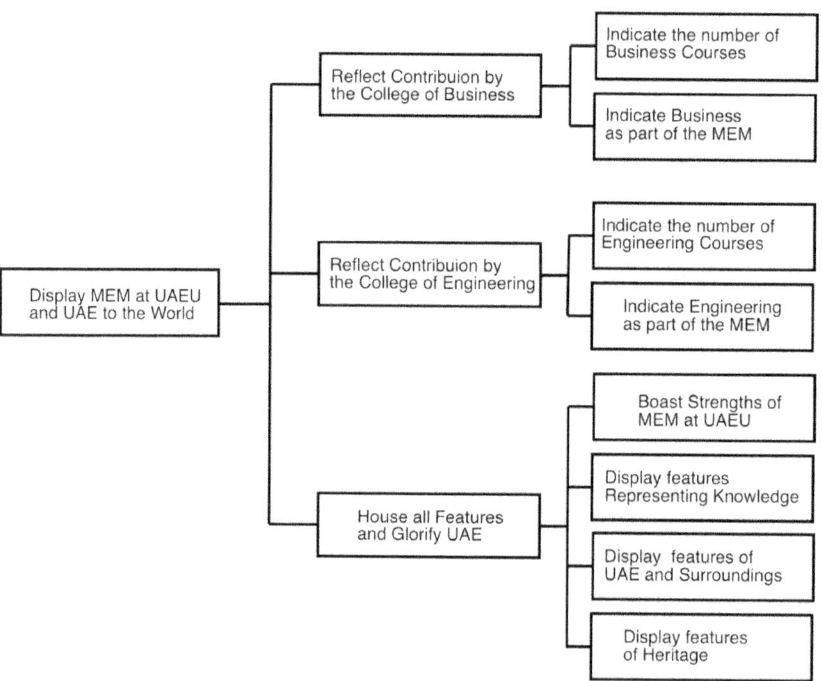

Figure 11.11 Function tree of a table-top memorabilia for the MEM program.

11.4.6 The need-metric matrix

The need-metric matrix translates the needs into metrics that have different levels of measurements. The need-metric matrix is shown in Table 11.20.

Table 11.20 Need-metric matrix

	Units	Yes/No	Yes/No	List	Number	List	List	Yes/No	Yes/No	n Dollars	Opinion Rating	kg	Yes/No	mm^4	
	Metric	Has Prismatic surf.	Free-form Surface	Local scenes	Number of Years	Include Land Marks	Included Values	Four Functions	Mention Colleges	Cost	Integrated Scenery	Weight of the item	Smooth surfaces	section of thin item	Importance Ratings
Number	Need														9
1	Include prismatic shapes														8
2	Include natural shapes														9
3	Can reflect the local scenes														7
4	Should last for long periods														9
5	Features of the landscape of UAE														8
6	Include UAE culture														8
7	Description of functions of EM														9
8	Include the program providers														9
9	Should cost 30 -50 US dollars														6
10	Integrated appearance														8
11	Should be harmonious														6
12	Light for easy to move around														7
13	Should be easy to clean														7
14	Should not be fragile														9

Table 11.21 Specifications for MEM memorabilia

No	Specification	Values
1	Prismatic shapes	Present
2	Free-form shapes	Present
3	Local scenes	Present
4	Show Functions of MEM	Yes
5	Technology and Business mentioned	Yes
6	Compact Size including many features	yes
7	Stepped Pyramid	yes
8	Strong structure	yes
9	Base size on the table	About 125 mm

11.4.7 Establishing the specifications

Considering the design brief, the requirements and metrics the specifications were drawn. The drawn specifications are shown in Table 11.21.

11.4.8 Criteria

The criteria for concept evaluation were determined by the design team. They are as follows:

1. Compact size
2. Represent many characteristics of the region
3. Represent MEM Program
4. Mention Business and Engineering
5. Show the number of courses
6. Easy and attractive geometry
7. Aesthetically pleasing
8. Functions of Engineering Management
9. Easy to manufacture

11.4.9 Conceptual designs in the first round of gallery method

Eight conceptual designs were proposed by a team of four. They are shown in Table 11.22.

a. Design (i) – A model airplane with the fuselage carrying MEM while the wings show Engineering and Business colleges. A degree hat is at the front of the plane.
b. Design (ii) – A degree hat on top of a rectangular platform having engineering and Business as the two sides.

Table 11.22 Conceptual designs in gallery method first phase

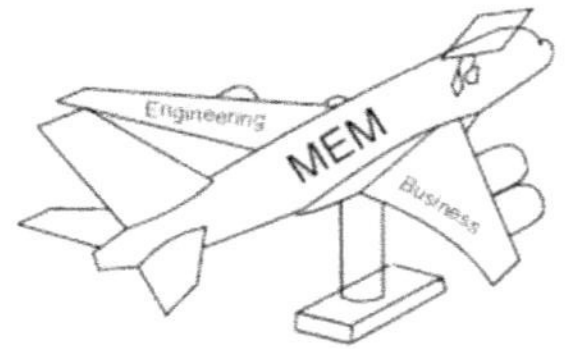 Engineering MEM Business (a)	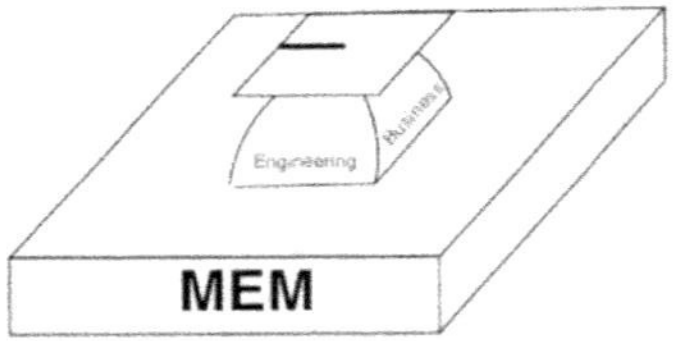 Engineering Business MEM (b)	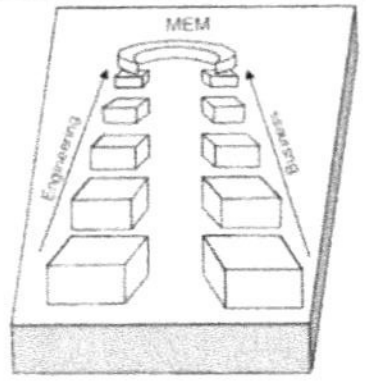 MEM Engineering Business (c)
 MEM $ (d)	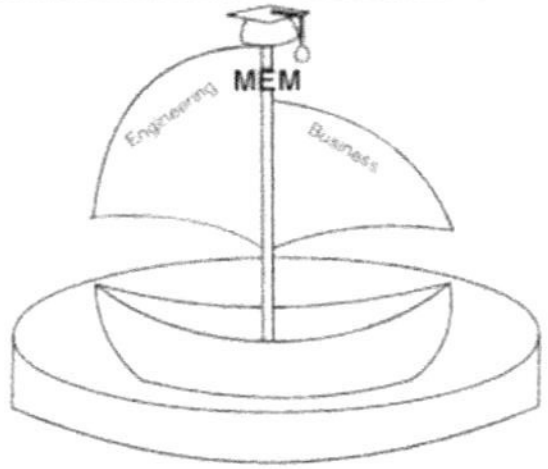 MEM Engineering Business (e)	 MEM MEM (f)
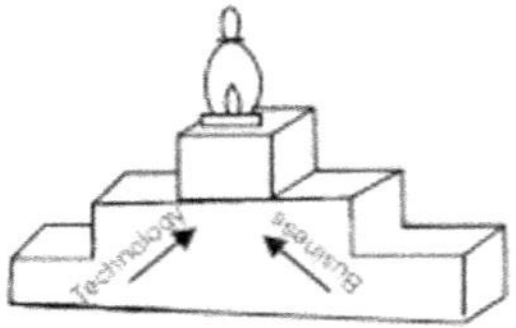 Technology Business (g)	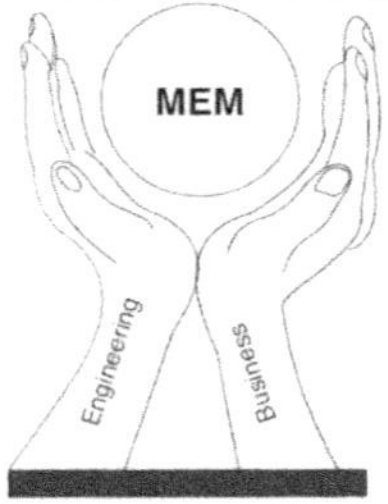 MEM Engineering Business (h)	

c. Design (iii) – A rectangular platform in which two rows of five steps representing the number of courses in Engineering and Business to achieve the MEM degree.
d. Design (iv) – A five stepped one side podium carrying the degree hat while the side showing a gear representing engineering and dollar sign representing business. MEM is also marked in the side.
e. Design (v) – A fishing boat with two sails representing engineering and business while the mast represents MEM.
f. Design (vi) – Showing MEM in a surrounding of castles and sand dunes.
g. Design (vii) – A two-sided podium with five positions carrying a torch. Two ascending arrows showing technology and business.
h. Design (viii) – Two hands representing Engineering and business preserving MEM inside their palms.

11.4.10 Pugh's matrix

All designs in the first round have many negatives. However, they had good features distributed among them. It appeared that when these good features are combined, they could result better designs. The team went for the second round and the designs they produced are given in Table 11.23. Pugh's matrix is shown in Table 11.24.

Table 11.23 Conceptual designs in the second round

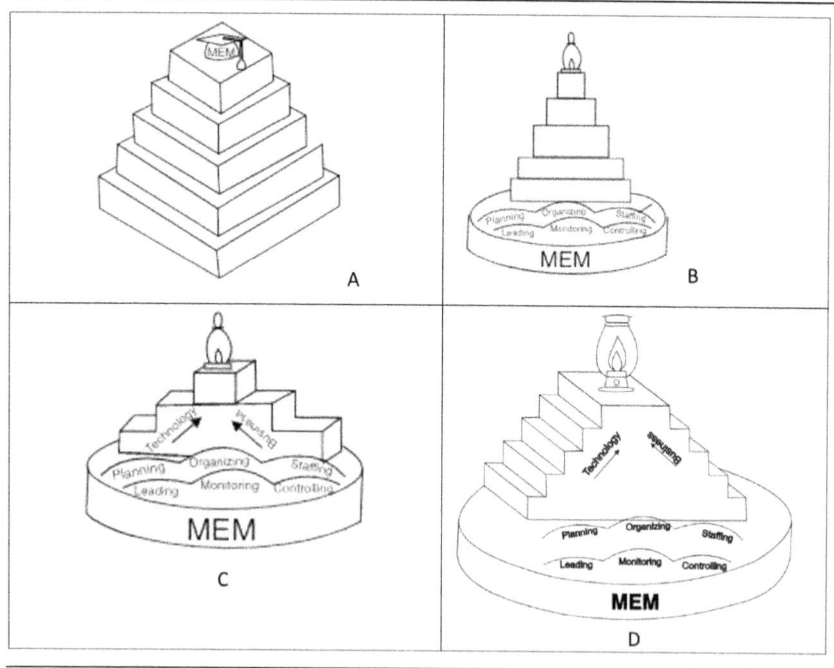

Table 11.24 Pugh's matrix first round

	Gallery Method – Round 1								Gallery Method – Round 2			
	Design (a)	Design (b)	Design (c)	Design (d)	Design (e)	Design (f)	Design (g)	Design (h)	Design A	Design B	Design C	Design D
Compact size	-	+	DATUM	S	+	-	-	+	+	+	+	+
Represent many characteristics Of the region	-	-		-	+	+	-	-	-	+	+	+
Represent MEM Program	S	S		S	S	S	-	+	S	S	S	S
Mention Business and Engineering	S	S		-	S	-	S	S	-	-	S	S
Show the Number of Courses	-	-		-	-	-	-	-	-	-	-	S
Easy and attractive Geometry	-	+		+	S	-	S	-	S	S	+	+
Aesthetically pleasing	+	S		S	+	+	S	+	S	S	+	+
Functions of Engineering Management	-	-		-	-	-	-	-	S	S	+	+
Easy to Manufacture	-	S		S	-	-	S	-	S	S	S	S
$\sum +$	1	2		1	3	2	0	3	1	2	5	5
$\sum -$	6	3		4	3	6	5	5	3	2	-	0
$\sum s$	2	4		4	3	1	4	1	5	4	3	4

11.4.11 Manufacturing and finished product

From Pugh's matrix for round 2, Design D was chosen to be the design for further development. It was manufactured using 3D printing.

The circular base was taken to be 125 mm in diameter and other features were given proportionate dimensions. The product was modelled in CATIA and 3D printed. The produced product is shown in Figure 11.12.

Figure 11.12 3D Printed product.

11.5 MANUAL PITH PRESS FOR A RESEARCH LABORATORY EXAMPLE

An entrepreneur wants to make briquettes for baking ovens using coconut pith. He thinks that there should be a particular density of the briquette to get the best burning performance. A manual pith press is needed to conduct experiments to achieve this purpose. In this assignment, you are required to design and fabricate such a press.

The conceptual design idea is to use a power screw to press the coconut pith inside a chamber. If the diameter of the compression chamber is 63 mm and the maximum pressure inside it is limited to 10 MPa continue the design of the press. From a height of 50 mm of pith, the minimum achievable thickness of the briquette is 10 mm. The design brief of the pith press is given in Table 11.25

Table 11.25 Manual pith press

	Manual press
Product description	A press for making Organic Fuel Briquettes (such as Coconut Pith Briquettes) with a thickness of 10 mm made by compressing 50 mm high material utilizing manual force that an average person can exert.
Product concept	An adaptation of the basic machine, screw, that will magnify the applied force several times to produce a pressure of up to 10 MPa.
Benefits to be delivered	Easy to operate (filling and removal) Can produce briquettes at different compression ratios Ability to read the height of the compressed material Minimum number of parts
Positioning and target price	Can support a development a price tag of \$300–\$500
Target market	Test laboratories
Assumptions and constraints	Square and ACME screws are easily manufacturable.
Stakeholders	Lab Technicians Researchers in Chemical Engineering Agricultural product researchers
possible features and attributes	Sequence of operations or process to produce briquettes
possible area for innovation	Cleanliness and safety

Drafted by UAEU mechanical engineers.

11.5.1 Design process planning

Design process is planned with the design model at the background. Activities, deliverables and needed knowledge at each stage are established in this activity. The plan is shown in Figure 11.13.

11.5.2 Needed types of requirements and stakeholder needs

Ullman's checklist [2] was consulted to identify the types of stakeholders' needs. After identifying the types one-to-one interviews were held with chosen customers and their verbatim were recorded. Since the product is small and requirements were few the verbatim were considered as same. Table 11.26 shows the requirement types and the needs.

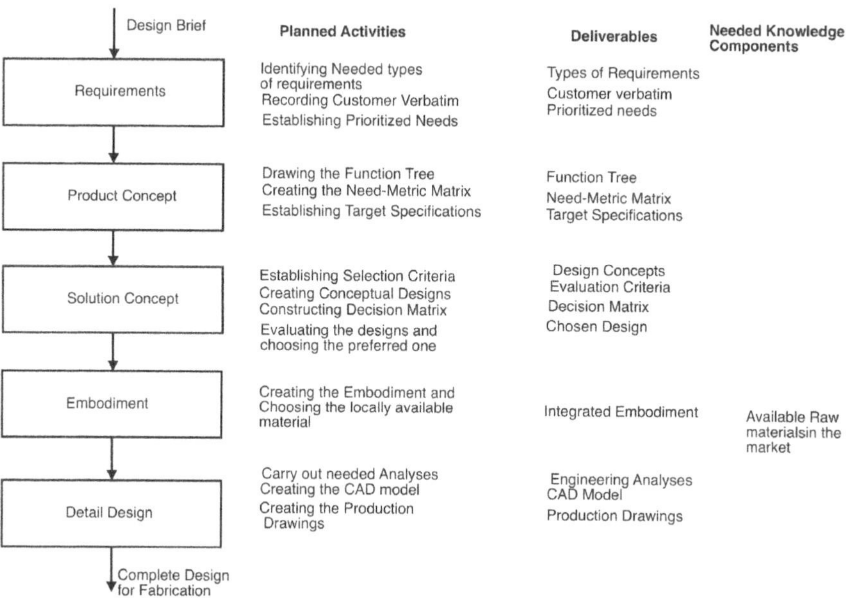

Figure 11.13 Design process plan.

Table 11.26 Types of requirements and stakeholder needs

Type	*Customer needs*
Flow of energy	Input effort less than 100 N
	Effort should be in the gravitational direction
Information flow	Height of material is always readable
Material flow	Easy to fill
	Easy to extract
Spatial envelope	Foot space less than 1 m × 0.5 m
Physical properties	Operational height around 0.6–0.065 m
	Operational height
Installability	Easy to install
Cost	Affordable
Human factors	Attractive

11.5.3 Prioritized stakeholder needs

Importance ratings were assigned to stakeholder needs by the interviewing members of the design team as shown in Table 11.27.

Table 11.27 Prioritized needs

No	*Customer needs*	*Importance*
1	Input effort less than 100 N	8
2	Effort preferably in gravitational direction	7
3	Height of pith is always readable	9
4	Easy to fill	7
5	Easy to extract	9
6	Foot space less than 1 m × 0.5m	7
7	Operational height around 0.6–0.65 m	6
8	Compression height from 50 to 10 mm	9
9	Affordable	7
10	Easy to Install	7

11.5.4 Function tree

The function tree is a description of the product in the functional domain. In order to draw the function tree, the goal and purpose functions are as follows:

Goal: Receive effort, convert into force and apply pressure on the pith

Purpose functions:

i. Transmit effort as torque on the screw
ii. Support the load from Ram as Thrust
iii. Receive torque and apply it as pressure
iv. Receive pressure from Ram and press the pith

The purpose functions are further sub-divided into action functions. The function tree for the pith press is shown in Figure 11.14.

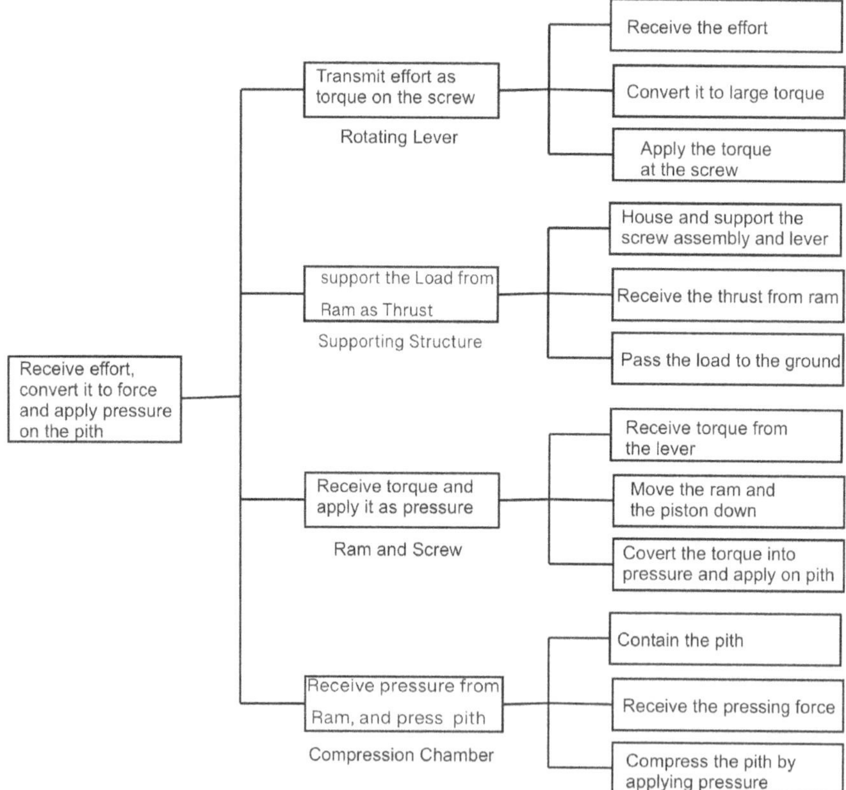

Figure 11.14 Function tree of the pith press.

11.5.5 Need-metric matrix

Table 11.28 Need-metric matrix

		Newton	Yes/No	Yes/No	mm	yes/No	Sq. mm	mm	mm	$	No
		Force	Vertical effort	Height read facility	Clearance at top	Extract Mechanism	Area of the foot	Operational Height	Compression height	cost	Number of steps
1	Input effort less than 100 N										
2	Effort preferably in gravitational direction										
3	Height of pith is always readable										
4	Easy to fill										
5	Easy to extract										
6	Foot space less than 1m x 0.5m										
7	Operational height around 0.6 to 0.65 m										
8	Compression height from 50 mm to 10 mm										
9	Affordable										
10	Easy to Install										

11.5.6 Specifications

Table 11.29 Specifications

	Specifications	*Units*
1	Diameter of the briquette be 63 mm	63 mm
2	Maximum compressing pressure 10MPa	10 MPa
3	Human effort needed is <150 N	<150 N
4	Removal mechanism for the briquette	Yes
5	Number of operation	5 steps
6	Should be at operable height from table-top	Yes

Table 11.30 Criteria and weights

	Criteria	*Value*	*Weight*
1	Easy removal of briquettes	7	7/28
2	Low friction to use	6	6/28
3	Ease of Manufacturing	5	5/28
4	Minimum number of parts	4	4/28
5	Make briquettes at varying compressions	3	3/28
6	Easy to fill	2	2/28
7	Ability to read height of briquette	1	1/28

11.5.7 Selection criteria

Criteria for selection are given below in the descending order of importance:

1. Easy removal of briquettes
2. Low friction to use
3. Ease of manufacturing
4. Minimum number of parts
5. Ability to make briquettes at different compressions
6. Easy to fill
7. Ability to read height of briquette

The criteria and the weights are given in Table 11.30.

11.5.8 Conceptual design

Table 11.31 Conceptual designs

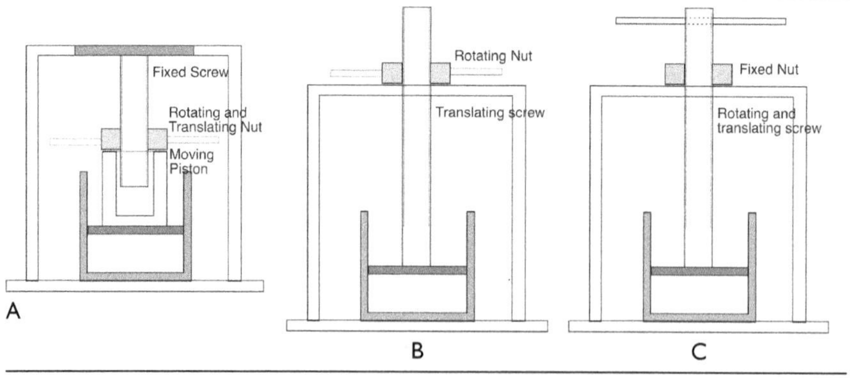

11.5.9 Decision matrix

	a	b	c	d	e	f	g	
	7/28	6/28	5/28	4/28	3/28	2/28	1/28	
A	10 / 70/28	10 / 60/28	8 / 40/28	8 / 32/28	10 / 30/28	8 / 16/28	4 / 4/28	252/28
B	6 / 42/28	6 / 36/28	5 / 25/28	6 / 24/28	10 / 30/28	6 / 12/28	3 / 3/28	172/28
C	8 / 56/28	8 / 48/28	8 / 40/28	8 / 32/28	10 / 30/28	8 / 16/28	4 / 4/28	226/28

Figure 11.15 Decision matrix.

11.5.10 Embodiment of the chosen concept

During the embodiment design, shape and dimensions have to be given to the following. Figure 11.16 shows them in the pith press concept.

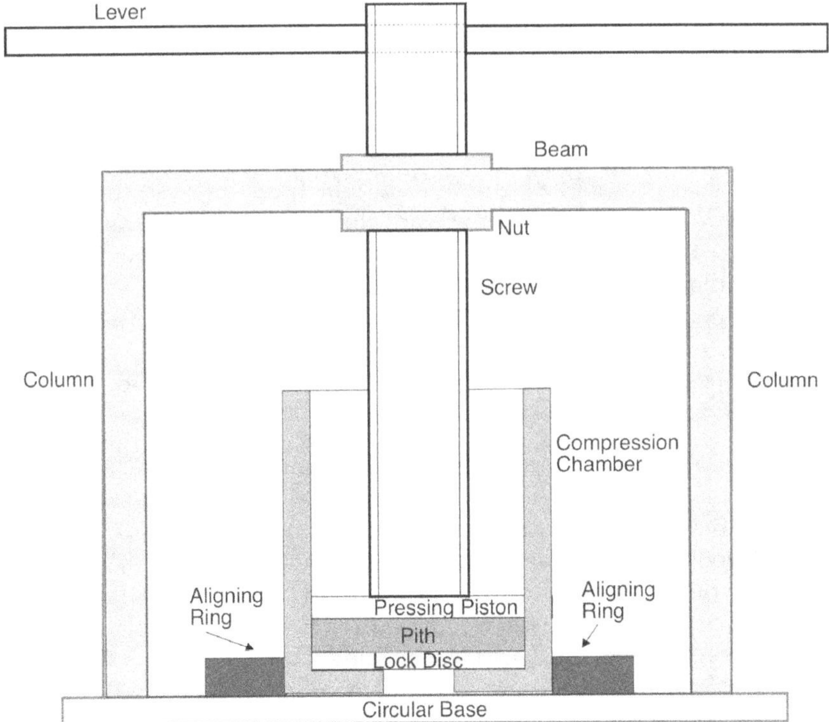

Figure 11.16 Embodiment of chosen conceptual design.

a. Support structure that holds the nut and screw
b. Interface between the nut and structure
c. A rough dimension for the lever
d. Compression chamber
e. Self-aligning mechanism
f. Screw design
g. Nut design

The embodiment design can be explained in the following way:

1. From the design brief and the requirements, compression chamber is the first thing that has to be given the geometry. It has to accommodate material that could fill to a height of 50mm to get compressed to a height of 10mm. In order to meet the requirement of easy to extract the briquette, a lock disc and a hole underneath in the chamber is needed. On top of it the piston comes, and therefore the height of the compression chamber can be 10mm for the base plate 10mm for the lock disc, 10mm for the piston and 50mm for the material making the height to be a minimum of 80mm.
2. The compression chamber should be taken out easily for that the piston should go clear of it by a good amount say 25mm at least.
3. The next one is the columns. These should be kept far apart for easy access to place the compression chamber and removing it. A clearance of 50mm on both sides is recommended as the minimum.
4. An aligning ring is the next one to embodied. It has its axis aligned with that of the screw so the screw can exert a uniform pressure on the pith, inside the chamber.
5. The next item is the beam. It should be firmly fixed to the columns as it would experience an upward thrust from the pith, through the screw and nut. It should also incorporate the nut as an integral part and the nut should have sufficient number of threads.
6. The next item is the screw. It should be sufficiently long to have clear space for the fingers between the lever and the beam.
7. The next item is the lever. It should be sufficiently long to magnify a small amount of effort by the operator.

The dimensions (i) can be determined by engineering calculations after choosing the material or (ii) can be given values to complete the embodiment and then proven to be adequate after choosing the material. The latter method would permit having a CAD model built to visualize the press and determine the engineering analyses needed to prove the design. In this project, the dimensions were assumed and proven later to be adequate.

The dimensions assumed are given in Table 11.32.

Other assumptions made are (i) the material selected 1,040 OQT condition with a Yield stress of 600MPa and (ii) Young's Modulus is 206GPa.

Assuming a safety factor of two permissible stress=300MPa.

Figure 11.17 shows the CAD model with the assumed values.

Table 11.32 Assumed values for the Parameters

No	*Parameter*	*Quantity*
1	Maximum chamber pressure	10 MPa
2	Chamber diameter (internal)	63 mm
3	Chamber thickness	5 mm
4	Bottom hole	25 mm
5	Screw length	300 mm
6	Screw outer diameter	30 mm
7	Screw pitch (single start)	8 mm
8	Coefficient of friction	0.12
9	Length of beam	210 mm
10	Cross section of beam + column	40 × 25 × 2.5 mm
11	Colum height (ground to top)	200 mm
12	Lever length from centre	175 mm
13	Nut Rectangular height × width	40 × 60 × 60 mm
14	Base diameter	300 mm
15	Base thickness	6 mm
16	Contact pressure for grease	89.6 MPa

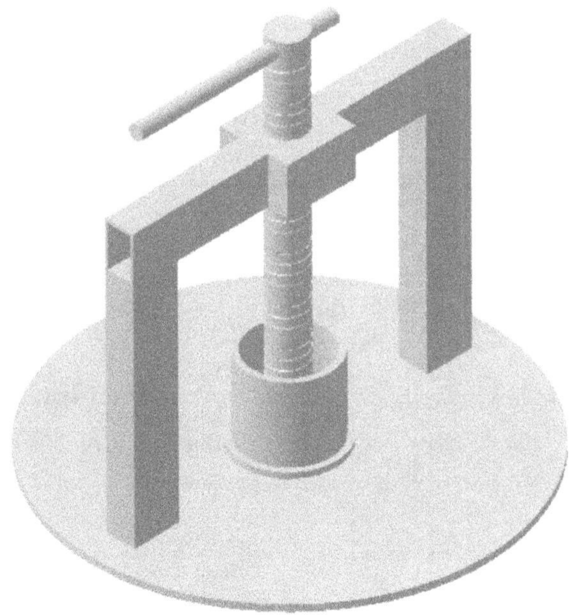

Figure 11.17 CAD Model with assumed values.

Table 11.33 Details of the proofs needed

No	Part	Details of the proofs needed
1	Compression Chamber	Axial and hoop stresses should be acceptable
2	Supporting columns	Tensile stresses should be acceptable
3	Beam	Bending stresses should be acceptable
4	Lever	Bending stress should be acceptable
5		Combined axial torsional stress acceptable
6	Screw	Contact pressure for grease in thread acceptable
7		Buckling load is greater than the compression
8	Nut	Number of threads sufficient to prevent tearing

11.5.11 Detailed design or engineering analysis

This section describes the engineering calculations that are carried out to prove the design. The needed proofs are identified by inspecting the CAD model and the identified ones are given in Table 11.33.

11.5.11.1 Compression chamber

Maximum pressure is realized when the piston reaches the bottom and the pith is pressed to the maximum, ie 10 mm. With the lock disc and allowance for filling the compression chamber will be 90 mm tall and 73 mm outer diameter. Full dimensions of the compression chamber are shown in Figure 11.18.

The compression chamber does not have axial stress as the load is taken up by the screw.

If the hoop stress is σ_h and the length and thickness of the chamber is l and t.

Tangential force is $= \sigma_h \times (l \times 2t)$.

This is balanced by the pressure force $= P \times lD$.

Equating the forces give $\sigma_h = \dfrac{PD}{2t}$.

Substituting the values give $\sigma_h = \dfrac{PD}{2t} = \dfrac{10 \times 63}{2 \times 5} = 63$ MPa.

This is very much less than the permissible value of 300 MPa and the chamber is safe.

11.5.11.2 Supporting columns

The supporting columns balance the upwards force by the screw on the beam.

The force on the on the beam by the screw $= 10 \times \left(\dfrac{\pi}{4} \times 63^2\right) = 31.18$ kN

Force on each column is 15.59 kN

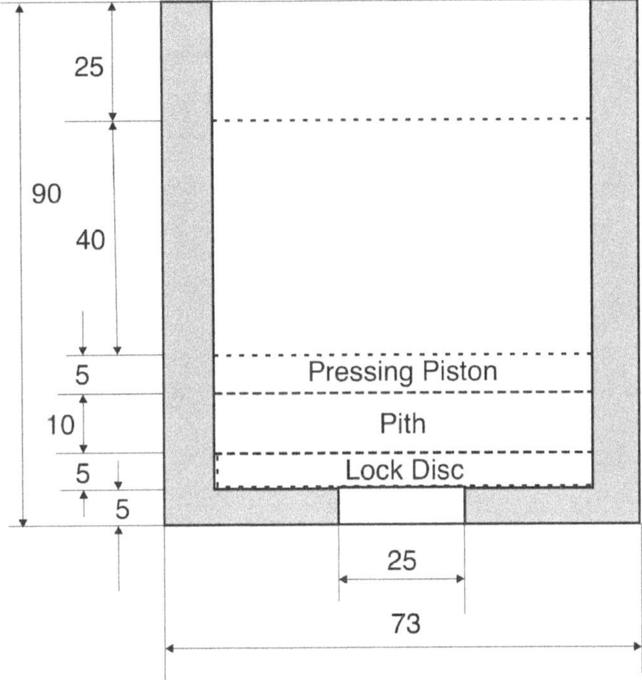

Figure 11.18 Compression chamber.

$$\text{Stress on the column} = \frac{15.59 \times 10^3}{40 \times 25 \times 2.5} = 6.24 \text{ MPa}$$

This again is very much less than the permissible 300 MPa and the columns are safe.

The load is tensile and hence there is no buckling (Figure 11.19).

11.5.11.3 The beam

The 210 mm long beam has a nut in between which is a 60 mm square block with 40 mm tall like the beam. This can accommodate five threads with 8 mm lead (Figure 11.20).

Second moment of area of the beam

$$= \left[\frac{1}{12}\left(25 \times 40^3 - 20 \times 35^3\right)\right] = \left[\frac{1}{12} 742,500\right] = 61,875 \text{ mm}^4$$

$$\text{Critical moment} = 15.59 \times 10^3 \times 55 = 857,450 \text{ Nmm}$$

Maximum bending stress at the joint between the nut and beam

$$= \frac{My}{I} = \frac{857,450}{61,875} \times 20 = 277.1 \text{ MPa}$$

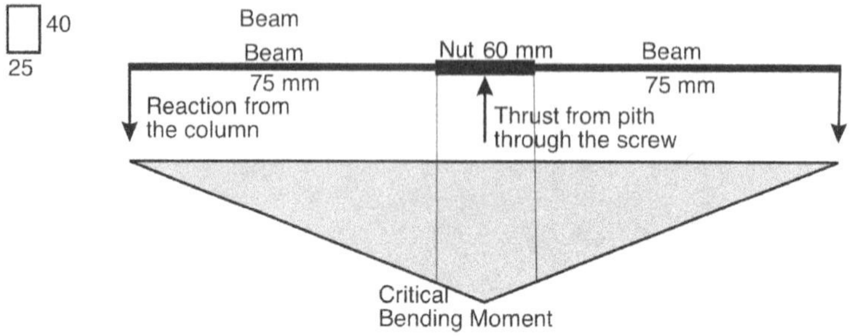

40
25
Beam
Beam
75 mm
Reaction from
the column
Nut 60 mm
Thrust from pith
through the screw
Beam
75 mm
Critical
Bending Moment

Figure 11.19 Bending moment in the beam.

This is less than the permissible 300 MPa. The beam therefore is safe.

11.5.11.4 Screw design

i. Safety against combined stresses

The force on the on the screw $= 10\ \text{MPa} \times \left(\frac{\pi}{4} \times 63^2\right) \text{mm}^2 = 31.18\,\text{kN}$

Torque required to lift this load $= F \times r_m \times \tan(\alpha + \lambda)$ where α is the lead angle and λ is the angle of friction and r_m is the mean radius.

Substituting the values $\alpha = \tan^{-1} \dfrac{\text{lead}}{\text{mean circumference}} = \tan^{-1} \dfrac{8}{26\pi}\ 5.6.$

Friction angle $\lambda = \tan^{-1} f = \tan^{-1} 0.12 = 6.84°$.

Torque is $= 31.18 \times 10^3 \times 13 \times \tan(5.6 + 6.84) = 89{,}416$ N mm.

This torque is applied on the core screw. The maximum torsional shear stress.

$$\tau = \frac{Tc}{J} = \frac{89{,}416 \times 11}{\frac{\pi}{32} \times 22^4} = 42.8\ \text{MPa}$$

Axial stress at the core screw $= \dfrac{31.18 \times 10^3}{\frac{\pi}{4} \times 22^2} = 82.02\text{MPa}.$

This is a situation of complex stresses. The radius of the Mohr's circle would be

$$= \sqrt{\left(42.8^2 + 41.01^2\right)} = 59.3\ \text{MPa}$$

The maximum shear stress of this complex stress is 59.3 MPa.

We may assume that the Yield shear strength as half of the yield stress. Therefore, the yield shear stress is 300MPa.

The maximum shear stress on the screw is 59.3 MPa which is much smaller than 300 MPa and hence the screw is safe from the complex stresses.

ii. safety against metal-to-metal contact

The nut is 40 mm tall and the lead of the screw is 8 mm. Therefore, the number of threads in contact is 5.

$$\text{Bearing pressure} = \frac{4W}{n\pi\left(D^2 - d^2\right)} = \frac{4 \times 31.18 \times 10^3}{5 \times \pi\left(30^2 - 22^2\right)} = 19.1 \text{ MPa}.$$

Permissible bearing pressure for normal grease is 89.6 MPa and the bearing pressure between threads is 19.1 MPa which is much less than the permissible one.

The screw is safe from metal-to-metal contact.

iii. Check for buckling

The screw is a circular column under axial load.

$$\text{Radius of gyration } r = \sqrt{I / A} = \frac{d}{4} = 5.5 \text{ mm}.$$

$$\text{Slenderness ratio} = \frac{kL}{r} = \frac{1 \times 180}{5.5} = 32.7.$$

$$\text{Column constant } C_c = \sqrt{\frac{2\pi^2 E}{S_y}} = \sqrt{\frac{2\pi^2 \times 206 \times 10^3}{600}} = 82.86.$$

The column is short because column constant is bigger than slenderness ration.

J. B. Johnson formula should be used.

$$P_{cr} = As_y\left[1 - \frac{s_y\left(KL / r\right)^2}{4\pi^2 E}\right]$$

$$\text{According to Johnson formula } P_{cr} = AS_y\left[1 - \frac{S_y\left(\frac{KL}{r}\right)^2}{4\pi^2 E}\right].$$

$$\text{Substituting the values give } P_{cr} = \frac{\pi}{4} \times 600\left[1 - \frac{600(32.7)^2}{4\pi^2 \times 206 \times 10^3}\right] =$$

$$\frac{\pi}{4} \times 600\left[1 - 0.078\right] = 434.5 \text{ MPa}$$

This is much more than the 82.02 MPa. Therefore, the screw will not buckle.

11.5.11.5 Nut

If the number of threads engaged is not sufficient, the threads may get sheared off at the root. This check is to make sure that the number of threads engaged is sufficient.

$$\tau_{root} = \frac{W}{n\pi D_{root}t} = \frac{31.18 \times 10^3}{5 \times \pi \times 22 \times 4} = 22.6 \text{ MPa}$$

This is much below than the permissible shear stress which is 300 MPa. The nut is therefore safe.

11.5.12 Production drawings

Production drawings are a set of drawings starting with the exploded view and the bill of quantities. The first sheet shows the exploded view and numbered parts. The following pages show drawings of individual parts. Each part is contained in a separate page with all necessary dimensions and details to manufacture the component. Figures 11.20–11.27 show the production drawings for the manual pith press.

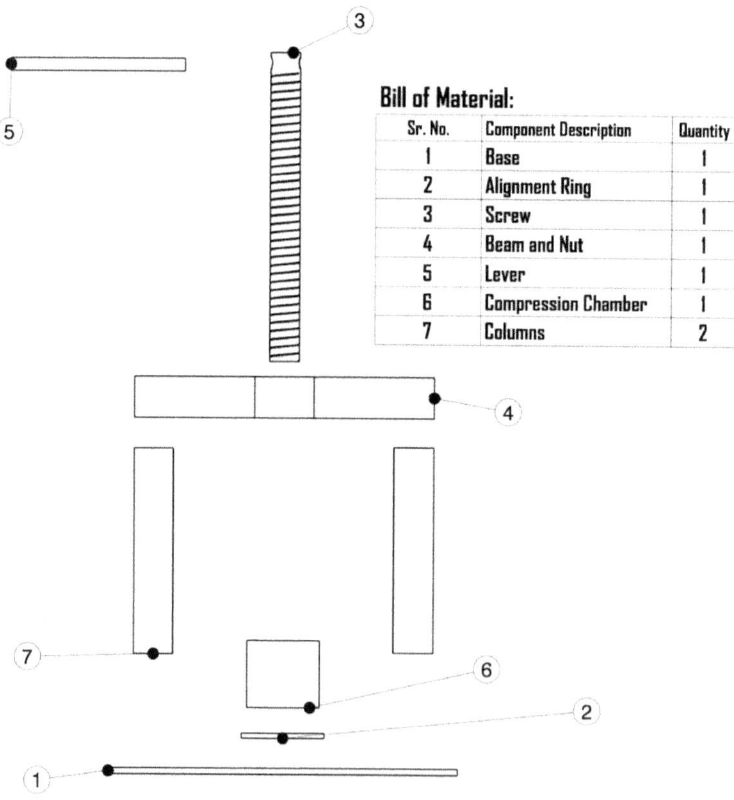

Bill of Material:

Sr. No.	Component Description	Quantity
1	Base	1
2	Alignment Ring	1
3	Screw	1
4	Beam and Nut	1
5	Lever	1
6	Compression Chamber	1
7	Columns	2

Figure 11.20 Exploded view of the assembled view.

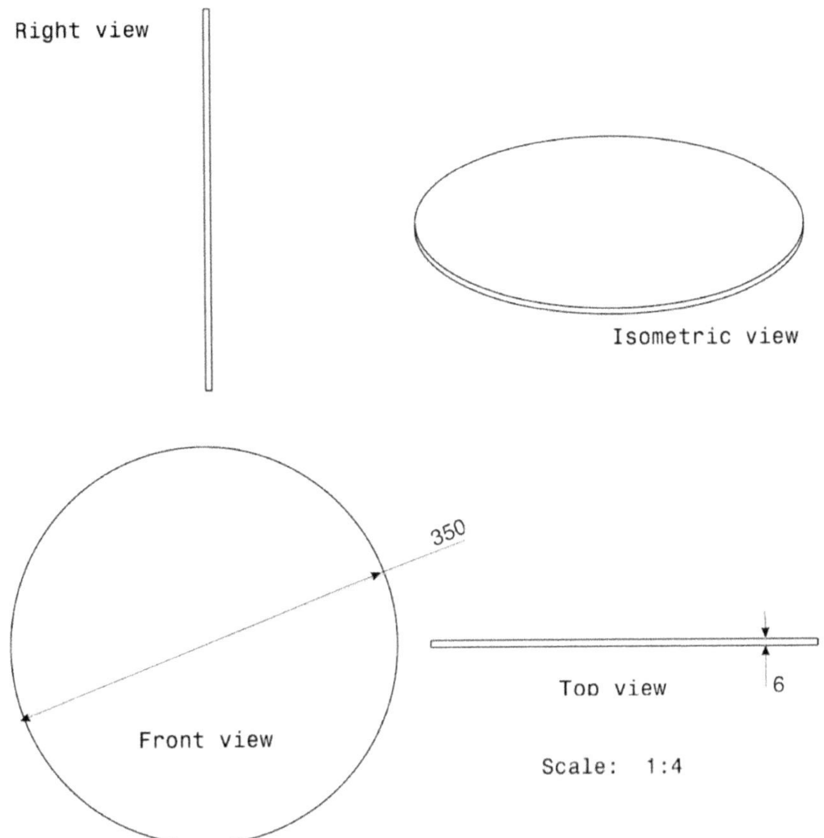

Figure 11.21 Base plate.

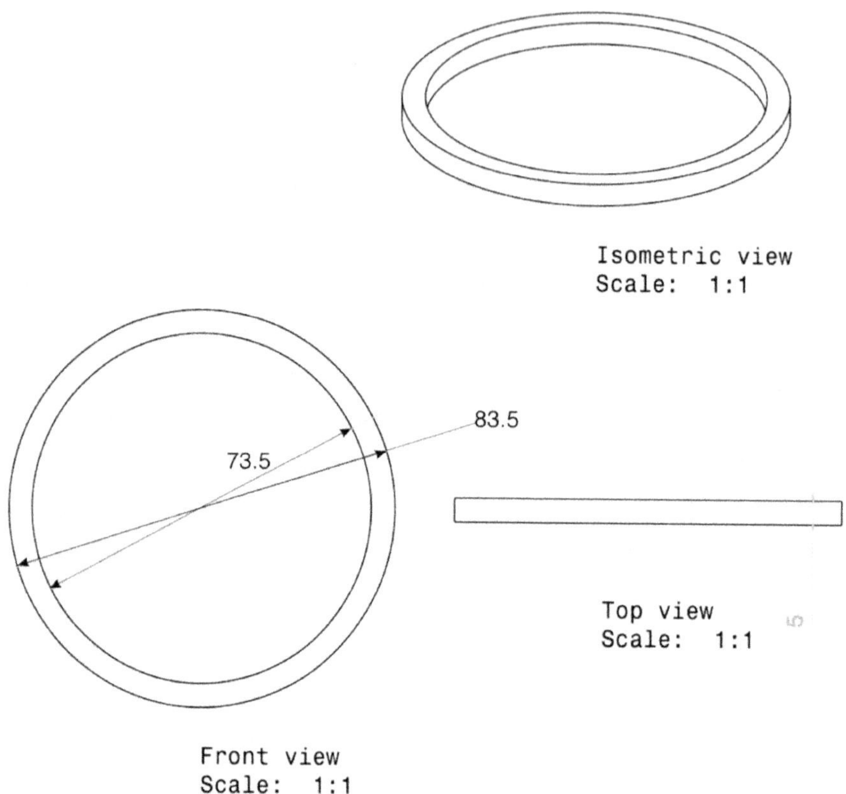

Figure 11.22 Alignment ring.

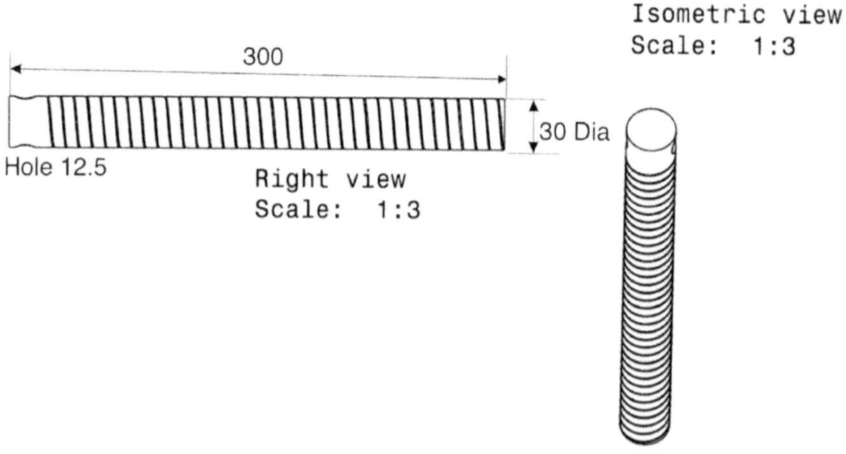

Figure 11.23 Screw.

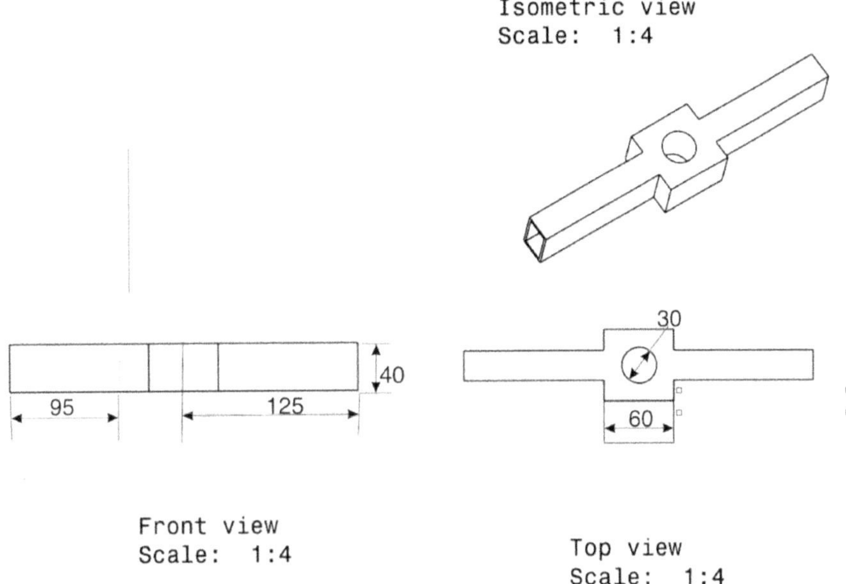

Figure 11.24 Beam and nut.

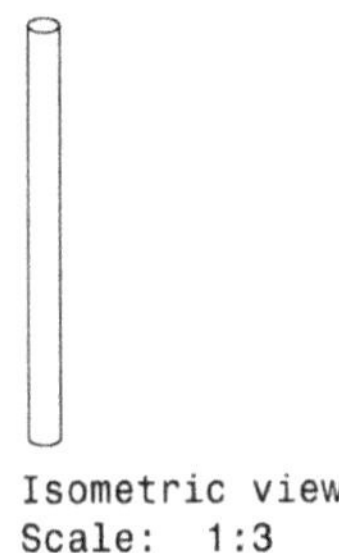

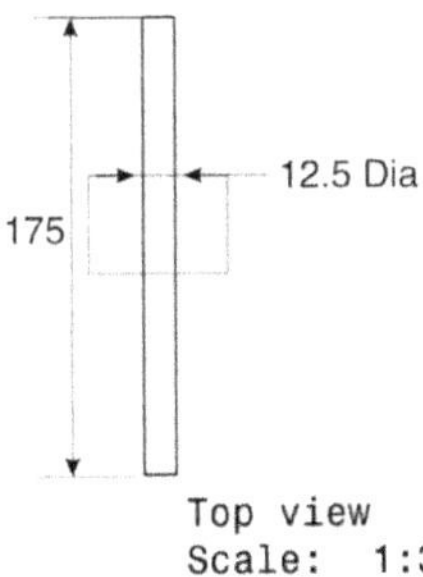

Figure 11.25 Lever.

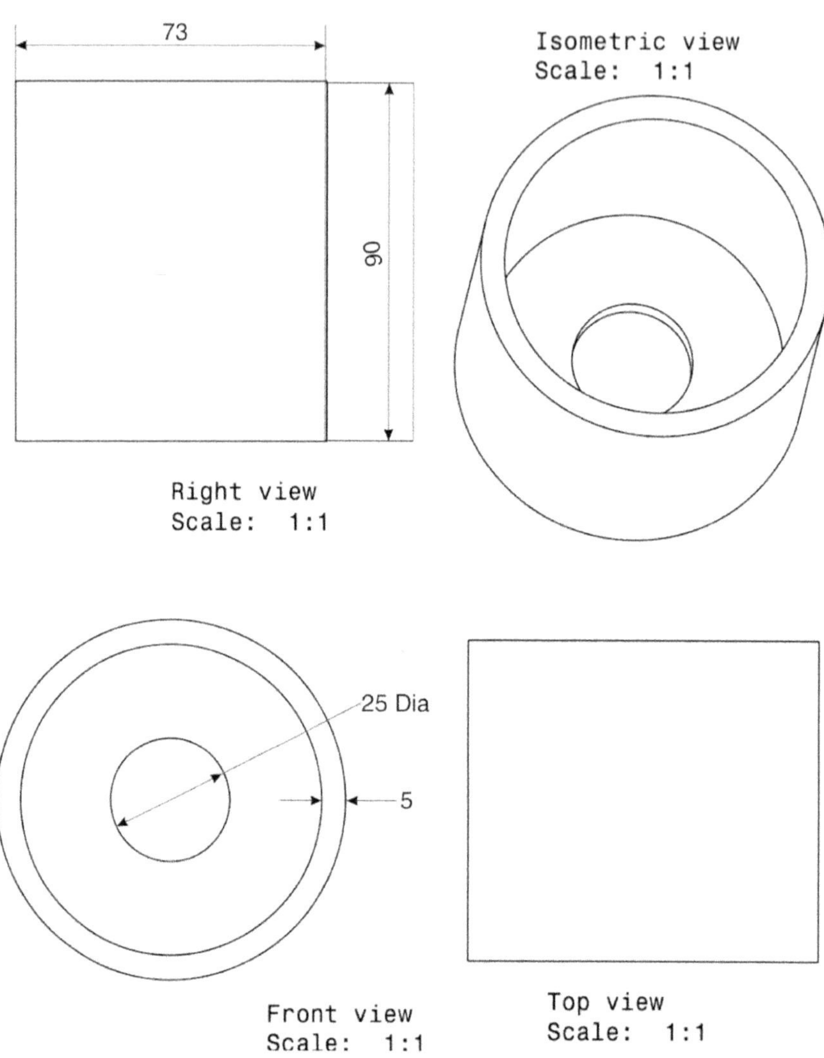

Figure 11.26 Compression chamber.

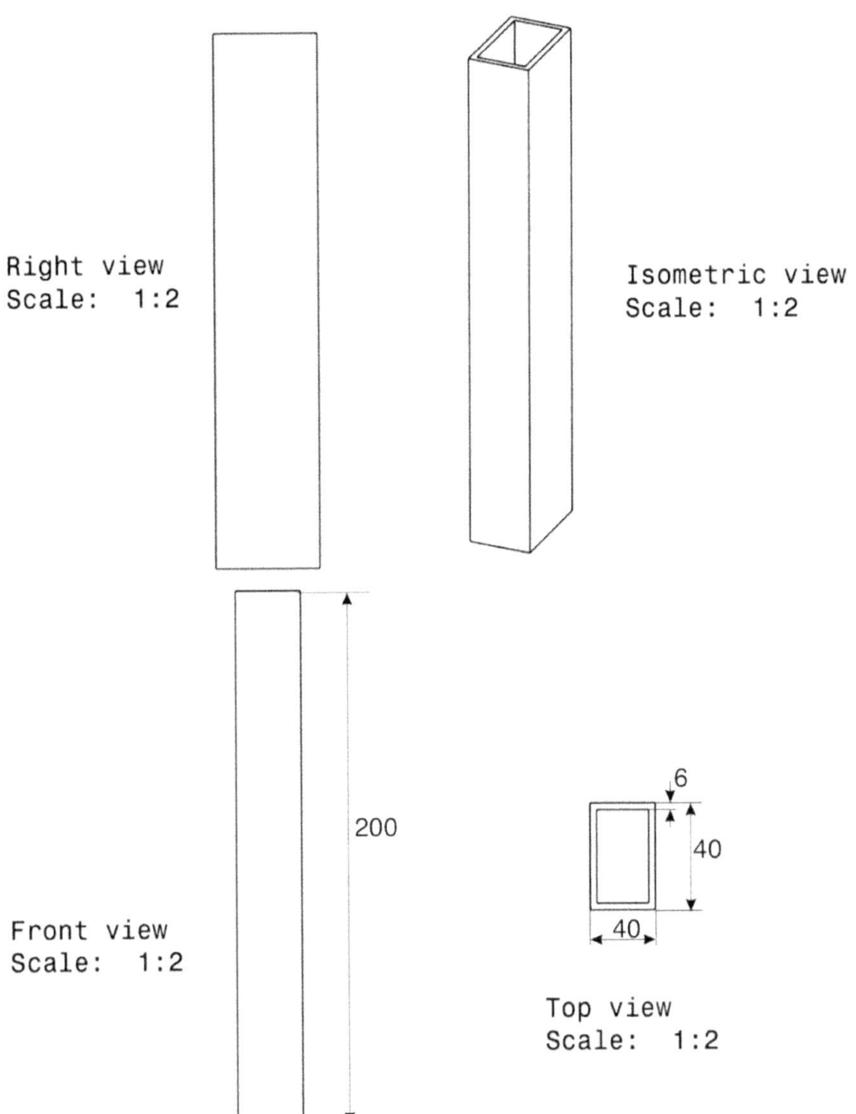

Figure 11.27 Column.

Chapter 12

Summary – closing note

This book started with an introduction to the nature of design, and systematic design process. Introduction continued with the descriptions of design model, design methods and an example that ran through the entire design process. Chapters 2–10 explained systematic design process from the start by receiving a design brief and finishing with the embodiment of a product. If it is a product design problem, the information at hand would be sufficient to produce the product. However, if it is an engineering product, further engineering analyses would be necessary to prove the design before its committal to manufacture. This book would have directed the reader in the right way of thinking during the design process. With the guidance and experience gained with the examples, the reader should be able to plan the design process and be able to implement the plan. This book would have enabled the reader to:

1. Plan the design process needed for the specific product under a systematic design model. It would have enabled the reader to plan the activities needed at each stage of the design model to choose the right and effective activities for the specific project – Explained in Chapter 2.
2. Understand the design brief well as a member of the design team and write a design brief as a member of the senior management team of a company – Explained in Chapter 3.
3. Plan and execute a stakeholder survey first by prior planning of the types of needed requirements and then by obtaining the prioritized requirements through one-to-one interviews – Explained in Chapter 4.
4. Represent the product in the functional domain as a function tree in a hierarchical manner or as a function structure representing the flows of material, energy and information – Explained Chapter 5.
5. Measure the desired outputs representing the required stakeholder requirements by defining the appropriate metrics and the level of measurement needed – Explained in Chapter 6.

DOI: 10.1201/9781003484950-15

6. Draw specifications as a set of measurable parameters with their permissible range, their level of measurement and units – Explained in Chapter 7.
7. Choose appropriate design methods and propose conceptual designs as a team and as an individual designer – Explained in Chapter 8.
8. Evaluate proposed conceptual designs using a set of criteria, improve them wherever possible and choose an optimal or near optimal design – Chapter 9.
9. Develop the physical form of the product by carrying out the Embodiment design – Explained in Chapter 10.

The reader can understand how a project is completed by reading Chapter 11 where examples are given in both product design and engineering design. The examples followed the steps explained in Chapters 2–10.

I sincerely believe that the simple examples given here would have helped the reader to understand the design process well and given the ability to handle much complicated design problems.

Index

For Product Safety Concerns and Information please contact our EU representative GPSR@taylorandfrancis.com
Taylor & Francis Verlag GmbH, Kaufingerstraße 24, 80331 München, Germany

www.ingramcontent.com/pod-product-compliance
Lightning Source LLC
LaVergne TN
LVHW020602110826
845149LV00002B/355
* 9 7 8 1 0 3 2 7 7 8 2 0 4 *